STUDIES IN ANALYTICAL CHEMISTRY 2

THE PRINCIPLES OF ION-SELECTIVE ELECTRODES AND OF MEMBRANE TRANSPORT

W. E. MORF

Swiss Federal Institute of Technology
Zürich, Switzerland

ELSEVIER SCIENTIFIC PUBLISHING COMPANY
Amsterdam–Oxford–New York 1981

The distribution of this book is being handled
by the following publishers

for the U.S.A. and Canada
Elsevier/North-Holland, Inc.
52 Vanderbilt Avenue
New York, New York 10017, U.S.A.

for the East European Countries
People's Republic of China, Democratic People's Republic of Korea,
Republic of Cuba, Socialist Republic of Vietnam, People's Republic of Mongolia
Akadémiai Kiadó, The Publishing House of the
Hungarian Academy of Sciences, Budapest

for all remaining areas
Elsevier Scientific Publishing Company
335 Jan van Galenstraat
P. O. Box 211, Amsterdam, The Netherlands

Library of Congress Cataloging in Publication Data

Morf, W E
    The principles of ion-selective electrodes and
of membrane transport

    (Studies in analytical chemistry; 2)
    Includes bibliographies and indexes.
    1. Ion-selective electrodes. 2. Membranes
(Technology) I. Title. II. Series.
QD565.M67   541.3'724   80-25386
    ISBN 0-444-99749-0 (Vol. 2)
    ISBN 0-444-41941-1 (Series)

Joint edition published by
Elsevier Scientific Publishing Company, The Netherlands and
Akadémiai Kiadó, The Publishing House of the Hungarian Academy
of Sciences, Budapest, Hungary

Printed in Hungary

CONTENTS

Preface   ix

1. Introduction and preliminary aspects   1
   1.1. Organization and EMF response of membrane electrode cells   2
   1.2. Membrane materials and their selectivity-determining principles   8
   1.3. Single-ion activities   13
   References   21

PART A – THEORY OF MEMBRANE POTENTIALS AND
MEMBRANE TRANSPORT

2. Description of the basic membrane model   27
   References   34
3. The phase-boundary potential (Donnan potential)   35
   References   43
4. The diffusion potential   44
   4.1. General formulation   44
   4.2. Practical solutions   50
   References   62

5. Calculation of liquid-junction potentials   64
   References   74

6. Solutions for the membrane potential   75
   References   86

7. Classical concepts of membrane transport   87
   7.1. The Nernst–Planck flux equation   87
   7.2. The Goldman–Hodgkin–Katz approximation   88
   7.3. Simple model for symmetrical membrane cells   91
   7.4. Schlögl's general theory and its applications   94
   7.5. Electrical properties and ion-transport selectivity of bulk membranes   101
   References   112

8. Free and carrier-mediated ion transport across bilayer membranes   113
   8.1. Description of the ion flux across the membrane interior   115
   8.2. Description of the ion flux across the interfaces   123
   8.3. Consequences of a closed-circuit flux of carriers   125
   8.4. Derivation of the general result   127

8.5. Bilayer model by Läuger and Stark   131
8.6. Bilayer model by Ciani, Eisenmann, and Krasne   134
8.7. Comparison with the bulk membrane model by Morf, Wuhrmann, and Simon   143
References   150

9. Summary of fundamental relationships   153

PART B – ION- SELECTIVE ELECTRODES

10. Solid-state membrane electrodes   165
    10.1. Characterization of membrane materials   166
    10.2. Basic theoretical aspects of solid-state membrane electrodes   168
    10.3. Potential response and detection limit of silver compound membranes in unbuffed
          solutions of the primary ions   171
    10.4. Potential response of silver halide membranes to different cations 183
    10.5. Selectivity of silver halide membranes towards different anions   186
    10.6. Potential reponse and selectivity of silver halide membranes towards different ligands 198
    References   207

11. Liquid-membrane electrodes based on liquid ion-exchangers   211
    11.1. Membrane materials and observed ion selectivities   212
    11.2. Implications of the general membrane theory   219
          11.2.1. Potential response of ion-exchange membranes to monovalent counterions 219
          11.2.2. Potential response of ion-exchange membranes to divalent counterions   229
          11.2.3. Mixed potential response to divalent and monovalent counterions (origin of
                  "potential dips")   233
    11.3. Theory of Sandblom, Eisenman, and Walker, and its extensions   236
    11.4. Interpretation of the apparent selectivity behavior of liquid membranes 246
    References   259

12. Liquid-membrane electrodes based on neutral carriers   264
    12.1. Characteristics of neutral carriers and reported selectivities for membrane electrodes 268
    12.2. Mechanism of cation specificity (permselectivity) of neutral carrier membranes 274
    12.3. Cation selectivity of carrier membrane electrodes   285
          12.3.1. Selectivity between cations of the same charge   287
          12.3.2. Monovalent/divalent cation selectivity   291
    12.4. Anion effects in carrier membrane electrodes   296
          12.4.1. Anion interference in conventional carrier membranes   296
          12.4.2. Properties of carrier-based liquid membranes with incorporated ion-ex-
                  changers   309
    12.5. Molecular aspects of cation-selective carriers   315
          12.5.1. Molecular basis of ion selectivity   315
          12.5.2. Design features of membrane-active complexing agents   322
    References   331

13. Glass electrodes   337
    13.1. Introduction   337
    13.2. Ion-exchange theories and n-type descriptions of glass membrane potentials 340
    13.3. Potential responses of $Na_2O-Al_2O_3SiO_2$ glasses   354
    13.4. Alternative approaches to heterogeneous-site glasses   359
    13.5. Further development of glass electrode theory (liquid-membrane concepts) 362
    References   372

14. Dynamic response behavior of ion-selective electrodes   375
    14.1. Electrical relaxation processes   377
    14.2. Kinetics of interfacial reactions   379
    14.3. Diffusion through a stagnant layer   382
    14.4. Diffusion within the ion-sensing membrane   384
    References   399

15. Special arrangements: gas-sensing electrodes and enzyme electrodes   401
    15.1. Gas-sensing electrodes   402
    15.2. Enzyme electrodes   406
    References   414

Author index   417
Subject index   425

PREFACE

Ion-selective membrane electrodes and separation techniques based on membranes have in recent years become a focus of attention. The development of such analytical devices and their application have progressed far, having been much stimulated by a variety of individual theories, membrane models, or more intuitive attempts aimed at a deeper understanding. Yet, there is no unified approach to the theory of potentiometric and electrodialytic membrane cells available.

The present work is intended to provide a comprehensive survey of the theory, the principles, and the fundamentals of ion-selective electrodes and of membrane transport. Virtually the whole treatise is based on the membrane model specified in Chapter 2. While risking some loss of universality, emphasis has been laid on simplicity of the derivations and on explicit results. Because of the complexity of the subject, the text has been subdivided into two parts.

Part A aims at a more general discussion of membrane potentials and membrane transport. Formulations of the interfacial potential contribution due to phase boundaries are given in Chapter 3. Chapter 4 contains a nearly encyclopedic treatment of the diffusion potential, taking into account the nonideality of diffusion layers or membrane phases. The extended Nernst, Planck, Teorell, and Schlögl solutions are derived in a new and simple way. Chapter 5 deals with the liquid-junction potentials arising in conventional potentiometric measuring cells. A comparison is given between different calculation procedures and experimental results. In Chapter 6 practical solutions for the membrane potential are derived and catalogued. Chapter 7 is meant to give a detailed analysis of the ion-transport and the electrical properties of bulk membranes. The discussions are based on the Nernst-Planck flux equation. The implications of Schlögl's theory and those of a simplified membrane model that leads to equivalent results are emphasized. A generalized theory of ion transport is set forth in Chapter 8; the treatment goes beyond the earlier models offered by Läuger's and Eisenman's groups. To account for the observed characteristics of lipid bilayer membranes, different shapes of the membrane-internal free energy profiles are considered. The given analysis permits rationalization of the free and carrier-mediated ion transport across membranes. For convenience, the key results of Part A are summarized in Chapter 9.

Part B deals in great detail with the fundamentals of ion-selective electrodes. Chapter 10 covers the principles of solid-state membrane electrodes. The response behavior, the ion selectivity, and the detection limits of such systems are discussed for the example of silver compound membranes. Special attention is paid to the role of dissolution and leaching processes, on the one hand, and of surface coverings, on the other. In Chapter 11 several important extensions and modifications of the Sandblom-Eisenman-Walker theory are presented. The new results contribute to a full understanding of the response phenomena observed for liquid ion-exchange membrane electrodes, so much the more since the chapter also includes an interpretation of the apparent selectivity of such sensors. Chapter 12 is considered to extensively cover the field of neutral carrier membrane electrodes. The theory of these systems has been revised in the light of new information concerning their response mechanism, and many practical examples are given to illustrate the results. The molecular basis of ion selectivity and the design features of electrically neutral ionophores are also briefly discussed. Chapter 13 contains two unified approaches to the theory of glass electrodes. The first of these is based on solid-state principles and encompasses the earlier ion-exchange theories and n-type descriptions of glass membrane potentials. The second one, based on liquid-membrane concepts, is of like import. Both treatments bridge the gaps that heretofore existed between the earlier, more specific theories. In Chapter 14 the electrical, chemical, and diffusional processes contributing to the time response of ion-selective electrodes are reviewed. Finally, Chapter 15 gives a short introduction into the principles of gas-sensing probes and enzyme electrodes. A simple model is presented that is successful in reconstructing the observed response of enzyme electrodes.

I am greatly indebted to Professor W. Simon for having given me the opportunity to prepare the present book, as well as for many valuable discussions on the subject. I acknowledge financial support by the Swiss National Science Foundation and the Swiss Federal Institute of Technology. I would also like to thank Dr. P. C. Meier for his critical reading of the English text, and Miss I. Port for her invaluable assistance in preparing the final manuscript.

Zürich, June 1979   Werner E. Morf

Chapter 1

Introduction and Preliminary Aspects

Since the end of the nineteen sixties, ion-selective electrodes have continued to be one of the most important developments in analytical chemistry. This is best documented by the large and still expanding literature on this subject. While the initial era saw an intensive search for novel electrode materials and new constructions, this has given way subsequently to more introspective studies on ion selectivity and electrode mechanisms, as well as to extensive practical applications of ion sensors, especially in clinical and environmental chemistry.

The field of ion-selective electrodes is perhaps one of the most eminent examples for interdisciplinary research in chemistry. Indeed, important books and reviews have been prepared or edited by the physiologists Eisenman [1 - 3] and Kessler et al. [4], the physicochemists Buck [5 - 8] and Covington [9 - 11], the electrochemists Koryta [12 - 15] and Cammann [16], the analytical chemists Pungor [17 - 21] and Simon and Morf [22 - 27], the membrane expert Lakshminarayanaiah [28 - 30], the extraction specialist Freiser [31], as well as authors more interested in practical applications and manufacturing of electrodes, such as Moody and Thomas [32 - 35], Durst [36, 37], Ross [38, 39], Rechnitz [40, 41], Bailey [42], and others. Most of these reviews cover both the practical and the fundamental aspects of different ion-selective electrodes.

The aim of the present work is to provide a unified, self-consistent approach to the theory and principles of ion-selective membranes and membrane electrodes. In Part A, a basic membrane model is formulated which encompasses a variety of

earlier theories on membrane potentials and membrane transport. The explicit key results of these general considerations
are summarized in Chapter 9. Part B, which is the major part,
is devoted to the fundamentals of ion-selective electrodes.
Although it is not intended here to produce another exhaustive
review, the history of ion-selective electrodes and the current
developments in the field are also covered in some detail.

1.1. ORGANIZATION AND EMF RESPONSE OF MEMBRANE ELECTRODE CELLS

Ion-selective electrodes are electrochemical sensors that
allow potentiometric determination of the activity of certain
ions in the presence of other ions; the sample under test is
usually an aqueous solution. Such an electrode constitutes a
galvanic half-cell, consisting of an ion-selective membrane,
an internal contacting solution (conventional construction,
see Figure 1.1) or a solid contact ("all-solid-state" con-
figuration), and an internal reference electrode. For practical convenience these elements are housed in a single body.
The other half-cell is given by an external reference electrode dipping into a reference electrolyte. The contact between the two half-cells is preferably maintained by an intermediate salt bridge which can be placed within the reference
electrode housing. The organization of a typical membrane
electrode cell may then be represented as follows (see also
Figure 1.1):

$$\underbrace{\underset{E_1}{Hg;}\ \underset{E_2}{Hg_2Cl_2;}\ KCl\ (satd)\ \bigg|\ \underset{E_3}{salt\ bridge}\ \bigg|\ \underset{E_J}{sample}}_{\text{reference electrode system}}$$

$$\underbrace{\bigg|\ \underbrace{membrane}_{E_M}\ \bigg|\ internal\ solution;\ AgCl;\ \underset{E_4}{\ }\ \underset{E_5}{Ag}}_{\substack{\text{ion-selective membrane electrode} \\ \text{(indicator electrode)}}} \qquad (1.1)$$

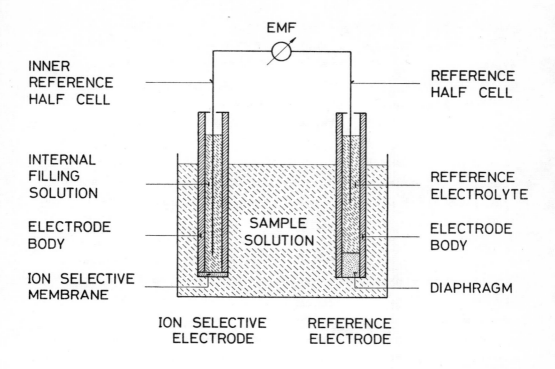

INNER
REFERENCE
HALF CELL

INTERNAL
FILLING
SOLUTION

ELECTRODE
BODY

ION SELECTIVE
MEMBRANE

EMF

REFERENCE
HALF CELL

REFERENCE
ELECTROLYTE

ELECTRODE
BODY

DIAPHRAGM

SAMPLE
SOLUTION

ION SELECTIVE
ELECTRODE

REFERENCE
ELECTRODE

Figure 1.1. Schematic diagram of a membrane electrode
measuring circuit and cell assembly.

The total electrical potential difference measured between
the two terminals of the cell is evidently composed of a considerable number of local potential differences E_n, arising
at solid-solid, solid-liquid, and liquid-liquid interfaces.
The sum of all these terms is observable as the emf-response
of the ion-selective electrode cell:

$$E = (E_1 + E_2 + E_3 + E_4 + E_5) + E_J + E_M = E_0 + E_J + E_M \tag{1.2}$$

where   E : cell potential (emf)

E_0: reference potential, encompassing the potential contributions E_1 to E_5 arising within the system of reference electrodes in cell (1.1)

E_J: liquid-junction potential

E_M: membrane potential

For a given membrane electrode assembly and a fixed temperature, E_0 is constant. Thus the emf of the cell reflects the
two electrical potential contributions that are influenced by
the sample solution.

The term E_J considers the potential difference generated
at the junction between the sample solution and the salt
bridge solution. Such liquid-junction potentials may cause
some problems in experimentation as they clearly interfere
with the intrinsic response of the ion-selective membrane.
For practical purposes, it is advisable to choose electrolyte
combinations where

$$E_J \approx 0 \qquad\qquad\qquad (1.3)$$

The convenient methods for reducing or eliminating the liquid-
junction potential will be discussed in Chapter 5.

The membrane potential E_M is found to be the fundamental
part since it clearly describes the whole performance of the
ion-selective membrane electrode. For a membrane which is
supposed to be ideally and exclusively selective for ions of
the sort I, the zero-current membrane potential is a direct
and specific measure of the respective activities in the contacting solutions on either side:

$$E_M = \frac{RT}{z_i F} \ln \frac{a_i'}{a_i''} \qquad\qquad (1.4)$$

The activity a_i' refers to the external solution (sample) and a_i'' to the internal solution; z_i is the charge of the ion, in units of the proton charge, R is the gas constant, T the absolute temperature, and F the Faraday equivalent. In this case, we may expect a <u>Nernstian response</u> of the membrane electrode cell since the composition of the internal filling solution is kept constant:

$$E = E_i^O + s \log a_i' \qquad (1.5)$$

As long as Eq. (1.3) applies, the intercept E_i^O of the linear response function represents a standard potential,

$$E_i^O = E_0 + E_J - s \log a_i'' \cong const, \qquad (1.6)$$

and s is identical to the Nernstian slope

$$s = 2.303 \ RT/z_i F = 59.16 \ mV/z_i \ (25^O C) \qquad (1.7)$$

The analytical technique of ion-selective electrodes obviously relies on the applicability of such potential-activity relationships (Figure 1.2). According to Eqs. (1.5) - (1.7) the basic response toward the primary ion I is derived as

$$1 \ mV \cong z_i \cdot 4 \ \% \ change \ in \ a_i' \qquad (1.5a)$$

The reproducibility of emf measurements conducted on modern ion-selective electrode cells is typically on the order of 0.1 mV (see Figure 1.3), which means that the precision in the determination of activities is seldom much better than 1%. Although direct potentiometry using ion-selective electrodes cannot therefore be considered an overly precise analytical technique, its advantages more than outweigh this limitation.

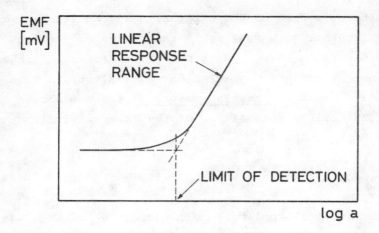

Figure 1.2. Schematic diagram of the response function of an
ion-selective electrode. The detection limit of ion-selective
electrodes is usually determined by interfering ions present
in the sample, contaminations stemming from impurities in the
reagents, leaching of ions out of the membrane, or intentionally introduced background electrolytes.

In practice an ideally specific electrode behavior such as
described by Eq. (1.5) can most often not be attained. Therefore, we generally have to consider additional contributions
to the total measured activity which result from the presence
of interfering species J in the sample solution. A semiempirical but rather successful approach to treat real membrane
electrode systems is offered by the extended Nicolsky equation
(or Eisenman equation) [1, 2, 44, 45]:

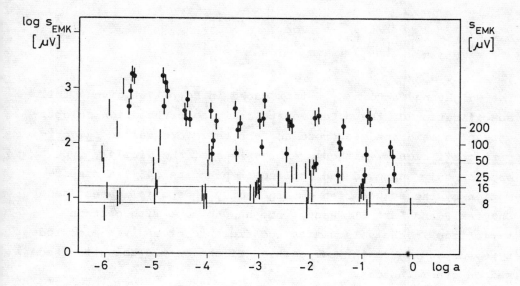

<u>Figure 1.3.</u> Errors associated with emf measurements [43].
Error bars with circles: standard deviations of routine measurements conducted on Na^+- and Ca^{2+}-selective liquid-membrane
electrodes. Error bars without circles: standard deviations of
precision measurements conducted on Li^+-, Na^+-, K^+-, Ca^{2+}-,
and Cl^--selective electrodes, special precautions being taken
to ensure freedom of drift, temperature fluctuations and contamination by the salt bridge electrolyte. The analytical
precision is found to increase with increasing activity of the
measured species. The theoretical limit is given by the overall noise of the electronic equipment (12 ± 4 µV).

$$E = E_i^o + s \log \left[a_i' + \sum_{j \neq i} K_{ij}^{Pot} (a_j')^{z_i/z_j} \right] \qquad (1.8)$$

The weighting factors K_{ij}^{Pot} introduced in Eq. (1.8) give a full specification of the potentiometrically observable ion selectivity of a membrane electrode and are therefore called selectivity coefficients. Although expressions of the Nicolsky type are commonly approved and applied in practice [45], a confirmation by theory has been obtained only in special cases (see Chapter 6 and Part B). Hence, one has to cope with certain variations in the selectivity coefficients of a given electrode type, depending on the composition of the sample solution used and on other factors.

Different methods have been proposed for the evaluation of selectivity factors [27, 32]. These are
a) the separate_solution_method,
b) the fixed_interference_method, and
c) the fixed_primary_ion_method.
The advantage of method a) is simplicity, but the selectivity data thus determined may be unrepresentative for mixed sample solutions. Hence, method b) was recommended by IUPAC [45]. It should be noted, however, that the uncertainty in the determination of selectivities by the mixed solution method seems to be higher than for the separate solution technique [46]. Method c) is less frequently used, except for illustrating the effect of pH on the emf-response to a given primary ion.

1.2. MEMBRANE MATERIALS AND THEIR SELECTIVITY-DETERMINING PRINCIPLES

A great number of ion-selective membrane electrodes have

been developed and recommended for analytical applications, and many of them have become commercially available (see Table 1.1 and Part B). These devices may be classified, according to the nature of the basic membrane material, into the following categories:

a) <u>Solid-state membrane electrodes</u>, based on various crystalline materials; the forms include single crystals, cast or sintered materials, pressed polycrystalline pellets, as well as heterogeneous combinations of precipitates held in hydrophobic polymer binders.

b) <u>Glass membrane electrodes</u>, usually formed from lithia, aluminosilicate or multi-component glasses.

c) <u>Liquid ion-exchanger membrane electrodes</u>, the membrane of which consists of an organic, water-immiscible liquid phase incorporating mobile ionic or ionogenic compounds, such as hydrophobic acids, bases, and salts.

d) <u>Neutral carrier liquid membrane electrodes</u>, where the membrane is usually formed from an organic solution of electrically neutral, ion-specific complexing agents (ion carriers, ionophores), held in an inert polymer matrix.

e) <u>Special arrangements</u>, such as gas-sensitive electrodes and enzyme electrodes, the potentiometric detection unit of which is based on conventional electrodes of the type a) - d).

f) <u>Ion-selective field effect transistors</u> (ISFET's), which are hybrids of ion-selective electrodes and metal-oxide field effect transistors (MOSFET's). In a conventional potentiometric measurement, the signal of the ion-selective electrode is transmitted by wire to the input MOSFET of the voltmeter and there modulates the drain current. In the ISFET, the metal gate of a MOSFET has been directly replaced by or contacted with a solid or liquid ion-sensitive membrane. For evident reasons, the response of such miniaturized sensors is linked to a current, in lieu of a potential [47 - 50].

9

<u>Table 1.1.</u> Commercially available membrane electrodes (from reference 37).

Cations	Anions	Neutral (gases)
Acetylcholine	Bromide	Ammonia
Arsenic[a]	Carbonate[a]	Carbon dioxide
Cadmium	Chloride	Chlorine
Calcium	Cyanide	Hydrogen sulfide
Chromium[a]	Fluoride	Oxygen[b]
Copper	Fluoroborate	Sulfur dioxide
Hydrogen	Iodide	Nitrogen oxides
Lead	Nitrate	
Mercury[a]	Perchlorate	
Potassium	Phosphate[a]	
Silver	Sulfate[a]	
Zinc	Sulfide	
Univalent	Thiocyanate	
Divalent		

[a] Electrode of limited analytical utility

[b] Amperometric (not potentiometric) sensor

Solid-state membrane electrodes are used primarily as sensors for those kinds of ions that are constituents of the insoluble salt forming the membrane. In addition, they make possible the detection of other species interacting with the ionic sites of the membrane material. Silver halide membranes, for example, can thus serve as sensors for silver ions, halide ions, sulfide ions, as well as for ligands that form stable complexes with the silver ion (e. g. cyanide). The theoretical selectivity of AgX-membranes for interfering anions Y^{z-} relative to the primary ions X^- is determined by the solubility products of the respective silver compounds (see Chapter

10). Hence, the selectivity sequence is basically the same for any membrane of this type:

$$S^{2-} >> I^- > Br^- \sim SCN^- > Cl^-$$

By the same principle, electrodes based on LaF_3 single crystals show a high preference for fluoride over other halide ions but are subject to some interference by hydroxyl ions.

The pH glass electrode was the first ion-selective electrode discovered and is still one of the most important standard laboratory devices. Its unsurpassed specificity for hydrogen ions is due to the strongly basic nature of the charged silica groups residing in the glass network. By replacing these components in part by alumina groups, which form ionic sites of much lower electric field strength, glass membrane electrodes with increased selectivities towards alkali and silver ions can be obtained (see also Chapter 13).

Liquid-membrane electrodes with electrically charged ion-exchange sites generally show permselectivity for oppositely charged counterions. In the case of nearly complete dissociation between sites and counterions, such as obtains for non-complexing ionic components in more polar membrane solvents, the selectivity between different counterions of the same charge is dictated mainly by the extraction behavior of the solvating membrane medium. The ionic extraction constants assume comparatively small values for counterions that are strongly hydrated in the aqueous phase, and large values for large, lipophilic organic ions R. Therefore, the following monotonic selectivity sequence is obtained for membrane electrodes based on dissociated cation-exchangers (e. g. tetraphenylborate in nitroaromatic solvents, see Chapter 11):

$$R^+ > Cs^+ > Rb^+ > K^+ > Na^+ > Li^+$$

and analogously for anion-exchangers (e. g. quaternary ammonium salts in appropriate solvents):

$$R^- > ClO_4^- > I^- > NO_3^- \sim Br^- > Cl^- > F^-$$

The discrimination of corresponding anion-sensitive electrodes between different anions (counterions) is less pronounced than for silver compound solid-state electrodes, however, which fact recommends them as sensors for various anions, such as nitrate, chloride, or perchlorate. For liquid membranes with almost complete association or complexation between ionic sites and counterions, the potentiometric selectivity depends in a rather complicated way on both the ion-extraction selectivity of the membrane solvent and the ion-binding specificity of the incorporated sites. Hence, the same ligand dialkylphosphate is used in certain calcium-selective electrodes (membrane solvent: dioctylphenyl-phosphonate) and in divalent-ion-sensors with comparable selectivities for calcium and magnesium ions (solvent: 1-decanol).

Neutral carrier membrane electrodes make use of the inherently outstanding cation specificity of certain natural and synthetic ionophores. The ion-binding selectivity of such electrically neutral complexing agents can be fully exploited in membranes, which is in contrast to the behavior reported for liquid ion-exchangers. For example, the selectivity of neutral carrier membranes among different cations of the same charge is virtually determined by the stability constants of the ion/carrier complexes involved. Accordingly, the natural ionophore valinomycin steadily induces the same selectivity sequence in biological and in artificial membrane systems (see Chapters 8 and 12):

$$Rb^+ \geqslant K^+ \geqslant Cs^+ >> Na^+ > Li^+$$

Such membranes constitute the working principle of the best potassium ion sensors available. Electrodes which show some preference for ammonium ions are obtained by using the macro-tetrolide antibiotics as carriers. The detailed study of the structure-selectivity relationship of natural ionophores and model compounds has led to the design of a respectable series of synthetic carrier ligands. At present, forms with con-siderable specificity for Ca^{2+}, Ba^{2+}, Na^+, Li^+, and other iohs are available and have found acceptance in ion-selective electrode applications (Chapter 12).

A comprehensive discussion of the potentiometric response behavior of the different membrane types is given in Part B.

1.3. <u>SINGLE-ION ACTIVITIES</u>

There is some evidence indicating that the usually employed "junction cells" shown in Figure 1.1 and Eq. (1.1) respond to single-ion activities rather than mean activities. Although these quantities can be formally defined by introducing single-ion activity coefficients γ_i (referring to molalities m_i) or y_i (referring to molarities c_i)

$$a_i = \gamma_i m_i = y_i c_i \qquad (1.9)$$

there is no possibility for assessing them individually on an exact thermodynamic basis. Thermodynamic methods, e. g. emf-measurements on cells without liquid junction of the type

anion-selective electrode|salt solution|cation-selective
electrode,

can rigorously yield information only on the mean activity coefficient $\gamma_\pm$, which is related to the single-ion activity coefficients γ_+ and γ_- as follows:

$$\log \gamma_\pm = \frac{|z_-|}{|z_+| + |z_-|} \log \gamma_+ + \frac{|z_+|}{|z_+| + |z_-|} \log \gamma_- \qquad (1.10)$$

The activity coefficients are well known to depend primarily on the ionic strength of the solution, defined as

$$I = 0.5 \sum_i z_i^2 c_i \qquad (1.11)$$

With increasing I, ionic interactions result in a characteristic variation of $\gamma_\pm$, as predicted by the Debye-Hückel theory and its extensions. The following law applies to aqueous 1:1 electrolytes up to $I \cong 0.1$ mol/l:

$$\log \gamma_\pm = \log f_{DH} = -\frac{A\sqrt{I}}{1 + B\,a\sqrt{I}} \qquad (1.12)$$

where a is the ion size parameter, and A = 0.509 and B = 0.328 $\overset{\circ}{A}^{-1}$ are constants (25°C). To allow for ions of higher valencies and ionic strengths of up to about 1 mol/l, a semi-empirical extension of Eq. (1.12) can be used [51, 52]:

$$\log \gamma_\pm = |z_+ z_-| \log f_{DH} + C\,I \qquad (1.13)$$

This relationship can be fitted to experimental data, and the values of the parameters a and C thus evaluated. A more elaborate approach was presented by Stokes and Robinson [53] who also accounted for effects arising from ionic hydration. The general result had the form:

$$\log \gamma_{\pm} = |z_+ z_-| \log f_{DH} - \frac{h}{\nu} \log a_w - \log [1-0.018 (h-\nu)m] \quad (1.14)$$

where m is the molality of the salt solution, a_w is the activity of water, h is the total hydration number

$$h = |z_-| h_+ + |z_+| h_- \quad (1.15)$$

and ν is the formal number of ions obtained from the salt

$$\nu = |z_+| + |z_-| \quad (1.16)$$

Except for extremely high molalities, Eq. (1.14) may be converted into a form analogous to (1.13) where a term C'm
appears instead of CI. This demonstrates the equivalence of
the two formulations which indeed lead to practically the same
correlations with experimental $\gamma_{\pm}$ values.

The predominant problem is then to split the mean activity
coefficients into individual ionic contributions, according
to Eq. (1.10). In the history of ion-selective electrodes,
the following procedures were called upon.

1) MacInnes convention [54, 55]. The single-ion activity coefficients for K^+ and Cl^- ions in aqueous solutions are set
equal to the mean activity coefficient of a KCl solution of
the same ionic strength:

$$\gamma_+(K^+) = \gamma_-(Cl^-) = \gamma_{\pm}(KCl) \quad (1.17)$$

2) pH-convention [56, 57]. The single-ion activity coefficient of Cl^- ions is assumed to obey the following Debye-Hückel relation:

$$\log \gamma_-(Cl^-) = -\frac{A\sqrt{I}}{1 + 1.5\sqrt{I}} \tag{1.18}$$

This allows to assess individual activity coefficients for cations in chloride solutions and, indirectly, those of other anions in solutions of the given cations.

3) Debye-Hückel convention [57, 58]. The single-ion and mean activity coefficients are interrelated most naturally, according to the Debye-Hückel theory, as follows:

$$\log \gamma_+ = |z_+/z_-| \log \gamma_\pm \tag{1.19a}$$

$$\log \gamma_- = |z_-/z_+| \log \gamma_\pm \tag{1.19b}$$

Accordingly, it is found that $\gamma_+ = \gamma_- = \gamma_\pm$ for 1:1 electrolytes, and $\gamma_+^{1/2} = \gamma_-^2 = \gamma_\pm$ for 2:1 electrolytes (see Figure 1.4).

4) Stokes-Robinson-Bates convention [53, 58, 59]. At ionic strengths $I \geqslant 1$ mol/l, methods 1) - 3) definitely fail, and a more precise procedure is then required. To this end, the single-ion activity coefficients may be formulated, in analogy to the Stokes-Robinson equation (1.14), as

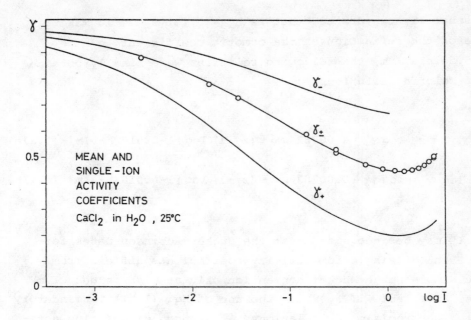

<u>Figure 1.4.</u> The mean and the single-ion activity coefficients
of calcium chloride as functions of the logarithm of the
ionic strength [43]. The circles denote measured mean activity coefficients $\gamma_\pm$. The curves were plotted using Eqs. (1.13)
and (1.19) with a = 5.0 Å and C = 0.04.

$$\log \gamma_+ = z_+^2 \log f_{DH} - h_+ \log a_w - \log [1-0.018(h-\nu)m] \quad (1.20a)$$

$$\log \gamma_- = z_-^2 \log f_{DH} - h_- \log a_w - \log [1-0.018(h-\nu)m] \quad (1.20b)$$

Bates et al. [59] suggested that anions such as chloride are

essentially unhydrated, that is $h_- = 0$ for $z_- = -1$. Making
use of the definition of the osmotic coefficient, $g \cong$
$-55.5 \ln a_w / \nu m$, the following relations were finally obtained
for chloride solutions:

$$\log \gamma_+ = z_+ \log \gamma_\pm + 0.00782\,hmg + (z_+ - 1) \log [1 - 0.018\,(h - \nu)m] \quad (1.21a)$$

$$z_+ \log \gamma_- = \log \gamma_\pm - 0.00782\,hmg - (z_+ - 1) \log [1 - 0.018\,(h - \nu)m] \quad (1.21b)$$

It may be recognized that the Stokes-Robinson-Bates formulation of single-ion activity coefficients differs from
the simple Debye-Hückel convention only at high values of m
(see also Table 1.2). While the use of Eq. (1.21) is mandatory
for high-precision calculations, e. g. when elaborating activity standards, the approximation (1.19) is more useful and
preferred for routine analytical work. In fact, the latter
relation allows ready calculation of single-ion activity coefficients from tabulated $\gamma_\pm$ data (see Figure 1.4) without
any knowledge of the experimental parameters g and h. Such
practical values for γ_+ (chloride salts) and γ_- (sodium
salts) are plotted in Figure 1.5.

Table 1.2. Single-ion activity coefficients (molal scale) for alkali and alkaline earth chlorides at 25°C.

Molality	Debye-Hückel convention[a]		Stokes-Robinson-Bates convention[b]	
m	γ_+	γ_-	γ_+	γ_-
	NaCl; a = 4.0 Å, C = 0.040			
0.1	0.777	0.777	0.783	0.773
0.2	0.732	0.732	0.744	0.726
0.5	0.681	0.681	0.701	0.661
1.0	0.660	0.660	0.697	0.620
	KCl; a = 3.65 Å, C = 0.015			
0.1	0.767	0.767	0.773	0.768
0.2	0.716	0.716	0.722	0.714
0.5	0.650	0.650	0.659	0.639
1.0	0.607	0.607	0.623	0.586
	CsCl; a = 2.6 Å, C = 0.014			
0.1	0.749	0.749	0.756	0.756
0.2	0.689	0.689	0.694	0.694
0.5	0.606	0.606	0.606	0.606
1.0	0.549	0.549	0.544	0.544
	$MgCl_2$; a = 5.2 Å, C = 0.060			
0.1	0.288	0.733	0.279	0.726
0.2	0.247	0.705	0.239	0.697
0.5	0.236	0.697	0.234	0.688
1.0	0.294	0.736	0.344	0.732
	$CaCl_2$; a = 5.0 Å, C = 0.040			
0.1	0.273	0.723	0.269	0.719
0.2	0.226	0.689	0.224	0.685
0.5	0.196	0.665	0.204	0.665
1.0	0.210	0.677	0.263	0.690
	$BaCl_2$; a = 4.4 Å, C = 0.040			
0.1	0.252	0.708	0.259	0.712
0.2	0.201	0.670	0.204	0.668
0.5	0.166	0.638	0.165	0.630
1.0	0.171	0.643	0.167	0.620

[a] Values were calculated from Eqs. (1.13) and (1.19).

[b] Values were derived from the hydration theory by Bates, Staples and Robinson [59].

19

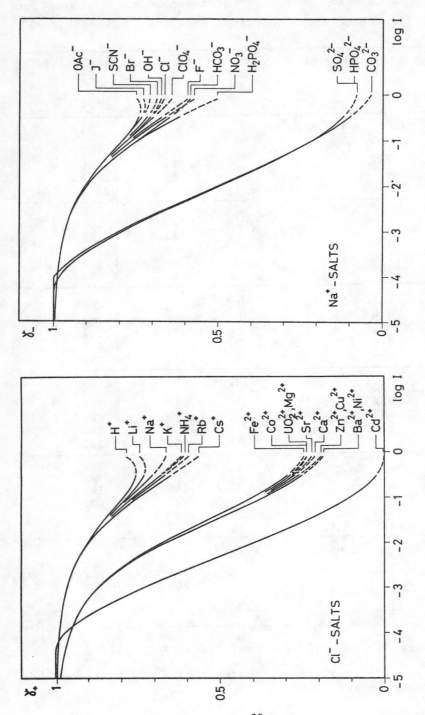

Figure 1.5. Single-ion activity coefficients as functions of the ionic strength; for cations (left) and anions (right). The curves were plotted using Eqs. (1.13) and (1.19). For details, see Reference 43.

20

REFERENCES

[1]   G. Eisenman, ed., Glass Electrodes for Hydrogen and
      Other Cations, M. Dekker, New York, 1967.

[2]   G. Eisenman, 'Theory of membrane electrode potentials:
      an examination of the parameters determining the selec-
      tivity of solid and liquid ion exchangers and of neutral
      sequestering molecules', chapter 1 of ref. [36].

[3]   G. Eisenman, ed., Membranes, Vol. 1 and 2, M. Dekker,
      New York, 1972/1973.

[4]   M. Kessler, L. C. Clark, Jr., D. W. Lübbers, I. A. Sil-
      ver, and W. Simon, eds., Ion and Enzyme Electrodes in
      Biology and Medicine, Urban & Schwarzenberg, Munich,
      1976.

[5]   R. P. Buck, 'Potentiometry, pH measurements and ion
      selective electrodes', chapter 2 of Physical Methods
      of Chemistry (A. Weissberger and B. W. Rossiter, eds.),
      part IIA, Interscience, New York, 1971.

[6]   R. P. Buck, Anal. Chem. 44, 270R (1972); 46, 28R (1974);
      48, 23R (1976); 50, 17R (1978).

[7]   R. P. Buck, Crit. Rev. Anal. Chem. 5, 323 (1975).

[8]   R. P. Buck, 'Theory and principles of membrane elec-
      trodes', chapter 1 of ref. [31].

[9]   A. K. Covington, 'Heterogeneous membrane electrodes',
      chapter 3 of ref. [36].

[10] A. K. Covington, 'Reference electrodes', chapter 4 of
      ref. [36].

[11] A. K. Covington, Crit. Rev. Anal. Chem. 3, 355 (1974).

[12] J. Koryta, Anal. Chim. Acta 61, 329 (1972).

[13] J. Koryta, Ion-Selective Electrodes, Cambridge Univ.
      Press, Cambridge, 1975.

[14] J. Koryta, Anal. Chim. Acta 91, 1 (1977).

[15] J. Koryta, ed., Medical and Biological Applications of
      Electrochemical Devices, Wiley, New York, in press.

[16] K. Cammann, Das Arbeiten mit ionenselektiven Elektroden,
      Springer-Verlag, Berlin, 1973.

[17] E. Pungor and K. Tóth, Pure Appl. Chem. 34, 105 (1973);
      36, 441 (1973).

[18] E. Pungor and I. Buzás, eds., Ion-Selective Electrodes
      (1st Symposium held at Mátrafüred, Hungary, 1972),
      Akadémiai Kiadó, Budapest, 1973.

[19] E. Pungor and I. Buzás, eds., Ion-Selective Electrodes
      (2nd Symposium held at Mátrafüred, Hungary, 1976), Aka-
      démiai Kiadó, Budapest, 1977.

[20] E. Pungor, ed., Ion-Selective Electrodes (Conference on
      Ion-Selective Electrodes held at Budapest, 1977), Aka-
      démiai Kiadó, Budapest, 1978.

[21] E. Pungor and K. Tóth, 'Precipitate-based ion-selective
      electrodes', chapter 2 of ref. [31].

[22] W. Simon, H.-R. Wuhrmann, M. Vašák, L. A. R. Pioda,
      R. Dohner, and Z. Štefanac, Angew. Chem. 82, 433 (1970);
      Angew. Chem. Intern. Ed. 9, 445 (1970).

[23] W. Simon, W. E. Morf, and P. Ch. Meier, Structure and
      Bonding 16, 113 (1973).

[24] W. E. Morf, D. Ammann, E. Pretsch, and W. Simon, Pure
      Appl. Chem. 36, 421 (1973).

[25] W. Simon, E. Pretsch, D. Ammann, W. E. Morf, M. Güggi,
      R. Bissig, and M. Kessler, Pure Appl. Chem. 44, 613
      (1975).

[26] W. E. Morf and W. Simon, 'Ion-selective electrodes
      based on neutral carriers', chapter 3 of ref. [31].

[27] W. Simon, D. Ammann, M. Oehme, and W. E. Morf, Ann.
      N. Y. Acad. Sci. 307, 52 (1978).

[28] N. Lakshminarayanaiah, Transport Phenomena in Membranes,
      Academic Press, New York, 1969.

[29] N. Lakshminarayanaiah, Electrochemistry, Spec. Period.
      Rep. 2, 203 (1972); 4, 167 (1974); 5, 132 (1975).

[30] N. Lakshminarayanaiah, Membrane Electrodes, Academic
      Press, New York, 1976.

[31] H. Freiser, ed., Ion-Selective Electrodes in Analytical
      Chemistry, Plenum Press, New York, 1978.

[32] G. J. Moody and J. D. R. Thomas, Selective Ion Sensitive
      Electrodes, Merrow, Watford, 1971.

[33] G. J. Moody and J. D. R. Thomas, <u>Selected Ann. Rev. Anal. Sci.</u> <u>3</u>, 59 (1973).

[34] G. J. Moody and J. D. R. Thomas, 'Poly(vinyl chloride) matrix membrane ion-selective electrodes', chapter 4 of ref. [31].

[35] G. J. Moody and J. D. R. Thomas, 'Applications of ion-selective electrodes', chapter 6 of ref. [31].

[36] R. A. Durst, ed., <u>Ion-Selective Electrodes</u>, Natl. Bur. of Standards Spec. Publ. <u>314</u>, Washington, 1969.

[37] R. A. Durst, 'Sources of error in ion-selective electrode potentiometry', chapter 5 of ref. [31].

[38] J. W. Ross, Jr., 'Solid-state and liquid membrane i n selective electrodes', chapter 2 of ref. [36].

[39] J. W. Ross, J. H. Riseman, and J. A. Krueger, <u>Pure Appl. Chem.</u> <u>36</u>, 473 (1973).

[40] G. A. Rechnitz, 'Analytical studies on ion-selective electrodes', chapter 9 of ref. [36].

[41] G. A. Rechnitz, <u>Pure Appl. Chem.</u> <u>36</u>, 457 (1973).

[42] P. L. Bailey, <u>Analysis with Ion-Selective Electrodes</u>, Heyden International Topics in Science, Heyden, London, 1976.

[43] P. C. Meier, D. Ammann, W. E. Morf, and W. Simon, 'Liquid-membrane ion-selective electrodes and their biomedical application', in ref. [15].

[44] B. P. Nicolsky, <u>Zh. Fiz. Khim.</u> <u>10</u>, 495 (1937).

[45] IUPAC Recommendations for Nomenclature of Ion-Selective Electrodes, <u>Pure Appl. Chem.</u> <u>48</u>, 127 (1976).

[46] P. Szepesváry and L. Naszódi, 'Design and evaluation of experiments for determination of the selectivity factor of ion-selective electrodes', in ref. [19].

[47] P. Bergveld, <u>IEEE Trans. Biomed. Eng.</u> <u>17</u>, 70 (1970); <u>19</u>, 342 (1972).

[48] S. D. Moss, J. Janata, and C. C. Johnson, <u>Anal. Chem.</u> <u>47</u>, 2238 (1975).

[49] R. G. Kelley, <u>Electrochim. Acta</u> <u>22</u>, 1 (1977).

[50] R. P. Buck and D. E. Hackleman, Anal. Chem. 49, 2315 (1977).

[51] E. A. Guggenheim, Phil. Mag. 19, 588 (1935).

[52] C. W. Davies, J. Chem. Soc., 2093 (1938).

[53] R. H. Stokes and R. A. Robinson, J. Am. Chem. Soc. 70, 1870 (1948).

[54] D. A. MacInnes, J. Am. Chem. Soc. 41, 1086 (1919).

[55] R. M. Garrels, 'Ion-sensitive electrodes and individual ion activity coefficients', chapter 13 of ref. [36].

[56] R. G. Bates and E. A. Guggenheim, Pure Appl. Chem. 1, 163 (1960).

[57] R. G. Bates and M. Alfenaar, 'Activity standards for ion-selective electrodes', chapter 6 of ref. [36].

[58] R. G. Bates, Pure Appl. Chem. 36, 407 (1973).

[59] R. G. Bates, B. R. Staples, and R. A. Robinson, Anal. Chem. 42, 867 (1970).

PART A

THEORY OF MEMBRANE POTENTIALS AND MEMBRANE TRANSPORT

CHAPTER 2

DESCRIPTION OF THE BASIC MEMBRANE MODEL

According to Schlögl's definition [1, 2], a membrane is a phase, finite in space, which separates two other phases and exhibits individual resistances to the permeation of different species. The membrane phases considered for electrode applications are solids, glasses, liquids, or gasses (e. g. air gap membranes in gas-sensing probes). To efficiently separate the outer phases, which are normally aqueous solutions, such membranes should preferably be nonporous and water-insoluble. Sufficient mechanical stability of liquid or gaseous membranes can be achieved by the introduction of supporting materials. In contrast to biological and artificial bilayer membranes, the active parts of ion-selective electrodes are obviously relatively thick, nearly electroneutral membranes.

The phenomenon of individual resistances mentioned above may be identified with the permeability selectivity of a membrane. The capability to differentiate between various permeating species is perhaps characteristic of any type of membrane, but it is much more pronounced for the membranes used in selective transport systems and electrochemical sensors. As will be shown later, there is a close relationship between the selectivity of ion transport (permeability selectivity) and the potentiometric ion selectivity of a given membrane. Since the permeation of a species involves distribution across the membrane/solution interfaces and translocation across the interior of the barrier, ion selectivity is frequently expressible in terms of ionic extraction parameters and diffusion coefficients or mobilities.

An almost encyclopedic account of membrane phenomena was

27

given by Lakshminarayanaiah [3]. He considered some 16 fundamental relationships between causative agents (driving forces)
and the net flows of matter, charge, and volume. In the present work, considerations will be restricted to the isothermal transport properties of nonporous membranes at constant
pressure. Hence the six causal relationships specified in
Figure 2.1 are of prime interest. The major driving forces
acting on the system are the transmembrane differences in concentrations (or activities), Δc_i, and in electrical potential,
$\Delta\phi$. The pivotal flows J_i indicated in Figure 2.1 are the mass
fluxes of chemical constituents. These quantities are basic to
the understanding of most membrane phenomena. The electric
current passing through a membrane of area A, for example, is
directly defined as $I = AF \sum_i z_i J_i$ and is seen to encompass the
fluxes of charge-carrying species. In fact, all the interrela-
tions shown in Figure 2.1 are connected with or mediated by
the transport of ionic and other solutes. Since transport in
artificial membranes is usually passive[*], the net fluxes
follow simply the direction of the driving forces acting on
the system. Now, selective membranes are capable of efficiently modifying or inhibiting the fluxes of different species.
This implies that such selectivity is expected to be observable for any of the cited membrane phenomena.

For a theoretical description of membrane potentials and
membrane transport, it is convenient to use the three-segmen-
ted membrane model set forth by Sollner [8], Teorell [9], and

---

[*] It should be noted that many features of active transport
are mimicked by carrier membranes. The introduction of electrically charged or neutral ionophores allows to couple the move-
ment of an ion to an energy source other than the chemical or
electrochemical potential gradient of this species. Thus
effects such as "ion pumping" and "uphill transport" can be
realized even for macroscopic model membranes [4 - 7].

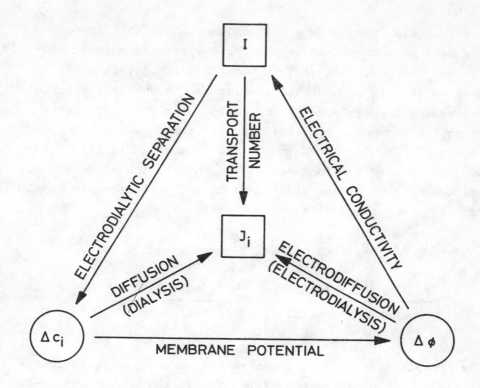

Figure 2.1. Schematic representation of the isothermal transport properties of nonporous membranes at constant pressure. The driving forces of concentration and electrical potential are shown encircled while the flows of matter and current are represented by squares. The relations indicated by arrows are associated with specific membrane phenomena.

Meyer and Sievers [10, 11]. Here the two boundary surfaces and the interior of the membrane are treated separately (see Figure 2.2). Although the location of the interfaces is not clear-cut because the physico-chemical properties, the electrical charge density, and the potential distribution vary

continuously from one phase to another [12], this approach
has many advantages and is appropriate for macroscopic membranes. The treatment of thin membranes ("bilayers", see
Chapter 8) will be modified insofar as the purely geometric
interfaces are replaced by realistic interfacial barriers.
Using the segmented membrane concept, the total membrane
potential E_M, introduced in Eqs. (1.1) and (1.2), is
subdivided into two fundamental components. These are the total boundary potential difference E_B and a membrane-internal
contribution called diffusion potential E_D:

$$E_M \equiv \phi'' - \phi'$$

$$= \underbrace{(\phi(0) - \phi') - (\phi(d) - \phi'')}_{\text{boundary potential } E_B} + \underbrace{(\phi(d) - \phi(0))}_{\substack{\text{diffusion} \\ \text{potential } E_D}} \qquad (2.1)$$

The symbols ϕ in Eq. (2.1) denote the local electrostatic potentials on the membrane surfaces and in the contacting solutions, respectively (see Figure 2.2).

The two potential contributions E_B and E_D can be evaluated
straightforwardly or expressed in simple terms when a series
of fundamental model assumptions is accepted [1 - 3, 7 - 16].
The general theoretical treatment set forth in the next chapters is based on the following assumptions concerning the
membrane.

I. There are no gradients of pressure and temperature across
   the membrane. The only driving forces to be considered
   are differences in concentrations and in electrical potential (see Figure 2.1).

II. A thermodynamic equilibrium exists between the membrane
    and each of the outside solutions at the respective phase
    boundaries.

30

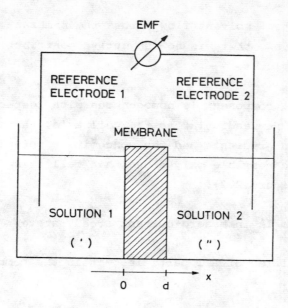

EMF

REFERENCE
ELECTRODE 1

REFERENCE
ELECTRODE 2

MEMBRANE

SOLUTION 1

(')

SOLUTION 2

(")

0   d   x

Figure 2.2. Schematic model of a membrane cell. The membrane
is interposed between two solutions, denoted by (') and (").
Three regions are discerned for the membrane, namely two boundary surfaces located at $x = 0$ and $x = d$, respectively, and
the intermediate bulk of the membrane.

III. The same solvent is used for the electrolyte solutions
      on either side of the membrane.

 IV. Within the membrane, the chemical standard potentials
      of all particles are invariant with space and time.

V. The effect of solvent flow across the membrane is negligible, i. e., there is no convective contribution to the flow of solutes.

VI. Every cell component is homogeneous with respect to a direction perpendicular to the cell axis; therefore, concentration gradients and the concomitant potential differences are possible only along this cell axis (x-coordinate in Figure 2.2).

VII. The system is in a zero-current steady-state.

VIII. The mobilities of all particles within the membrane are invariant.

IXa. The local activity coefficients are the same for all ions in the membrane (MacInnes convention), or

IXb. The individual activity coefficients in the membrane are the same for all cations and for all anions, respectively (Debye-Hückel convention).

While assumptions I - VIII were explicitly stipulated in earlier treatments [13 - 15], assumptions IXa and IXb are introduced here the first time to allow for the nonideal behavior of the membrane phase. They replace the more restrictive ideality assumption, $\gamma_i = 1$, which was imposed in virtually all membrane models presented so far. It will be demonstrated in the following chapters, however, that this extension has no dramatic effect on the form of the derived results, except that concentrations are replaced by activities throughout. Nevertheless, the model assumptions I - IX are consistently observed in the present work, and nearly all the fascinating implications of membrane theory are based on these. Exceptions will be discussed in Chapter 3 (description of non-equilibrium steady-state), Chapter 7 (current different from zero), Chapter 8 (deviations from equilibrium and zero-current conditions; barrier-type profiles of the chemical standard potentials), as well as in Chapters 10 and 13 ("n-type" description of activity coefficients in solid ion-exchangers). The dynamics of equilibration processes in membrane electrode cells are examined in Chapter 14.

REFERENCES

[1] R. Schlögl, 'Zum Materietransport durch Porenmembranen',
     Habilitationsschrift, Georg-August-Universität,
     Göttingen, 1957.

[2] R. Schlögl, Stofftransport durch Membranen, Steinkopff,
     Darmstadt, 1964.

[3] N. Lakshminarayanaiah, Transport Phenomena in Membranes,
     Academic Press, New York, 1969.

[4] E. L. Cussler, AIChE J. 17, 1300 (1971).

[5] E. M. Choy, D. F. Evans, and E. L. Cussler, J. Am. Chem.
     Soc. 96, 7085 (1974).

[6] H.-K. Wipf, 'Komplexbildung von Antibiotika der Valinomy-
     cin- und Nigericin-Gruppe mit Alkalikationen sowie ionen-
     spezifischer Transport in Modellmembranen', Dissertation
     ETH No. 4492, Juris, Zürich, 1970.

[7] D. Erne, W. E. Morf, S. Arvanitis, Z. Cimerman,
     D. Ammann, and W. Simon, Helv. Chim. Acta 62 (4), 994
     (1979).

[8] K. Sollner, Z. Elektrochemie 36, 36 (1930).

[9] T. Teorell, Trans. Faraday Soc. 33, 1053 (1937).

[10] K. H. Meyer and J. F. Sievers, Helv. Chim. Acta 19, 649,
     665 (1936).

[11] K. H. Meyer and J. F. Sievers, Trans. Faraday Soc. 33,
     1073 (1937).

[12] R. P. Buck, in Ion-Selective Electrodes in Analytical
     Chemistry (H. Freiser, ed.), Plenum Press, New York,
     1978, p. 1.

[13] G. Eisenman, in Ion-Selective Electrodes (R. A. Durst,
     ed.), Natl. Bur. of Standards Spec. Publ. 314,
     Washington, 1969, p. 1.

[14] H.-R. Wuhrmann, W. E. Morf, and W. Simon, Helv. Chim.
     Acta 56, 1011 (1973).

[15] W. E. Morf, D. Ammann, E. Pretsch, and W. Simon, Pure
     Appl. Chem. 36, 421 (1973).

[16] R. P. Buck, Crit. Rev. Anal. Chem. 5, 323 (1975).

Chapter 3

The Phase-Boundary Potential (Donnan Potential)

An interface between two liquid or solid phases that each constitutes a partly ionic conductor represents a potential-generating system. The phase-boundary potential or interfacial potential difference arises mainly from the non-uniform distribution of electrically charged species between the two phases; this involves differences in the single-ion chemical standard potentials. In a more general sense, the electrical boundary potentials are related to or exert a controlling influence on the charge transfer reactions at the interfaces. This implies that, generally, chemical and electrical potential contributions must be taken into account in descriptions of ion transport or ion distribution. Both terms are combined in the electrochemical potential $\tilde{\mu}_i$ [1]:

$$\tilde{\mu}_i \equiv \mu_i + z_i F \phi \tag{3.1}$$

The symbol μ_i denotes the chemical potential of a species I (charge z_i), ϕ is the local electrostatic potential, and F is the Faraday equivalent. Using the simplifying assumption I of Chapter 2, the purely chemical contribution is given in the well known form (3.2):

$$\mu_i = \mu_i^o + RT \ln a_i \tag{3.2}$$

where μ_i^o is the chemical standard potential and a_i the activity of a species, R is the gas constant, and T the absolute temperature.

In the following we consider more specifically the interfaces between a membrane positioned at $x = 0 \ldots d$ and two
external contacting solutions (') and (") (see Figure 2.2).
If a thermodynamic equilibrium is a priori assumed to exist
at each interface (assumption II), this has the mathematical
consequence

$$\tilde{\mu}_i' = \tilde{\mu}_i(0) \tag{3.3a}$$

$$\tilde{\mu}_i'' = \tilde{\mu}_i(d) \tag{3.3b}$$

Insertion of the former definitions (3.1) and (3.2) for the
two aqueous solutions and the interposed membrane leads to
the expressions:

$$\mu_{i,aq}^o + RT \ln a_i' + z_i F \phi' = \mu_{i,m}^o + RT \ln a_i(0) + z_i F \phi(0) \tag{3.4a}$$

$$\mu_{i,aq}^o + RT \ln a_i'' + z_i F \phi'' = \mu_{i,m}^o + RT \ln a_i(d) + z_i F \phi(d) \tag{3.4b}$$

The chemical standard potentials $\mu_{i,m}^o$ and $\mu_{i,aq}^o$, referring to
the membrane and the external solutions, are the same for
each interface (assumptions III and IV). They can be included
in a thermodynamic parameter k_i, called the distribution coefficient of species I:

$$k_i \equiv \exp \left[-(\mu_{i,m}^o - \mu_{i,aq}^o)/RT \right] \tag{3.5}$$

Hence the following fundamental relationship holds between
the interfacial potentials and the local distribution of ions
at equilibrium:

$$\phi(0) - \phi' = \frac{RT}{z_i F} \ln \frac{k_i\, a_i'}{a_i(0)} \qquad (3.6a)$$

$$\phi(d) - \phi'' = \frac{RT}{z_i F} \ln \frac{k_i\, a_i''}{a_i(d)} \qquad (3.6b)$$

An alternative procedure for assessing the phase-boundary potentials is based on a kinetic approach. A theory of reaction rates at electrode surfaces was founded by Butler [2], Erdey-Gruz, and Volmer [3], and was later extended to membranes by Eyring's school [4 - 6] (see also Chapter 8). Accordingly, the rate of transfer of exchangeable species across the membrane/solution interfaces (in x-direction, see Figure 2.2) may be formulated as

$$J_i = \vec{k}_i a_i' \exp\left[-\alpha \frac{z_i F}{RT}(\phi(0)-\phi')\right] - \overleftarrow{k}_i a_i(0) \exp\left[(1-\alpha)\,\frac{z_i F}{RT}(\phi(0)-\phi')\right]$$

$$(3.7a)$$

and

$$J_i = \overleftarrow{k}_i a_i(d) \exp\left[(1-\alpha)\frac{z_i F}{RT}(\phi(d)-\phi'')\right] - \vec{k}_i a_i'' \exp\left[-\alpha\frac{z_i F}{RT}(\phi(d)-\phi'')\right]$$

$$(3.7b)$$

J_i is the total flux density (mol cm^{-2} s^{-1}) for the species I, $\vec{k}_i$ and $\overleftarrow{k}_i$ are the rate constants of the two partial transfer reactions at each interface, and $\alpha \approx 0.5$ is the transfer coefficient. At equilibrium, the same rates are established for the forward and backward reactions, and the total mass flux therefore approximates zero:

$$J_i = 0 \qquad (3.8)$$

Hence, Eqs. (3.7a,b) immediately reduce to the following relationships:

$$a_i(0) = k_i a_i' \exp\left[-\frac{z_i F}{RT} (\phi(0)-\phi')\right] \qquad (3.9a)$$

$$a_i(d) = k_i a_i'' \exp\left[-\frac{z_i F}{RT} (\phi(d)-\phi'')\right] \qquad (3.9b)$$

These expressions are equivalent to the former results (3.6a,
b) except that the distribution parameter k_i is here defined
kinetically:

$$k_i \equiv \frac{\vec{k}_i}{\overleftarrow{k}_i} \qquad (3.10)$$

Quasi-thermodynamic formulations according to Eqs. (3.8)
and (3.9a,b) are strictly valid only for the hypothetical
situation in which charge transfer is accomplished by a single
species (ion or electron) and the electrical current density
is ideally zero. In real electrode cells, however, it is conceivable that various charge carriers are simultaneously involved in the interfacial processes. The condition of "zero-
current" then does not guarantee a state of "equilibrium" for
each and every reaction because of the likely differences in
reaction rates. This implies that competing reactions may
proceed irreversibly even in the absence of a net current
flow. In such cases, a mixed potential will be established
which differs from the reversible potential in Eq. (3.9) by
an amount η_i. The steady-state current density j_i of any
species is then determined by the Butler-Volmer equation
(3.11) as a function of the corresponding overpotential η_i
and the exchange-current density $j_{i,o}$:

$$j_i \equiv z_i F J_i = j_{i,o} \exp\left[\alpha \frac{z_i F}{RT} \eta_i\right] - j_{i,o} \exp\left[-(1-\alpha)\frac{z_i F}{RT} \eta_i\right] \qquad (3.11)$$

$$j_{i,o} = z_i F \, (\vec{k}_i a_i')^{1-\alpha} \, (\overleftarrow{k}_i a_i(0))^{\alpha}$$

It may be recognized that reversibility of the i-th reaction, equivalent to a rapid local equilibrium, will generally be fulfilled only if the corresponding exchange-current density is very high. In this case, one gets $\eta_i \cong 0$, which justifies the use of Eqs. (3.6) and (3.9) as excellent approximations.

It is a well known fact underscored by experimental evidence that virtually all ion-selective membrane electrodes exhibit reversible behavior, which means that ions traverse the membrane surfaces in rapid equilibrium. One reason for the applicability of purely thermodynamic principles lies in the selectivity of such electrodes, which ensures that the transfer reactions are dominated by certain ions. In addition, the exchange-current densities of the selected ions are sufficiently high[*] so that a net transfer of material does not disturb the interfacial equilibrium. Consequently, the ion distribution and the potential difference established across each interface are expressible in thermodynamic terms. The following relationship for the total boundary potential of reversible membrane electrodes is in the end justified (see Eqs. (2.1) and (3.6a,b)):

_____

[*] A direct relationship seems to exist between the ion selectivities and the apparent exchange-current densities of different solid-state and liquid membranes [7, 8].

$$B = \frac{RT}{z_i F} \ln \frac{k_i\, a_i'}{a_i(0)} - \frac{RT}{z_i F} \ln \frac{k_i\, a_i''}{a_i(d)}$$

$$= \frac{RT}{z_i F} \ln \frac{a_i'\, a_i(d)}{a_i''\, a_i(0)} \tag{3.12}$$

This important result is valid, in the framework of the present model, for any cationic or anionic species that is capable of freely moving across the phase boundaries. It is, of course, not applicable to ionogenic groups that are permanently trapped in one phase, e. g., the fixed sites in solid ion-exchangers.

Equation (3.12) demonstrates that the boundary potential generated by reversible membrane electrodes is a clear-cut function of the activity ratios of exchangeable ions between the membrane surfaces and the bathing solutions. If the membrane-internal activity of an ion could be fixed by a specific mechanism, i. e.

$$a_i(0) = a_i(d) \tag{3.13}$$

the emf-response would represent a direct measure of the ion activities in the outside solutions:

$$E \cong E_0 + E_B = E_0 + \frac{RT}{z_i F} \ln \frac{a_i'}{a_i''} \tag{3.14}$$

This gives a simplified picture of the function of ion-selective electrodes. In practice, however, an ideal ion specifi-

40

city of sensors according to Eq. (3.14) can scarcely be realized and contributions by more than one ion must be included in any realistic descriptions of the potential terms. A more general result for the boundary potential difference E_B may be obtained from expressions of the type (3.9) by summarizing ions of the same charge z_i and rearranging [9, 10]:

$$E_B = \frac{RT}{z_i F} \ln \frac{\Sigma w_i k_i a_i'}{\Sigma w_i a_i(0)} - \frac{RT}{z_i F} \ln \frac{\Sigma w_i k_i a_i''}{\Sigma w_i a_i(d)} \qquad (3.15)$$

Here the symbol w_i represents any additional weighting factor. Equation (3.15) turns out to be very useful in the treatment of permselective membranes that are permeable for one class of ions. Another fundamental relationship may be derived for fixed-site membranes and two classes of counterions, I^{2+} and J^+ (or I^{2-} and J^-). If the total activity X of anionic (or cationic) sites is assumed to be constant, one can write:

$$2 \Sigma a_i(0) + \Sigma a_j(0) = X \qquad (3.16)$$

A second interrelation between the unknown terms $\Sigma a_i(0)$ and $\Sigma a_j(0)$ follows from Eq. (3.9a), respectively (3.15):

$$\frac{\Sigma a_i(0)}{\Sigma k_i a_i'} = \left(\frac{\Sigma a_j(0)}{\Sigma k_j a_j'}\right)^2 = \exp\left[-\frac{2F}{RT}(\phi(0)-\phi')\right] \qquad (3.17)$$

A quadratic equation results, which yields the following solution for the interfacial potential difference:

$$\phi(0) - \phi' = \frac{RT}{F} \ln \frac{\sqrt{8X \Sigma k_i a_i' + (\Sigma k_j a_j')^2} + \Sigma k_j a_j'}{2X} \qquad (3.18)$$

An analogous relation holds for the other membrane surface. Finally, the total boundary potential E_B assumes the form [10]:

$$E_B = \frac{RT}{F} \ln \frac{\sqrt{8X\ \Sigma k_i a_i' + (\Sigma k_j a_j')^2} + \Sigma k_j a_j'}{\sqrt{8X\ \Sigma k_i a_i'' + (\Sigma k_j a_j'')^2} + \Sigma k_j a_j''} \qquad (3.19)$$

Expressions of this type can be used for assessing the response of ion-selective electrodes to mixed solutions of monovalent and divalent ions [10 - 12].

The boundary potential difference E_B is often called the "Donnan term" [13, 14]. In fact, it was Donnan [15, 16] who first formulated the equilibrium between two electrolyte solutions separated by a (porous) membrane having the capability to completely prevent the permeation of at least one kind of ion (e. g. because its size exceeds the diameter of the pores). The Donnan potential established between the two solutions at equilibrium is of the form (3.14) where the index i refers to any permeating cation or anion. It should be noted that the ion activities at equilibrium are in this case unequal to the initial values since extensive diffusion processes take place across the "indifferent" membrane before an equilibrium is reached. In contrast, diffusion becomes negligible for ideally homogeneous, compact membranes under zero-current conditions. Solid-state membranes, for example, establish an equilibrium with the contacting solution films by dissolution of a small amount of crystalline material. Since Eq. (3.13) is applicable to such systems, the emf-response is again dictated by the Donnan potential, Eq. (3.14), where i denotes an ionic component of the membrane. Liquid membranes, when equilibrated with electrolyte solutions, are subject to salt extraction and ion-exchange reactions at the surfaces. This leads to inhomogeneities in the interior of the membrane which normally give rise to a diffusion potential. An extensive discussion of this fundamental potential contribution is presented in the next chapter.

REFERENCES

[1] E. A. Guggenheim, J. Phys. Chem. 33, 842 (1929); 34, 1540 (1930).

[2] J. A. V. Butler, Trans. Faraday Soc. 19, 729 (1924).

[3] T. Erdey-Gruz and M. Volmer, Z. Phys. Chem. (Leipzig) 150, 203 (1930).

[4] B. J. Zwolinski, H. Eyring, and C. E. Reese, J. Phys. Colloid Chem. 53, 1426 (1949).

[5] F. H. Johnson, H. Eyring, and M. J. Polissar, The Kinetic Basis of Molecular Biology, Wiley, New York, 1954.

[6] R. B. Parlin and H. Eyring, in Ion Transport across Membranes (H. T. Clarke, ed.), Academic Press, New York, 1954.

[7] K. Cammann and G. A. Rechnitz, Anal. Chem. 48, 856 (1976).

[8] K. Cammann, Conference on Ion-Selective Electrodes - Budapest, 1977 (E. Pungor, ed.), Akadémiai Kiadó, Budapest, 1978.

[9] W. E. Morf, Anal. Chem. 49, 810 (1977).

[10] W. E. Morf and W. Simon, in Ion-Selective Electrodes in Analytical Chemistry (H. Freiser, ed.), Plenum, New York, 1978.

[11] W. E. Morf, D. Ammann, E. Pretsch, and W. Simon, Pure Appl. Chem. 36, 421 (1973).

[12] R. P. Buck and J. R. Sandifer, J. Phys. Chem. 77, 2122 (1973).

[13] R. Schlögl, Z. Phys. Chem. (Frankfurt am Main) 1, 305 (1954).

[14] J. Koryta, Anal. Chim. Acta 61, 329 (1972).

[15] F. G. Donnan, Z. Elektrochem. 17, 572 (1911); Chem. Rev. 1, 73 (1924).

[16] F. G. Donnan and E. A. Guggenheim, Z. Phys. Chem. (Leipzig) 162, 346 (1932).

Chapter 4

The Diffusion Potential

While the Donnan equilibrium plays an important role in
the regulation of interfacial processes at the phase boundaries, such equilibria are usually not attained in the interior
of membranes. Thus the free energies of the membrane components
will undergo variations with space and time, although the membrane may still be considered to be a uniform phase in the
sense of assumption IV. The corresponding electrochemical potential gradients give rise to diffusional fluxes of ions
within the membrane. Because the intrinsic rates of diffusion
would not be the same for all species, a membrane-internal
diffusion potential is generated in order to maintain a zero-
current steady-state.

4.1. GENERAL FORMULATION

The flux $\vec{J}_i$ of any species within a phase, e. g. a membrane, can be described most correctly by the generalized
Nernst-Planck equation:

$$\vec{J}_i = c_i \vec{v}_i = \frac{a_i}{\gamma_i} \vec{v}_i \qquad (4.1)$$

where c_i, a_i, and γ_i represent the concentration, the activity, and the activity coefficient of species I, respectively.
The flow velocity $\vec{v}_i$ is composed of the velocity $\vec{v}$ of the
local center of mass and contributions given by the individual

mobility u_i of the species[*)] and the forces acting on them. The total force has been identified with the negative of the local gradient of the electrochemical potential $\tilde{\mu}_i$, hence:

$$\vec{v}_i = -u_i \text{ grad } \tilde{\mu}_i + \vec{v} \qquad (4.2)$$

According to assumptions V and VI of the membrane model proposed in Chapter 2, we may restrict the following considerations to the unidimensional flux J_i in x-direction of the membrane coordinate system (Figure 2.2), giving in place of Eq. (4.1)

$$J_i = -u_i c_i \frac{\partial \tilde{\mu}_i}{\partial x} = -u_i \frac{a_i}{\gamma_i} \frac{\partial \tilde{\mu}_i}{\partial x} \qquad [0 \leqslant x \leqslant d] \qquad (4.3)$$

Equation (4.3) is the usual form of the Nernst-Planck equation. Since the electrochemical potential was approximated by Eqs. (3.1) and (3.2), using assumptions I and IV, one can write

$$J_i = -u_i c_i RT \frac{\partial \ln a_i}{\partial x} - z_i u_i c_i F \frac{\partial \phi}{\partial x} \qquad (4.4)$$

[*)] The absolute mobility u_i (diffusional mobility, electrochemical mobility) has the dimension $cm^2 \ s^{-1} \ (J/mol)^{-1}$. Other mobility parameters referred to in the literature are the electrical mobility $u_i^* = u_i |z_i| F$ (physical mobility, in $cm^2 \ s^{-1} \ v^{-1}$), the equivalent ionic conductivity $\lambda_i = u_i |z_i| F^2$, the diffusion coefficient $D_i = u_i RT$ ($cm^2 \ s^{-1}$), and the friction coefficient $f_i = 1/u_i$.

or

$$J_i \gamma_i = -u_i RT \frac{\partial a_i}{\partial x} - z_i u_i a_i F \frac{\partial \phi}{\partial x} \qquad (4.5)$$

Accordingly, the driving forces acting on each species are given by the gradient of its chemical potential (activity gradient) and that of the electrical potential, whereas a frictional force is included implicitly in the phenomenological coefficient u_i. An alternative useful formulation of Eqs. (4.4) and (4.5) is obtained by summarizing the driving forces:

$$J_i \gamma_i \exp(z_i F\phi/RT) = -u_i RT \frac{\partial}{\partial x} [a_i \exp(z_i F\phi/RT)] \qquad (4.6)$$

It becomes evident that the mass flux J_i in a membrane can be neglected only if (a) the concentration of the species is $c_i = 0$, (b) the mobility is $u_i = 0$, or (c) the species assumes a Boltzmann equilibrium distribution with $a_i \cdot \exp (z_i F\phi/RT) = \mathrm{const}(x)$.

The Nernst-Planck flux equation represents an extension of Fick's diffusion law; the second term in Eqs. (4.4) and (4.5) considers the interaction of charged species with the electrical potential. Another interrelation between the fluxes of ions is given by the electrical current density j, which is equal to zero for membranes in potentiometric measurements (assumptions VI and VII):

$$j = F \Sigma z_i J_i = 0 \qquad (4.7)$$

Combination of Eqs. (4.4) and (4.7) leads to the following universal integral equation which describes the diffusion potential within a membrane or plane diffusion layer of thickness d:

$$E_D = \int_0^d \frac{\partial \phi}{\partial x}\, dx = -\frac{RT}{F} \int_0^d \frac{\Sigma\, z_i u_i c_i \dfrac{\partial \ln a_i}{\partial x}}{\Sigma\, z_i^2 u_i c_i}\, dx \qquad (4.8)$$

The sums in Eqs. (4.7) and (4.8) include all cations M and all anions X within the membrane, except nonpermeating species with $J_i = 0$ (e. g., fixed or stationary ions that are confined to the membrane phase). For all permeating species one can define an electrical transference number t_i (see also Chapter 7):

$$t_i = \frac{z_i^2 u_i c_i}{\Sigma\, z_i^2 u_i c_i} \qquad ; \qquad \Sigma\, t_i = 1 \qquad (4.9)$$

Hence Eq. (4.8) may readily be converted into the classical form [1]:

$$E_D = -\frac{RT}{F} \int_{(0)}^{(d)} \Sigma\, \frac{t_i}{z_i}\, d \ln a_i = -\frac{1}{F} \int_{(0)}^{(d)} \Sigma\, \frac{t_i}{z_i}\, d\mu_i \qquad (4.10)$$

This fundamental relationship is well known in the literature and was deduced many years ago from conventional thermodynamic arguments. A more recent derivation is based on the thermodynamics of irreversible processes [2] (see also [3]). The terms t_i/z_i were replaced by reduced mass transference numbers t_i^r, and possible contributions by electrically neutral species (e. g. water in porous membranes) were also included. The interrelation between the electrical potential and the

flow of solvent is essential to the understanding of electro-
kinetic phenomena, such as electroosmosis or streaming potentials [3]. In the case of compact membranes, these solvent
effects may be neglected, according to assumption V.

The last expressions clearly show that, generally, a
thorough knowledge of the concentration profiles of all species is required for the exact evaluation of diffusion potentials. In practice, however, it is at best the boundary values
at $x = 0$ and $x = d$ that are known or expressible in thermodynamic terms. The integrals in Eqs. (4.8) and (4.10) must
therefore be solved in order to arrive at practical descriptions of E_D. To this end, additional restrictions or approximations must be incorporated, beside those already used
(assumptions I - VII). All of the classical approaches, given
by Nernst [4], Planck [5], Johnson [6], Henderson [7],
Pleijel [8], Goldman [9], Teorell [10, 11], Meyer and
Sievers [12], Schlögl [13, 14] and Helfferich [14, 15], as
well as most of the theories developed by Eisenman et al.
[16 - 20] and others [21 - 27] stipulated ideality of the
diffusion layer or membrane. In addition, the mobilities of
all diffusing species were assumed to be constant (assumption
VIII) and the electroneutrality assumption was usually imposed.

Only a few of the available theories were based on less
idealized membrane models. Single-ion activity coefficients
were accounted for in the case of free diffusion of a single
salt [1] (Section 4.2.b) or in "n-type" descriptions of the
interdiffusion of two counterions in a solid ion-exchanger
[16, 17] (Chapter 13). Mean activity coefficients were considered for systems with 1:1 electrolytes [28]. However,
virtually all the aforementioned definite solutions for the
diffusion potential may be obtained in a more general form.
The following derivation reveals that integration of Eqs.(4.8)

and (4.10) can be carried out straightforwardly by using, instead of the restrictive ideality condition $\gamma_i = 1$, either assumption IXa:

$$\gamma_i(x) = \gamma_{\pm}(x) \text{ (for all ions within the diffusion layer)}, \quad (4.11a)$$

or the even more convincing assumption IXb:

$$\gamma_m(x) = \gamma_+(x) \quad \text{(for all cations M)}$$
$$\hspace{10cm} (4.11b)$$
$$\gamma_x(x) = \gamma_-(x) \quad \text{(for all anions X)}$$

When using assumptions VIII and IXa, Eq. (4.8) may be re-written as

$$E_D = -\frac{RT}{F} \int_0^d \frac{\frac{\partial}{\partial x}\left[\Sigma |z_m| u_m a_m(x) - \Sigma |z_x| u_x a_x(x)\right]}{\Sigma z_m^2 u_m a_m(x) + \Sigma z_x^2 u_x a_x(x)} \, dx \quad (4.12)$$

Solutions based on this formulation will be discussed in Sections 4.2.d-f. Assumptions VIII and IXb, on the other hand, lead to the alternative description[*]:

$$E_D = -\frac{RT}{F} \int_{(0)}^{(d)} \Sigma \frac{t_m}{|z_m|} \cdot d \ln \Sigma |z_m| u_m a_m$$
$$\quad + \frac{RT}{F} \int_{(0)}^{(d)} \Sigma \frac{t_x}{|z_x|} \cdot d \ln \Sigma |z_x| u_x a_x \quad (4.13)$$

---

[*] The validity of Eq. (4.13) is less obvious. For a proof, the integrals may be rearranged into $\int \Sigma (1/\gamma_i) z_i u_i da_i / \Sigma z_i^2 u_i c_i$, which is equivalent to the form given in Eq. (4.8).

Integration of the last expression is accomplished in Sections 4.2.a-c. It will be shown that the complexity of the solution depends on how many valency classes [13] of diffusing ions are present in the system.

4.2. PRACTICAL SOLUTIONS

a) Solution for permselective membranes (one class of ions)

The simplest case is realized for membranes that are permeable for only one class of counterions, e. g. cations of the same charge. Such an ideal permselectivity is met if the concentration or mobility of all other ions (e. g., co-ions) is negligible within the membrane, or if these ions are confined to the membrane phase (dissociated ion-exchange sites). The explicit condition for cation permselectivity reads:

$$\Sigma \ t_m = 1 \quad ; \quad \Sigma \ t_x = 0 \qquad\qquad (4.14a)$$

Since all the particles to be considered are here of the same charge z_m, Eqs. (4.12) and (4.13) reduce immediately to

$$E_D = \frac{RT}{z_m F} \ \ln \ \frac{\Sigma \ u_m a_m(0)}{\Sigma \ u_m a_m(d)} \qquad\qquad (4.15a)$$

An analogous result is obtained for ideal anion-exchangers:

$$\Sigma \ t_x = 1 \quad ; \quad \Sigma \ t_m = 0 \qquad\qquad (4.14b)$$

hence

$$E_D = \frac{RT}{z_x F} \ln \frac{\Sigma \, u_x a_x(0)}{\Sigma \, u_x a_x(d)} \tag{4.15b}$$

Expressions of this type were first suggested by Lark-Horovitz [29] and were later made public by the work of Eisenman's group [16 - 20]. The basically new aspect of Eqs. (4.15a,b) is that nonidealities of the membrane phase are evidently taken into account. A modification of the present results will be discussed in Chapter 13, where another description of activity coefficients is preferred.

b) Extension of Nernst's solution (two ion classes)

In contrast to ideally permselective membranes, free electrolyte solutions contain at least two classes of mobile ions, namely cations and anions, which may be involved in diffusion processes. As early as in 1889, Nernst [4] offered a formulation of diffusion potentials arising within the liquid junction between two dilute solutions of one and the same salt. An extension of this pioneering theory is based on the following assumptions: (1) all cations M within the diffusion layer have the same charge z_m, the same activity coefficient γ_m, and the same mobility u_m, (2) all anions have the same z_x, γ_x, and u_x, and (3) electroneutrality holds. Substitution of these assumptions in Eq. (4.9) yields:

$$\Sigma \, \frac{t_m}{|z_m|} = \frac{u_m}{|z_m| u_m + |z_x| u_x} = \text{const} \tag{4.16a}$$

$$\Sigma \, \frac{t_x}{|z_x|} = \frac{u_x}{|z_m| u_m + |z_x| u_x} = \text{const} \tag{4.16b}$$

51

Integration of Eq. (4.13) is now easily accomplished. Finally one gets an extended version of Nernst's equation:

$$E_D = \frac{u_m}{|z_m|u_m + |z_x|u_x} \frac{RT}{F} \ln \frac{\Sigma a_m(0)}{\Sigma a_m(d)} - \frac{u_x}{|z_m|u_m + |z_x|u_x} \frac{RT}{F} \ln \frac{\Sigma a_x(0)}{\Sigma a_x(d)}$$

$$(4.17)$$

Evidently, this result represents an intermediate form, bridging the gap between the limiting cases realized for permselective membranes. This description may easily be modified to include membrane systems or liquid junctions with ions of differing individual mobilities (see below).

c) Rederivation of Planck's solution (two ion classes)

An important contribution to the theory of diffusion potentials was made in 1890 by Planck [5]. He offered an exact solution to the problem of diffusion layers with more than one electrolyte. The principal drawbacks of the classical Planck theory are the rather voluminous derivation as well as the unwieldy implicit form of the result (see also MacInnes [30]). A new and less circuitous derivation of Planck's relation, and its conversion into a more transparent form, was presented only recently by Morf [25]. Here, the earlier treatment will be revised insofar as activity coefficients are no longer neglected.

The Planck theory of diffusion potentials is based on the following restrictions concerning the diffusion layer: (1) assumption of a steady-state, i. e., $J_i = \text{const}(x)$ for all ions, (2) assumption of electroneutrality, and (3) restriction to one class of mobile cations and one class of mobile anions (i. e., ions of the same charge z_m and z_x and the same activity coefficients γ_m and γ_x, respectively; the original paper [5] restricts itself to monovalent ions in ideal solutions). With these assumptions it is possible to introduce so-called mean mobilities, $\bar{u}_i$, characteristic of each ion class [25]:

$$\bar{u}_m = \frac{\Sigma \, J_m}{\Sigma \, (J_m/u_m)} = \text{const}(x) \qquad (4.18a)$$

$$\bar{u}_x = \frac{\Sigma \, J_x}{\Sigma \, (J_x/u_x)} = \text{const}(x) \qquad (4.18b)$$

Using this pivotal substitution, Eq. (4.7) can be rewritten in the form:

$$j/F = |z_m| \, \bar{u}_m \, \Sigma \, (J_m/u_m) - |z_x| \, \bar{u}_x \, \Sigma \, (J_x/u_x) = 0 \qquad (4.19)$$

After insertion of the fluxes according to Eq. (4.4), one arrives at relationships similar to Eqs. (4.8)-(4.13) where, however, all individual ionic mobilities u_i are replaced by mean mobilities $\bar{u}_i$. Thus the problem of integration becomes basically the same as in the preceding section, where identical mobilities were a priori inserted for all cations and all anions, respectively. In analogy to Eq. (4.16), we can therefore define integral transference numbers, τ_m and τ_x:

$$\Sigma \frac{t_m}{|z_m|} \doteq \frac{\tau_m}{|z_m|} = \frac{\bar{u}_m}{|z_m|\bar{u}_m + |z_x|\bar{u}_x} \qquad (4.20a)$$

$$\Sigma \frac{t_x}{|z_x|} \doteq \frac{\tau_x}{|z_x|} = \frac{\bar{u}_x}{|z_m|\bar{u}_m + |z_x|\bar{u}_x} \qquad (4.20b)$$

and the solution for the diffusion potential is formally
equivalent to Eq. (4.17):

$$E_D = \frac{\bar{u}_m}{|z_m|\bar{u}_m + |z_x|\bar{u}_x} \frac{RT}{F} \ln \frac{\Sigma a_m(0)}{\Sigma a_m(d)} - \frac{\bar{u}_x}{|z_m|\bar{u}_m + |z_x|\bar{u}_x} \frac{RT}{F} \ln \frac{\Sigma a_x(0)}{\Sigma a_x(d)}$$

(4.21)

This result corresponds to Planck's exact solution of the
problem (originally for $|z_m| = |z_x| = 1$ and $\gamma_m = \gamma_x = 1$) but
is obtained here in a new and more practical form that impressively shows the relationship with other approaches. The
mean mobilities are found, from Eqs. (4.6) and (4.18), to be
given as

$$\bar{u}_i = \frac{\Sigma\, u_i a_i(d) \cdot e^{z_i FE_D/RT} - \Sigma\, u_i a_i(0)}{\Sigma\, a_i(d) \cdot e^{z_i FE_D/RT} - \Sigma\, a_i(0)}$$

(4.22)

Exceptions aside, these mobility parameters depend on E_D.
Thus, the Planck solution does not generally yield the diffusion potential explicitly but has to be evaluated for E_D by
iterative methods [25].

A significant reduction of Planck's result is obtained
for liquid junctions formed by equimolar electrolyte solutions.
For a constant total ion concentration

$$\Sigma\, c_m(0) = \Sigma\, c_m(d) \quad ; \quad \Sigma\, c_x(0) = \Sigma\, c_x(d)$$

(4.23)

it holds as an excellent approximation that

$$\gamma_m = \gamma_+ = \text{const}(x) \quad ; \quad \gamma_x = \gamma_- = \text{const}(x) \qquad (4.24)$$

In this case the logarithmic terms in Eq. (4.21) become zero. As a consequence, it must also hold that $|z_m|\bar{u}_m + |z_x|\bar{u}_x = 0$. Hence, using Eqs. (4.22)-(4.24), we may derive an explicit solution for this special type of liquid-junction potential which, as a matter of fact, corresponds to Goldman's equation [9] for the diffusion potential of biological membranes[*] (see also Chapter 7 and [3, 25, 31]):

$$E_D = \frac{RT}{F} \ln \frac{\Sigma\, u_m c_m(0) + \Sigma\, u_x c_x(d)}{\Sigma\, u_m c_m(d) + \Sigma\, u_x c_x(0)} \qquad (4.25)$$

for $|z_m| = |z_x| = 1$.

If this relation is applied to the simplest case, namely a liquid junction formed by two single electrolytes of the same concentration and with either the same anion or cation, it further reduces to the well-known formula of Lewis and Sargent [32]:

$$E_D = \pm \frac{RT}{F} \ln \frac{\Lambda(0)}{\Lambda(d)} \qquad (4.26)$$

Here, the only parameters to be inserted are the equivalent conductivities of the two electrolyte solutions, $\Lambda = \Lambda_m + \Lambda_x$, with

---

[*] Goldman's equation was originally derived from a different model, based on the assumption of a constant electric field. This condition is apparently met for the present case.

$$\Lambda_i = |z_i| u_i F^2 \qquad\qquad (4.27)$$

If the two solutions are identical in every respect, one gets the trivial result:

$$E_D = 0, \text{ for } c_i(0) = c_i(d) \qquad\qquad (4.28)$$

This is consistent with the obvious fact that net diffusion within ideally homogeneous systems remains negligible.

d) Teorell's_solution_and_its_extension_(two_classes_of_mobile
   ions_and_fixed_sites)

The theory of permselective membranes which are easily permeable for ions with a certain charge (counterions) but poorly permeable for the oppositely charged ions (coions) was developed mainly by Teorell [10, 11]. He considered electroneutral systems containing a constant number of fixed charged sites. Ideal permselectivity, as stipulated in Section 4.2.a, can then be rationalized simply by an exclusive uptake of counterions and rejection of coions by the "charged" membrane. For general situations, however, the permeation of more than one class of ions must be taken into account. Hence Eq. (4.15a,b) has to be replaced by a more complete description, based on Eq. (4.12) and the electroneutrality condition (4.29)

$$\Sigma z_i a_i(x) = -\omega X = \text{const}(x) \qquad\qquad (4.29)$$

where ω is the charge and X the mean activity of fixed sites in the membrane.

The original theory of Teorell [11] was restricted to membranes interposed between solutions of the same 1:1 electrolyte. Since the cation M^+ and the anion X^- were the only permeating species to be considered, Teorell's result for the diffusion potential has the form:

$$E_D = \frac{u_m - u_x}{u_m + u_x} \frac{RT}{F} \ln \frac{u_m a_m(0) + u_x a_x(0)}{u_m a_m(d) + u_x a_x(d)} \tag{4.30}$$

A generalized version of Teorell's solution may be applied to any system with two ion classes for which Eq. (4.29) is obeyed. After introducing mean mobilities $\bar{u}_k$ and $\bar{u}_l$ for all ions of class k and l, respectively, integration of Eq. (4.12) leads to:

$$E_D = \frac{\bar{u}_k - \bar{u}_l}{z_k \bar{u}_k - z_l \bar{u}_l} \frac{RT}{F} \ln \frac{\Sigma \, z_k^2 \bar{u}_k a_k(0) + \Sigma \, z_l^2 \bar{u}_l a_l(0)}{\Sigma \, z_k^2 \bar{u}_k a_k(d) + \Sigma \, z_l^2 \bar{u}_l a_l(d)} \tag{4.31}$$

or, with $q = (\bar{u}_k - \bar{u}_l)/(z_k \bar{u}_k - z_l \bar{u}_l)$

$$E_D = q \frac{RT}{F} \ln \frac{\Sigma \, z_k^2 a_k(0) + \Sigma \, z_l^2 a_l(0) + q \, z_k z_l \omega X}{\Sigma \, z_k^2 a_k(d) + \Sigma \, z_l^2 a_l(d) + q \, z_k z_l \omega X} \tag{4.32}$$

Whereas Teorell's case corresponds to the situation with k = m (cations) and l = x (anions), another important case covered by Eq. (4.31) or (4.32) is that of a permselective ion-exchange membrane with monovalent and divalent counterions. For simplicity, it is often reasonable to assume that

57

$\bar{u}_k = \bar{u}_l = u$ and hence

$$q \cong 0 \quad \text{and} \quad E_D \cong 0 \tag{4.33}$$

The membrane potential can then be approximated by the interfacial potential contribution alone, Eq. (3.19) [23, 26, 33].

e) <u>Schlögl's solution (N classes of mobile ions and fixed sites)</u>

General integrated solutions for the Nernst-Planck flux equations and the diffusion potential at steady-state were worked out by Pleijel [8] and by Schlögl [13]. These derivations represent mathematical masterpieces but suffer from their complicated, perplexing form. Therefore, a simplified derivation of Schlögl's diffusion potential has been attempted. To this end, flux equations of the type (4.5) are multiplied with a constant factor p_i/u_i and added subsequently:

$$RT \frac{d}{dx} \Sigma \, p_i a_i + \Sigma \, p_i z_i a_i \cdot F \frac{d\phi}{dx} = - \Sigma \, \frac{p_i J_i}{u_i} \cdot \gamma_{\pm} \tag{4.34}$$

Recalling the electroneutrality condition, Eq. (4.29), we can write:

$$RT \frac{d}{dx} \Sigma \, z_i a_i = 0 \tag{4.35}$$

Addition of the two expressions leads to

$$RT \frac{d}{dx} \Sigma \, (p_i + z_i) a_i + \Sigma \, p_i z_i a_i \cdot F \frac{d\phi}{dx} = - \Sigma \, \frac{p_i J_i}{u_i} \cdot \gamma_{\pm} \tag{4.36}$$

58

Now, the constant terms p_i are chosen in such a way that the following relations be satisfied:

$$\Sigma \ (p_i J_i/u_i) = 0 \qquad\qquad (4.37)$$

and

$$p_i + z_i = p_i z_i \cdot q \qquad\qquad (4.38)$$

where q is a new constant. Hence Eq. (4.36) reduces to

$$q \cdot RT \ \frac{d}{dx} \ \Sigma \ p_i z_i a_i + \Sigma \ p_i z_i a_i \cdot F \ \frac{d\phi}{dx} = 0 \qquad\qquad (4.36a)$$

which can easily be integrated to yield Schlögl's result:

$$E_D = q \ \frac{RT}{F} \ \ln \ \frac{\Sigma \dfrac{z_i a_i(0)}{q - 1/z_i}}{\Sigma \dfrac{z_i a_i(d)}{q - 1/z_i}} \qquad\qquad (4.39)$$

If the fluxes J_i of all ions are known or expressible in terms of readily accessible quantities (see Chapter 7), Eq. (4.37) allows determination of the parameter q:

$$\Sigma \ \frac{J_i/u_i}{q - 1/z_i} = 0 \qquad\qquad (4.37a)$$

For systems with N classes of diffusing ions, there exist N-1 real or complex values of q that are compatible with Eqs.(4.37a) and (4.39). The limiting case for two ion classes is, of course, given by the generalized Teorell equation (4.32), which can be verified by using Eqs. (4.19), (4.37a), and (4.39).

f) Henderson's approximation

An ingenious method for the calculation of diffusion potentials (liquid-junction potentials) was introduced in 1907 by Henderson [7]. The Henderson approximation is probably the most frequently used in practice, although the exact Planck and Schlögl solutions are more convincing from the theoretical standpoint. Henderson's general approach relies upon the arbitrary assumption of linear concentration profiles for all ions within the diffusion layer. In the framework of the present model, this corresponds to the approximation:

$$\frac{\partial}{\partial x} a_i(x) \cong \frac{a_i(d) - a_i(0)}{d} = \frac{\Delta a_i}{d} \tag{4.40}$$

Insertion into Eq. (4.12) leads to the well-known formula:

$$E_D \cong \frac{\Sigma |z_m| u_m \Delta a_m - \Sigma |z_x| u_x \Delta a_x}{\Sigma z_m^2 u_m \Delta a_m + \Sigma z_x^2 u_x \Delta a_x} \times$$

$$\times \frac{RT}{F} \ln \frac{\Sigma z_m^2 u_m a_m(0) + \Sigma z_x^2 u_x a_x(0)}{\Sigma z_m^2 u_m a_m(d) + \Sigma z_x^2 u_x a_x(d)} \tag{4.41}$$

Evidently, the Henderson equation constitutes an explicit general description, incorporating as special cases many of the solutions discussed before. This allows an easy and rather close characterization of the diffusion potential in terms of boundary concentrations and mobilities of diffusing ions. Calculations of liquid-junction potentials according to Henderson's approach and comparison with other methods are presented in Chapter 5.

REFERENCES

[1] E. A. Guggenheim, Thermodynamics, North Holland Publ. Co., Amsterdam, 1950.

[2] A. J. Staverman, Trans. Faraday Soc. 48, 176 (1952).

[3] N. Lakshminarayanaiah, Transport Phenomena in Membranes, Academic Press, New York, 1969.

[4] W. Nernst, Z. Phys. Chem. 4, 129 (1889).

[5] M. Planck, Ann. Phys. 39, 161 (1890); 40, 561 (1890).

[6] K. R. Johnson, Ann. Phys. (Leipzig) 14, 995 (1904).

[7] P. Henderson, Z. Phys. Chem. 59, 118 (1907); 63, 325 (1908).

[8] H. Pleijel, Z. Phys. Chem. 72, 1 (1910).

[9] D. E. Goldman, J. Gen. Physiol. 27, 37 (1943).

[10] T. Teorell, Z. Elektrochem. 55, 460 (1951); Prog. Biophys. Biophys. Chem. 3, 305 (1953).

[11] T. Teorell, Proc. Nat. Acad. Sci. U.S.A. 21, 152 (1935); Proc. Soc. Exp. Biol. Med. 33, 282 (1935).

[12] K. H. Meyer and J. F. Sievers, Helv. Chim. Acta 19, 649, 665, 987 (1936).

[13] R. Schlögl, Z. Phys. Chem. (Frankfurt am Main) 1, 305 (1954).

[14] F. Helfferich and R. Schlögl, Disc. Faraday Soc. 21, 133 (1956).

[15] F. Helfferich, Ion Exchange, McGraw-Hill, New York, 1962.

[16] F. Conti and G. Eisenman, Biophys. J. 5, 247, 511 (1965).

[17] G. Eisenman, ed., Glass Electrodes for Hydrogen and Other Cations, M. Dekker, New York, 1967.

[18] J. P. Sandblom, G. Eisenman, and J. L. Walker, Jr., J. Phys. Chem. 71, 3862 (1967).

[19] S. M. Ciani, G. Eisenman, and G. Szabo, J. Membrane Biol. 1, 1 (1969).

[20] G. Eisenman, in Ion-Selective Electrodes (R. A. Durst, ed.), National Bureau of Standards, Spec. Publ. 314, Washington, 1969.

[21] W. F. Pickard, Math. Biosciences 13, 113 (1972).

[22] O. K. Stephanova and M. M. Shults, Vestnik Leningrad. Univ. 1972, No. 4, 80.

[23] R. P. Buck and J. R. Sandifer, J. Phys. Chem. 77, 2122 (1973).

[24] H.-R. Wuhrmann, W. E. Morf, and W. Simon, Helv. Chim. Acta 56, 1011 (1973).

[25] W. E. Morf, Anal. Chem. 49, 810 (1977).

[26] W. E. Morf and W. Simon, in Ion-Selective Electrodes in Analytical Chemistry (H. Freiser, ed.), Plenum, New York, 1978.

[27] R. Gaboriaud, J. Chim. Phys. 72, 347 (1975).

[28] H. Linderholm, Acta Physiol. Scand. Suppl. 27, 97 (1952).

[29] K. Lark-Horovitz, Naturwiss. 19, 397 (1931).

[30] D. A. MacInnes, The Principles of Electrochemistry, Dover, New York, 1961.

[31] R. P. Buck, Crit. Rev. Anal. Chem. 5, 323 (1975).

[32] G. N. Lewis and L. W. Sargent, J. Am. Chem. Soc. 31, 363 (1909).

[33] W. E. Morf, D. Ammann, E. Pretsch, and W. Simon, Pure Appl. Chem. 36, 421 (1973).

Chapter 5

Calculation of Liquid-Junction Potentials

The classical reference electrodes used in potentiometric measuring cells are the mercury/calomel electrode and the silver/silver chloride electrode, each in contact with an aqueous solution of fixed chloride activity (e. g., a saturated KCl solution). Frequently an additional electrolyte solution, called salt bridge, is interposed between the external reference electrolyte and the sample in order to inhibit interactions of these two solutions. The liquid junction between the salt bridge and the sample is usually established within a porous diaphragm or defined by a streaming boundary.

Since the liquid junction represents an interface where one electrolyte diffuses into the other, it is the origin of an electrical potential difference contributing to the emf of the cell. The liquid-junction potential E_J, for evident reasons, corresponds to a diffusion potential:

$$E_J \triangleq E_D \tag{5.1}$$

It can therefore be characterized straightforwardly by the relations given in Chapter 4 when the "membrane" or "diffusion layer" is identified here with the mixture region formed between the salt bridge (at $x = 0$) and the sample solution (at $x = d$). In the event, the classical approaches by Planck [1, 2] and Henderson [3], treated in Sections 4.2.c and f, were originally devoted to the calculation of liquid-junction potentials[*]), and were only later applied to diffusional membranes.

Although it is often assumed that the liquid-junction potential arising in ion-selective electrode cells is independent of the composition of the sample solution, or can even be set equal to zero, this contribution to the emf does in fact change considerably, as an evaluation of the Henderson equation (4.41) or the Planck equations (4.21) and (4.22) demonstrates. A change of the liquid-junction potential has, however, the same effect as a change in the activity measured by the ion-selective electrode. Thus, estimations of E_J are of practical interest in view of a systematical elimination or numerical correction of erroneous responses (see Figure 5.1). It is clear that the calculation of E_J hinges on numerical values for the ion mobilities u_i. As the single-ion or mean activity coefficients were explicitly accounted for in all expressions for E_D given in Chapter 4, these mobilities must refer to infinitely diluted aqueous solutions[**]). They can be calculated, according to Eq. (4.27), from the corresponding equivalent ionic conductivities tabulated in handbooks. The mobility data used in this work are compiled in Table 5.1.

---

[*]) A third method, differing from Planck's or Henderson's procedures, was suggested by Pickard [4]. His derivation of the diffusion potential relies on the assumption that the transference numbers (and not the concentrations) of all ions vary linearly with x.

[**]) A different procedure (see also [6]) for calculating E_J values uses concentrations instead of activities but takes into account the nonideality of the aqueous solutions by inserting the individual mobilities or conductivities observed for these ions at the given concentration. According to the Debye-Hückel theory, both approaches are equivalent.

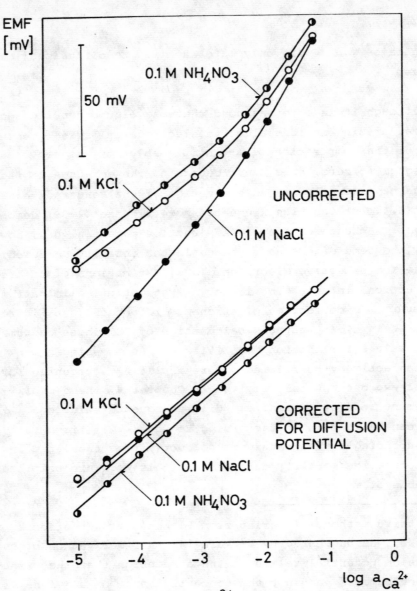

Figure 5.1. The response of a Ca^{2+}-selective electrode (see Chapter 12, Table 12.4) towards CaCl$_2$ solutions for three bridge electrolytes purposely chosen to be non-ideal: 0.1M NH$_4$NO$_3$, KCl, and NaCl. The raw emf values are plotted at the top. After correction for E$_J$ using Eq. (4.41), the points were replotted (bottom) and linear regressions were carried out over the activity ranges indicated [5]. For clarity, the data for raw and corrected emf were pulled apart.

Table 5.1. Mobility parameters of ions in aqueous solution at 25°C

Ion	Equivalent ionic conductivity at infinite dilution [7] Λ_i [$\Omega^{-1}cm^2$ equiv.$^{-1}$]	Absolute mobility, according to Eq. (4.27) $u_i \cdot 10^9$ [$cm^2 s^{-1} J^{-1} mol$]
H^+	350	37.6
Li^+	39.5	4.24
Na^+	50.9	5.47
K^+	74.5	8.00
NH_4^+	74.5	8.00
Ag^+	63.5	6.82
Mg^{2+}	54	2.90
Ca^{2+}	60	3.22
Ba^{2+}	65	3.49
OH^-	192	20.6
Cl^-	75.5	8.11
I^-	76.0	8.16
NO_3^-	70.6	7.58
$OAc^{-*)}$	40.8	4.38
SO_4^{2-}	79	4.24

*) Abbreviation for acetate anion.

To provide a test for the applicability of the calculation procedures to experimental situations (see also Figure 5.1), liquid-junction potentials measured by MacInnes and Yeh [8] are correlated in Table 5.2 with computed E_J values. The agreement between theory and experiment is found to be surprisingly good, the mean absolute deviation between the two sets of data being only around 0.5 mV. The calculations in Table 5.2 were based on the simplified formula of Lewis and Sargent [9], Eq. (4.26), which is valid for the discussed examples of liquid junctions formed by two equimolar chloride solutions. For more general cases, the complete formalism of the Planck theory or the Henderson approximation must be used for the evaluation of liquid-junction potentials. An iterative calculation method has been devised [2] to facilitate the rigorous application of the Planck relation. On the other hand, the more practical Henderson equation as a rule leads to similar or even the same values for E_J. Some representative results in Table 5.3 clearly demonstrate that there is a satisfactory agreement between the two modes of calculation. Deviations exceed 10% only in the case where HCl is used as sample.

A more detailed inspection of both theoretical expressions and numerical values reveals that there are two possibilities for a minimization of the liquid-junction potential in ion-selective electrode cells.

a) A solution which perfectly corresponds to the sample solution (i. e. $c_i(0) \approx c_i(d)$) is used as salt bridge.

b) A so-called "equitransferent solution" [10] (the exact condition is $\Sigma \, |z_m| \, u_m c_m(0) \cong \Sigma \, |z_x| \, u_x c_x(0)$ or $u_m \cong u_x$) has to be used as salt bridge, its concentration being much higher than that of the sample solution. An idealized example is given in Table 5.3 and is based on the fact that $u_K = 0.8 \, u_{Cl} + 0.2 \, u_{NO_3}$ (see Table 5.1).

Table 5.2. Comparison between calculated and measured liquid-junction potentials ($25^{\circ}C$)

Solutions		E_J in mV	
at x = 0	at x = d	calculated from Eq.(4.26)	observed MacInnes & Yeh [8]
0.01M HCl	0.01M KCl	26.79	25.73
0.01M HCl	0.01M NaCl	31.18	31.16
0.01M HCl	0.01M LiCl	33.62	33.75
0.01M HCl	0.01M NH_4Cl	26.79	27.02
0.01M KCl	0.01M NaCl	4.39	5.65
0.01M KCl	0.01M LiCl	6.83	8.20
0.01M KCl	0.01M NH_4Cl	0.00	1.30
0.01M NaCl	0.01M LiCl	2.44	2.63
0.01M NaCl	0.01M NH_4Cl	-4.39	-4.26
0.01M LiCl	0.01M NH_4Cl	-6.83	-6.89

Although procedure a) is often recommended for practical applications it turns out to be deceptive. Indeed, the liquid-junction potential may approximate zero for ideal situations but it generally remains very sensitive to changes in the concentration or ionic composition of the sample, as is shown in Table 5.4. On the other hand, procedure b) may lead to values different from zero but it guarantees the highest-possible constancy of the liquid-junction potential. The commonly used salt bridge solution is saturated or 3M KCl, which leads to favorable results in that uncertainties in the reference electrode potential can be reduced drastically (Tables 5.3 - 5.5). Other "equitransferent solutions" suited for salt bridges are KNO_3, NH_4NO_3, and LiOAc (see Figure 5.2);

Table 5.3. Liquid-junction potential values at 25°C, calculated from the Planck theory and the Henderson approach [2]. These diffusion potentials are generated in the aqueous diffusion layer ($0 \leqslant x \leqslant d$) between the sample solution (at $x=d$) and the electrolyte of the reference electrode (at $x=0$). A mixed solution of KCl and KNO_3 (4:1) is used as salt bridge or reference electrode solution

Sample solution	Relative activity of sample solution, $\Sigma a_i(d)/\Sigma a_i(0)$	E_J in mV according to Planck Eqs.(4.21), (4.22), and (4.25)*)		E_J in mV according to Henderson Eq. (4.41)
KCl	10^{-4}	0.00	(1)	0.00
	10^{-3}	0.00	(1)	0.00
	10^{-2}	0.01	(1)	0.01
	10^{-1}	0.05	(1)	0.05
	1	0.18	–	0.18
	10^{1}	0.45	(1)	0.45
	10^{2}	0.82	(1)	0.82
	10^{3}	1.21	(1)	1.21
NaCl	10^{-4}	0.00	(1)	0.00
	10^{-3}	0.03	(1)	0.03
	10^{-2}	0.20	(1)	0.20
	10^{-1}	1.11	(2)	1.14
	1	4.60	–	4.60
	10^{1}	12.45	(3)	12.11
	10^{2}	23.13	(2)	22.45
	10^{3}	34.52	(2)	33.72
HCl	10^{-4}	−0.04	(1)	−0.04
	10^{-3}	−0.32	(1)	−0.28
	10^{-2}	−2.07	(3)	−1.73
	10^{-1}	−9.40	(10)	−8.31
	1	−26.73	–	−26.77
	10^{1}	−52.84	(27)	−57.58
	10^{2}	−84.32	(7)	−94.06
	10^{3}	−118.81	(5)	−131.95
NaOH	10^{-4}	0.02	(1)	0.02
	10^{-3}	0.17	(1)	0.16
	10^{-2}	1.11	(2)	1.02
	10^{-1}	5.66	(4)	5.27
	1	19.35	–	18.85
	10^{1}	43.54	(10)	44.33
	10^{2}	73.24	(5)	76.42
	10^{3}	105.24	(4)	110.35

*) The value in brackets gives the number of iteration steps that are needed to come within ±0.01 mV of the final result [2].

Table 5.4. Liquid-junction potentials generated between two solutions of the same salt (25°C). According to Eq.(4.17), E_J varies linearly with increasing logarithm of the activity on the sample side at $x = d$

"Equitransferent solutions"	$\Delta E_J/\Delta \log a$ in mV	Other solutions	$\Delta E_J/\Delta \log a$ in mV
KCl	0.40	NaCl	11.5
KNO_3		$CaCl_2$	19.9
NH_4NO_3	-1.59	HCl	-38.2
KI	0.59	NaOH	34.3
NaF	1.48	$NaNO_3$	9.6
LiOAc	0.96	NaOAc	-6.5
Li_2SO_4 *)	0.00	K_2SO_4 *)	-13.5

*) K_2SO_4 is an example of an equitransferent electrolyte (with $\Lambda_m \cong \Lambda_x$ or $t_m \cong t_x$) which is not suited for salt-bridge applications. In contrast, Li_2SO_4 meets the basic requirement $u_m = u_x$ very exactly. Thus, the term "equitransferent" can be misleading when it is used for characterizing attractive salt bridge solutions.

the latter are preferable when a contamination of the sample by potassium or chloride ions is to be avoided.

In some cases, emf measurements can also be carried out on cells without liquid junction, i. e. without transference of ions. This is accomplished by replacing the junction-type reference electrode system by a second ion-selective electrode that responds specifically to some ion in the sample solution other than the primary ion. Thereby, the activity of this species is introduced as a reference level. Such cells have

Table 5.5. Liquid-junction potentials for different sample solutions and KCl bridge electrolytes at $25^{\circ}C$.
Values were calculated according to the Henderson (H) and Planck (P) methods

Sample activity [M]	E_J in mV, for a salt bridge consisting of					
	3.5M KCl[a]		1M KCl[a]		0.1M KCl[a]	
	H	P	H	P	H	P
KCl 1	-0.2	-0.2	0.0	0.0	0.4	0.4
0.1	-0.6	-0.6	-0.4	-0.4	0.0	0.0
0.01	-1.0	-1.0	-0.8	-0.8	-0.4	-0.4
0.001	-1.4	-1.4	-1.2	-1.2	-0.8	-0.8
NaCl 1	1.9	1.9	4.4	4.4	12.0	12.4
0.1	-0.2	-0.2	0.7	0.7	4.4	4.4
0.01	-1.0	-1.0	-0.6	-0.6	0.7	0.7
0.001	-1.4	-1.4	-1.2	-1.2	-0.6	-0.6
CaCl$_2$ 0.5	3.7	3.5[b]	8.0	8.0[b]	21.3	22.4[b]
0.05	0.2	0.2[b]	1.6	1.5[b]	8.0	8.0[b]
0.005	-0.9	-0.8[b]	-0.5	-0.4[b]	1.6	1.5[b]
0.0005	-1.4	-1.3[b]	-1.2	-1.1[b]	-0.5	-0.4[b]
HCl 1	-15.1	-16.2	-26.8	-26.8	-57.5	-52.8
0.1	-4.2	-4.9	-8.6	-9.7	-26.8	-26.8
0.01	-1.5	-1.8	-2.4	-2.8	-8.6	-9.7
0.001	-1.3	-1.5	-1.3	-1.5	-2.4	-2.8
NaOH 1	9.6	10.2	18.6	19.1	44.2	43.4
0.1	1.7	1.8	4.9	5.2	18.6	19.1
0.01	-0.6	-0.6	0.3	0.3	4.9	5.2
0.001	-1.3	-1.4	-1.0	-1.0	0.3	0.3
H$_2$O (pH 7)	-2.8	-3.0	-2.6	-2.8	-2.2	-2.4

[a] The same E_J values are expected for NH_4Cl salt bridges of the corresponding activities; Table 5.1 shows that $u_K = u_{NH_4}$.

[b] For systems with more than two ion classes, the extended theory of Schlögl [11] has to be consulted (see Sections 4.2.e and 7.4).

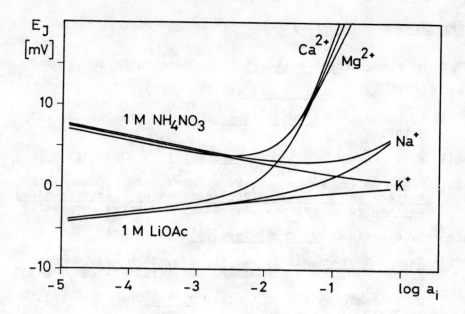

Figure 5.2. The liquid-junction potential E_J as a function of the single-ion activity of the sample cation (Na^+, K^+, Mg^{2+}, Ca^{2+}; chloride salts) for 1M ammonium nitrate and 1M lithium acetate bridge electrolytes. Calculations according to Eq. (4.41) using single-ion activities [5].

been used, for example, for the monitoring of the Cl^-/F^- activity ratio on a fluorocarbon plant [12], for the measurement of the Na^+/K^+ ratio in biomedical applications [13], or for measurements in complex biological mixtures by spiking the sample with a reference ion [14]. The same principle of reference electrodes is generally applied in gas-sensing electrodes (Chapter 15) for which the liquid-junction problem does not exist at all.

REFERENCES

[1] M. Planck, Ann. Phys. 39, 161 (1890); 40, 561 (1890).

[2] W. E. Morf, Anal. Chem. 49, 810 (1977).

[3] P. Henderson, Z. Phys. Chem. 59, 118 (1907); 63, 325 (1908).

[4] W. F. Pickard, Math. Biosciences 13, 113 (1972).

[5] P. C. Meier, D. Ammann, W. E. Morf, and W. Simon, in Medical and Biological Applications of Electrochemical Devices (J. Koryta, ed.), Wiley, New York, in press.

[6] R. Gaboriaud, J. Chim. Phys. 72, 347 (1975).

[7] Handbook of Chemistry and Physics, 56th ed., Chemical Rubber Publ. Co., Cleveland, Ohio, 1975-1976, p. D-153.

[8] D. A. MacInnes and Y. L. Yeh, J. Am. Chem. Soc. 43, 2563 (1921).

[9] G. N. Lewis and L. W. Sargent, J. Am. Chem. Soc. 31, 363 (1909).

[10] K. V. Grove-Rasmussen, Acta Chem. Scand. 2, 937 (1948).

[11] R. Schlögl, Z. Phys. Chem. (Frankfurt am Main) 1, 305 (1954).

[12] P. O. Kane, 'Some on-line applications of ion-selective electrodes', International Reference and Ion-Selective Electrodes Conference, Newcastle upon Tyne, Jan. 7-9, 1976.

[13] M. Oehme, Dissertation ETH No. 5953, Juris, Zürich, 1977.

[14] M. S. Mohan and R. G. Bates, Clin. Chem. 21, 864 (1975).

CHAPTER 6

SOLUTIONS FOR THE MEMBRANE POTENTIAL

A critical examination of the theoretical results in Chapters 3 and 4 reveals that these descriptions will be of practical significance only if all unknown activities referring to the membrane phase can be successfully replaced by outside values. Such evaluations often bear serious algebraic problems or otherwise conflicting situations, however. For example, the following relationship holds for the diffusion potential of ideally ion-specific membranes, which are permeable for the primary ions I only (see Section 4.2.a):

$$E_D = \frac{RT}{z_i F} \ln \frac{a_i(0)}{a_i(d)} \qquad (6.1)$$

If it is now attempted to eliminate the membrane activities $a_i(0)$ and $a_i(d)$ with the help of Eqs. (3.9.a,b), one immediately finds that another unknown term is thereby introduced, namely the boundary potential E_B:

$$E_D = \frac{RT}{z_i F} \ln \frac{a_i'}{a_i''} - E_B \qquad (6.2)$$

Evidently, the diffusion potential and the boundary potential of ion-selective membranes cannot be assessed independently, unless additional specifications are given (e. g., $z_i a_i(0) = z_i a_i(d) = -\omega X$, for fixed-site membranes).

In contrast, the formulation of the total membrane potential turns out to be less problematic. A respectable number of practical solutions for E_M is accessible by simply combining appropriate expressions for E_B and E_D. For that reason, a detailed catalogue of alternative descriptions was offered in the preceding chapters. The above example of ion-specific membranes directly leads to the following elementary result:

$$E_M = E_B + E_D = \frac{RT}{z_i F} \ln \frac{a_i'}{a_i''} \tag{6.3}$$

It was already shown in Chapter 1 that a Nernstian emf-response function results in this case (for $a_i'' = $ const):

$$E = E_i^o + \frac{RT}{z_i F} \ln a_i' \tag{6.4}$$

While the last equations have been explicitly derived for ideally permselective membranes, where I is the only permeating species, analogous results are obtained for ideally homogeneous membranes. Here the diffusion potential approximates zero (see Eq. (4.28)) and the total membrane potential is dictated by the Donnan term (3.14). Hence, the response behavior of solid-state membranes, formed from a poorly soluble salt MX, will be characterized by the following relationships:

$$E = E_m^o + \frac{RT}{z_m F} \ln a_m' \tag{6.5a}$$

respectively

$$E = E_x^o + \frac{RT}{z_x F} \ln a_x' \tag{6.5b}$$

In this case, the primary ions sensed by the electrode cell
are identical to the constituents of the membrane material.

An ideally ion-specific behavior of membrane electrodes
according to Eq. (6.4) must usually be considered an imaginary
standard because a perfect exclusion of interfering species
cannot be realized in practice. There are permselective membranes available that are nearly impermeable for certain
classes of ions (coions), but these systems generally will
extract and transport more than one sort of counterions. If
considerations are restricted to cation-permselective membranes
and counterions of the same charge z_m, Eqs. (3.15) and (4.15a)
can be combined to yield:

$$E_M = \frac{RT}{z_m F} \ln \frac{\Sigma u_m k_m a_m'}{\Sigma u_m k_m a_m''} \tag{6.6a}$$

An analogous result follows for anion-exchange membranes:

$$E_M = \frac{RT}{z_x F} \ln \frac{\Sigma u_x k_x a_x'}{\Sigma u_x k_x a_x''} \tag{6.6b}$$

Both expressions are consistent with the classical Nicolsky
equation describing the emf of membrane cells:

$$E = E_i^o + \frac{RT}{z_i F} \ln \left[a_i' + \sum_{j \neq i} K_{ij}^{Pot} a_j' \right] \tag{6.7}$$

From the theoretical standpoint, the choice between a primary
ion I and interfering ions J is arbitrary. Evidently, the
potentiometric selectivity between any two ions of the same
charge is here given by the ratio of the respective individual

mobilities u and single-ion extraction constants k, each of
these parameters being characteristic of the membrane material:

$$K_{ij}^{Pot} = \frac{u_j k_j}{u_i k_i} \tag{6.8}$$

Equations of the type (6.6) - (6.8) were pioneered by Eisenman
and coworkers in their treatments of solid [1 - 4] and liquid
ion-exchangers [4, 5], as well as of neutral carrier membranes
[4, 6, 7] (see also Chapters 8 and 11 - 13).

Combination of expressions for E_D and E_B is generally more
difficult when two classes of diffusing ions are involved.
Nevertheless, we may arrive at useful descriptions of the
membrane potential. An interesting example, which was not
treated in the older literature, is the case of permselective
membranes with two classes of permeating cations (respectively
anions). Here we focus on electroneutral membranes that contain divalent and monovalent cations, I^{2+} and J^+, as well as
dissociated anionic sites (fixed or mobile) of a constant activity X. The boundary potential of such systems has been characterized previously by Eq. (3.19), while the diffusion potential
is determined by Eq. (4.32). If there exist no significant
differences in cation mobilities, Eq. (4.33) applies and the
resulting solution for the membrane potential simply reads:

$$E_M = \frac{RT}{F} \ln \frac{\sqrt{8 X \Sigma k_i a_i' + (\Sigma k_j a_j')^2} + \Sigma k_j a_j'}{\sqrt{8 X \Sigma k_i a_i'' + (\Sigma k_j a_j'')^2} + \Sigma k_j a_j''} \tag{6.9}$$

For cases with only two sorts of cations present, this expression may be converted into the following emf relation-

ship [8, 9]*):

$$E = E_i^o + \frac{RT}{F} \ln \left[\sqrt{a_i' + \frac{1}{4} K_{ij}^M (a_j')^2} + \sqrt{\frac{1}{4} K_{ij}^M (a_j')^2} \right] \qquad (6.10)$$

where the selectivity factor is given by

$$K_{ij}^M = \frac{(k_j)^2}{2 \times k_i} \qquad (6.11)$$

This theoretical result represents a form intermediate between the two admissible empirical expressions of the Nicolsky type (see also Chapter 12):

$$E = E_i^o + \frac{RT}{2F} \ln \left[a_i' + K_{ij}^{Pot} (a_j')^2 \right] \qquad (6.12)$$

and

$$E = E_j^o + \frac{RT}{F} \ln \left[a_j' + K_{ji}^{Pot} (a_i')^{\frac{1}{2}} \right] \qquad (6.13)$$

It will be demonstrated below that the latter formulations are rigorously valid only for more hypothetical cases. Although significant differences are found to persist between Eqs. (6.10), (6.12), and (6.13), the use of either of the two Nicolsky equations can often be tolerated in practice and has been sanctioned by IUPAC [11].

---

*) An equivalent expression was derived by Buck and Sandifer [10] but these authors did not reduce their result to the simple form given by Morf et al. [8, 9].

An extension of Eq. (6.9), allowing for counterions I^{2+} and J^+ of widely different mobilities in the cation-exchanger phase, is still based on Eqs. (3.19) and (4.32). After lengthy but trivial algebra, the following new relationships are obtained:

$$E_M = (1-q) \frac{RT}{F} \ln \frac{\sqrt{8 X k_i a_i' + (k_j a_j')^2} + k_j a_j'}{\sqrt{8 X k_i a_i'' + (k_j a_j'')^2} + k_j a_j''}$$

(6.14)

$$+ q \frac{RT}{F} \ln \frac{(1-q) \sqrt{8 X k_i a_i' + (k_j a_j')^2} - q\, k_j a_j'}{(1-q) \sqrt{8 X k_i a_i'' + (k_j a_j'')^2} - q\, k_j a_j''}$$

and

$$E = E_i^o + (1-q) \frac{RT}{F} \ln \left[\sqrt{a_i' + \frac{1}{4} K_{ij}^M (a_j')^2} + \sqrt{\frac{1}{4} K_{ij}^M (a_j')^2} \right]$$

(6.15)

$$+ q \frac{RT}{F} \ln \left[\sqrt{a_i' + \frac{1}{4} K_{ij}^M (a_j')^2} - \frac{q}{1-q} \sqrt{\frac{1}{4} K_{ij}^M (a_j')^2} \right]$$

with

$$q = \frac{u_i - u_j}{2u_i - u_j} \quad ; \quad 1-q = \frac{u_i}{2u_i - u_j}$$

Evidently, the thermodynamic factor K_{ij}^M as well as the mobility ratio u_j/u_i enter in Eq. (6.15) as the selectivity-determining parameters. This fact is corroborated by a study of the limiting cases. For $u_i \gg u_j$, Eq. (6.15) clearly approximates the conventional Nicolsky equation (6.12), the potentiometric selectivity factor being derived as

$$K_{ij}^{Pot} = \frac{u_j}{2u_i} K_{ij}^{M} \qquad\qquad (6.16)$$

The form of this result bears resemblance to the former Eq. (6.8). As might be expected, the alternative Nicolsky-type relation (6.13) follows in the other limit, namely for $u_j \gg u_i$. Here the potentiometric selectivity comes out to be given by

$$K_{ji}^{Pot} = \frac{2u_i}{u_j} (K_{ij}^{M})^{\frac{1}{2}} \qquad\qquad (6.17)$$

Although the presumed large variance of cation mobilities is conceivable for certain solid ion-exchangers (e. g. glass membranes), the assumption of nearly identical mobilities for all species is often more realistic, especially for liquid membranes. The use of Eq. (6.10), which exactly holds for $u_i = u_j$, is therefore more convincing from the theoretical point of view. It should finally be noted that Eqs. (6.9) - (6.15) can be transcribed for anion-selective systems by simply changing the sign of the logarithmic terms.

An analytical solution for the membrane potential is also known for systems that undergo deviations from permselectivity. Such failure of coion exclusion ("Donnan exclusion") is often observed for porous membranes, e. g. conventional ion-exchangers, and for liquid membranes in the presence of highly extractable coions. A thorough theoretical treatment of porous membranes was offered many years ago by Teorell [12], Meyer and Sievers [13], and Schlögl [14, 15]. As was shown in Chapter 4, the key results of these theories are also applicable to compact membranes. The classical Teorell-Meyer-Sievers theory is devoted to fixed-site membranes containing cations M^+ and anions X^- as the only permeating species. It relies on the applicability of relations of the type (3.9), (3.12), (4.29), and (4.30). The final result for E_M assumes the form[*]:

$$E_M = \frac{RT}{F} \ln \frac{a'_m}{a''_m} - \frac{RT}{F} \ln \frac{\sqrt{1+Q'^2} \pm 1}{\sqrt{1+Q''^2} \pm 1} + q \frac{RT}{F} \ln \frac{\sqrt{1+Q'^2} \pm q}{\sqrt{1+Q''^2} \pm q} \qquad (6.18)$$

where $q = (u_m - u_x)/(u_m + u_x)$, and

$$Q'^2 = \frac{4 \, k_m a'_m \, k_x a'_x}{(\omega X)^2} \; ; \qquad Q''^2 = \frac{4 \, k_m a''_m \, k_x a''_x}{(\omega X)^2} \qquad (6.19)$$

The positive sign in the logarithmic terms of Eq. (6.18) applies to membranes with negative sites, and vice versa. The quantities Q represent a direct measure of permselectivity. For ideally permselective systems it holds that $Q' \ll 1$ and $Q'' \ll 1$, and hence the Teorell-Meyer-Sievers equation simply reduces to the Donnan equation (6.3). A loss of permselectivity, on the other hand, is signaled by values of $Q' \gtrsim 1$ and/or $Q'' \gtrsim 1$. In this case, coions infiltrate from the outer electrolyte solutions and the membrane becomes easily permeable for both cations and anions. This effect, failure of Donnan

---

[*] In the version given by Lakshminarayanaiah [16], reciprocals of activity coefficients referring to the membrane phase appear instead of the distribution coefficients introduced by Meyer and Sievers [13].

exclusion, is manifested by the fact that now the membrane potential reflects cationic as well as anionic contributions (for Q' and $Q'' \gg 1$):

$$E_M = \frac{u_m}{u_m + u_x} \frac{RT}{F} \ln \frac{a'_m}{a''_m} - \frac{u_x}{u_m + u_x} \frac{RT}{F} \ln \frac{a'_x}{a''_x} \qquad (6.20)$$

Such a mixed response is frequently observable as a near-zero slope of the emf vs. log (activity) function:

$$E \cong E^o_{mx} + \frac{u_m - u_x}{u_m + u_x} \frac{RT}{F} \ln a'_{\pm} \qquad (6.21)$$

In order to exclude coions more effectively, a high site density of the membrane, sufficiently low external activities, and the absence of highly extractable species must be ensured. These requirements for permselectivity are readily deducible from Eq. (6.19).

The basic difference between the so-called porous membranes and typical representatives of liquid membranes is that the latter, by definition, do not incorporate fixed sites in the sense of chemically bound ion-exchanger groups. The ionic species present in such systems are essentially mobile. Consequently, the Teorell-Meyer-Sievers concept may be inappropriate for liquid membranes and should preferably be replaced by an analysis along the lines of Planck's liquid-junction theory. A useful extension of the theory was accomplished only recently by Morf [17]. This new description of the membrane potential was obtained by combining the transformed Planck relation for the diffusion potential, Eq. (4.21), and the generalized Donnan potential term, Eq. (3.15). The final solution for E_M reads:

$$E_M = (1 - \tau_x) \frac{RT}{z_m F} \ln \frac{\Sigma k_m a_m'}{\Sigma k_m a_m''} + \tau_x \frac{RT}{z_x F} \ln \frac{\Sigma k_x a_x'}{\Sigma k_x a_x''} \qquad (6.22)$$

where τ_x is the integral anionic transference number

$$\tau_x = \frac{|z_x| \bar{u}_x}{|z_m| \bar{u}_m + |z_x| \bar{u}_x} \qquad (6.23)$$

The mean ionic mobilities entering Eq. (6.23) may be expressed as follows (see Eqs. (4.22) and (3.15)):

$$\bar{u}_i = \frac{\Sigma u_i k_i a_i'' \cdot e^{z_i F E_M / RT} - \Sigma u_i k_i a_i'}{\Sigma k_i a_i'' \cdot e^{z_i F E_M / RT} - \Sigma k_i a_i'} \qquad (6.24)$$

The system of Eqs. (6.22) - (6.24) offers an implicit but well-defined solution for the membrane potential. Since evaluation for E_M is not nearly as cumbersome as would be estimated [17], the use of this formalism is to be encouraged. For simplicity, it is often legitimate to insert the same mobilities for all cations, $u_m = \bar{u}_m$, and for all anions, $u_x = \bar{u}_x$. This approximation leads to an explicit solution, analogous to Eq. (6.20), which still allows a rather universal description of the electrical potentials arising over liquid membranes (see Chapters 11 and 12).

In the above treatment, emphasis has been laid on explicit formulations. All the catalogued practical solutions for the membrane potential or the emf are, of course, subclasses of a more general description, given by Eqs. (3.12) and (4.8). The special cases discussed here have substantiated that the po-

tentiometric ion selectivity of membranes is usually expressible in terms of ionic extractabilities and mobilities
(i. e. permeabilities) or transference numbers. A more detailed
analysis of the transport-selectivity relationships is presented in the next chapters.

REFERENCES

[1] G. Eisenman, D. O. Rudin, and J. U. Casby, Science 126,
     871 (1957).

[2] F. Conti and G. Eisenman, Biophys. J. 5, 247, 511 (1965).

[3] G. Eisenman, ed., Glass Electrodes for Hydrogen and Other
     Cations, M. Dekker, New York, 1967.

[4] G. Eisenman, in Ion-Selective Electrodes (R. A. Durst,
     ed.), Natl. Bur. of Standards Spec. Publ. 314, Washing-
     ton, 1969.

[5] J. P. Sandblom, G. Eisenman, and J. L. Walker, Jr., J.
     Phys. Chem. 71, 3862, 3871 (1967).

[6] S. Ciani, G. Eisenman, and G. Szabo, J. Membrane Biol.
     1, 1 (1969).

[7] G. Eisenman, ed., Membranes, Vol. 2, M. Dekker, New York,
     1973.

[8] W. E. Morf, D. Ammann, E. Pretsch, and W. Simon, Pure
     Appl. Chem. 36, 421 (1973).

[9] W. E. Morf and W. Simon, in Ion-Selective Electrodes in
     Analytical Chemistry (H. Freiser, ed.), Plenum Press,
     New York, 1978.

[10] R. P. Buck and J. R. Sandifer, J. Phys. Chem. 77, 2122
     (1973).

[11] IUPAC Recommendations for Nomenclature of Ion-Selective
     Electrodes; Pure Appl. Chem. 48, 127 (1976).

[12] T. Teorell, Proc. Soc. Exp. Biol. Med. 33, 282 (1935).

[13] K. H. Meyer and J. F. Sievers, Helv. Chim. Acta 19, 649,
     665, 987 (1936).

[14] R. Schlögl, Z. Phys. Chem. (Frankfurt am Main) 1, 305
     (1954).

[15] R. Schlögl, Stofftransport durch Membranen, Steinkopff,
     Darmstadt, 1964.

[16] N. Lakshminarayanaiah, Transport Phenomena in Membranes,
     Academic Press, 1969.

[17] W. E. Morf, Anal. Chem. 49, 810 (1977).

CHAPTER 7

CLASSICAL CONCEPTS OF MEMBRANE TRANSPORT

7.1. THE NERNST-PLANCK FLUX EQUATION

The primary aim of a theoretical analysis of the ion transport through membranes is to describe the ionic fluxes as a
closed function of ion activities in contact with the membrane
and the applied voltage. Such descriptions can be sought, in
principle, among the following categories:

a) Theories on the basis of the Nernst-Planck flux equation.
b) Formulations based on Eyring's theory of reaction rates.
c) Applications of the laws of irreversible thermodynamics.

It has been established that the different approaches often
lead to very similar or even identical results (see Chapter 8
and References 1-3). However, the formal descriptions belonging
to class c) are usually cast in terms of phenomenological coefficients, a thorough interpretation of which also requires
the knowledge of the equivalent Nernst-Planck formalism. To
avoid a purely phenomenological treatment of membrane transport, we shall preferably restrict the following discussion
to modifications of the classical Nernst-Planck flux equation,
which indeed allows diffusion or migration of ions to be
characterized in terms of conventional experimental parameters, such as diffusion coefficients, distribution coefficients, etc.

Suitable solutions to the Nernst-Planck flux equation can
be best obtained for the steady-state, where the ionic
fluxes J_i have become constant throughout the membrane. Accordingly, the basic relationship (4.4) applies here in the
time-independent version:

$$J_i = -\frac{D_i}{\gamma(x)} \left[\frac{da_i(x)}{dx} + z_i a_i(x) \frac{F}{RT} \frac{d\phi(x)}{dx} \right] = \text{const}(x) \qquad (7.1)$$

The notation of symbols is the same as in Chapter 4, except that the term $u_i RT$ has been identified with the diffusion coefficient D_i (Nernst-Einstein relation), and the individual activity coefficients have been replaced by the mean activity coefficient γ (assumption IXa). Equation (7.1) can be directly integrated to yield

$$J_i = -D_i \frac{a_i(d) \exp(z_i F\phi(d)/RT) - a_i(0) \exp(z_i F\phi(0)/RT)}{\int_0^d \gamma(x) \exp(z_i F\phi(x)/RT) \, dx} \qquad (7.2)$$

Obviously, in addition to the boundary conditions a thorough knowledge of the concentration profiles (activity coefficients) and the potential profile is required for the evaluation of Eq. (7.2). Although a general solution to the problem was formulated by Schlögl [4] (see Section 7.4), practical descriptions of electrodiffusion across membranes are obtained only if further simplifying assumptions or approximations are invoked.

7.2. THE GOLDMAN-HODGKIN-KATZ APPROXIMATION

An ingenious approach to the theory of membrane transport was introduced in 1943 by Goldman [5]. He assumed that the electrical potential varies linearly from $x = 0$ to $x = d$, that is

$$\frac{d\phi}{dx} \approx \frac{\phi(d) - \phi(0)}{d} = \frac{\Delta\phi}{d} \qquad (7.3)$$

In addition, the ideality assumption $\gamma = 1$ was imposed. These approximations and the resulting formulae seem to be realistic for very thin membranes, such as biological membranes. Insertion of the constant field approximation (7.3) in Eq. (7.2) and subsequent integration leads to the well-known Goldman flux equation:

$$J_i \cong - z_i D_i \, \frac{F}{RT} \, \frac{\Delta\phi}{d} \, \frac{c_i(d) \, \exp(z_i F \Delta\phi / RT) - c_i(0)}{\exp(z_i F \Delta\phi / RT) - 1} \tag{7.4}$$

In their pioneering work on nerve cell potentials, Hodgkin and Katz [6] made use of the following simplifying assumptions:

$$c_i(0) \approx k_i a_i' \quad ; \quad c_i(d) \approx k_i a_i'' \tag{7.5}$$

$$\Delta\phi \approx \phi'' - \phi' \equiv - V \tag{7.6}$$

This permits replacement in Eq. (7.4) of all terms referring to the membrane boundaries by the corresponding outside values. Hence

$$J_i \cong P_i \, \frac{z_i FV/RT}{\exp(z_i FV/RT) - 1} \, (a_i' \, \exp(z_i FV/RT) - a_i'') \tag{7.7}$$

where

$$P_i = \frac{D_i k_i}{d} \tag{7.8}$$

89

The outstanding feature of the Goldman-Hodgkin-Katz flux equation, in addition to simplicity, is that the extraction capacity of the membrane and its resistance to diffusion are characterized by a single parameter, namely the so-called permeability P_i (with the dimension of a velocity). On the other hand, the ion permeabilities thus determined often do not represent real constants, a fact which manifests the approximative nature of the basic formulae. Several extensions of Eqs. (7.1)-(7.8) have been advanced to refine the interpretation of experimental data obtained on bimolecular membranes (see Chapter 8).

Nevertheless, the Goldman-Hodgkin-Katz theory leads to some very interesting conclusions. Equation (7.7) suggests that the net flux of permeating species can generally be considered as a difference of two partial fluxes, namely a component $\vec{J}_i$ in the direction of the x-axis and a component $\overleftarrow{J}_i$ in the opposite direction:

$$J_i = \vec{J}_i - \overleftarrow{J}_i \tag{7.9}$$

The following fundamental relationship may then be obtained (Ussing's equation of passive and independent transport [7]):

$$\frac{\vec{J}_i}{\overleftarrow{J}_i} = \frac{a_i'}{a_i''} \exp(z_i FV/RT) \tag{7.10}$$

Non-validity of this equation has been widely used as a criterion for active, coupled, or saturated transport in biological membrane systems [8].

One of the most important and frequently applied formulae in membrane biophysics is the Goldman-Hodgkin-Katz equation for the zero-current membrane potential:

$$V_0 = - E_M \cong - \frac{RT}{F} \ln \frac{\Sigma P_m a_m' + \Sigma P_x a_x''}{\Sigma P_m a_m'' + \Sigma P_x a_x'} \qquad (7.11)$$

This relation, derived from flux equations of the form (7.7), is valid for monovalent ions M^+ and X^-. To rationalize the observed electrical behavior of nerve cells, Hodgkin and Katz [6] took into account the permeation of potassium, sodium, and chloride ions. For Loligo axons, the permeability ratios $P_K : P_{Na} : P_{Cl}$ were found to be 1.0 : 0.04 : 0.45 in the resting state, and 1.0 : 20 : 0.45 during activity.

7.3. SIMPLE MODEL FOR SYMMETRICAL MEMBRANE CELLS

While the Goldman-Hodgkin-Katz model constitutes a reasonable approach to the ion transport behavior of very thin membranes, the underlying assumptions are commonly not met for bulk membranes as they are used for ion-selective electrodes. For example, Eq. (7.5) does not account for the phase-boundary potentials which, however, can no longer be neglected in the case of electroneutral membranes (see also Chapter 3). The correct formulation of the Donnan equilibria must therefore read:

$$a_i(0) = k_i a_i' \exp[-z_i F(\phi(0) - \phi')/RT] \qquad (7.12a)$$

$$a_i(d) = k_i a_i'' \exp[-z_i F(\phi(d) - \phi'')/RT] \qquad (7.12b)$$

91

By the same argument, one has to replace the former approximation (7.6) by the rigorous relation

$$\Delta\phi = -V + (\phi(d) - \phi'') - (\phi(0) - \phi')$$ (7.13)

In addition, the Goldman assumption of a constant electric field, Eq. (7.3), can be justified only for certain special cases. So the general description of ion fluxes in bulk membrane phases turns out to be much more involved than the one used for microscopic barriers.

In order to simplify the exposé of the theory of thick cation-permselective membranes, Morf, Wuhrmann, and Simon [9, 10] made a series of elementary assumptions. The simplest version of the theory is restricted to symmetrical systems where the membrane is interposed between two aqueous solutions of identical composition, i. e.

$$a'_i = a''_i$$ (7.14)

A thermodynamic equilibrium is assumed to exist between the membrane boundaries and the adjoining solutions (Eq. (7.12)). Then, the membrane composition evidently becomes constant throughout:

$$c_i = const(x)$$ (7.15)

$$\gamma = const(x)$$ (7.16)

Using the last relations, one can write the Nernst-Planck equation (7.1) in the following convenient form, as might be gleaned from the Goldman equation (7.4):

$$J_i = - z_i D_i c_i \frac{F}{RT} \frac{\Delta\phi}{d} \qquad (7.17)$$

The basic assumption of a symmetrical cell implies that the phase-boundary potential differences established at the two membrane/solution interfaces have exactly the same value:

$$\phi(0)-\phi' = \phi(d)-\phi'' \qquad (7.18a)$$

hence

$$\Delta\phi = - V \qquad (7.18b)$$

Accordingly, when a voltage is applied across a homogeneous membrane that is in equilibrium with two identical solutions, the whole potential drop lies within the membrane phase. This leads to the following relationship for the ion fluxes:

$$J_i = z_i D_i c_i \frac{F}{RT} \frac{V}{d} \qquad (7.19)$$

from which the electric current density is found to be given by Ohm's law. For the evaluation of the membrane-internal concentrations c_i, according to Eq. (7.12), the assumptions of electroneutrality and of a fixed concentration c of anionic sites were used [9, 10]:

$$\Sigma z_i c_i = c \qquad (7.20)$$

This permits explicit transport equations to be derived for a series of important cases. Several extensions of the model demonstrate the applicability of the relations given for cation-permselective membranes [9, 10] (see also Sections 7.4 and 7.5 and Chapter 8).

7.4. SCHLÖGL'S GENERAL THEORY AND ITS APPLICATIONS

In 1954, Schlögl [4] solved the algebraically difficult problem of electrodiffusion in thick electroneutral membranes; the general theoretical analysis allowed for any number of permeating ions as well as for fixed ionic sites (charge ω, site density X). He recognized that the ensemble of diffusing ions must be subdivided into valency classes, and that the complexity of the theoretical description depends on what number N of such ion classes are present in the system. The following version of the Nernst-Planck equation (7.1) was used to describe the steady-state flux of the i-th ion of class k:

$$J_{ik} = -D_{ik} \frac{A_{ik}}{A_k} \frac{A_k}{\sum_k A_k} \quad \sum_k A_k = \text{const}(x) \qquad (7.21)$$

where

$$A_{ik} \equiv -J_{ik}/D_{ik} = \frac{1}{\gamma} \left[\frac{da_{ik}}{dx} + z_k a_{ik} \frac{F}{RT} \frac{d\phi}{dx} \right] \qquad (7.22)$$

$$A_k \equiv \sum_i A_{ik} \qquad\qquad\qquad (7.23)$$

In the original work by Schlögl [4], Eq. (7.22) was written in the ideal-solution form, whereas here activity coefficients are taken into account according to assumption IXa in Chapter 2. Equation (7.21) is found to be composed of three fundamental parts. The first and the last of these terms may be readily determined, recalling Eqs. (7.2) and (7.22):

$$\frac{A_{ik}}{A_k} = \frac{a_{ik}(d) \; \exp(z_k F\Delta\phi/RT) \; - \; a_{ik}(0)}{a_k(d) \; \exp(z_k F\Delta\phi/RT) \; - \; a_k(0)} \qquad\qquad (7.24)$$

and

$$\sum A_k = \frac{\sum a_k(d) \; - \; \sum a_k(0) \; - \; \omega \, X \cdot F\Delta\phi/RT}{\bar{\gamma}d} \qquad\qquad (7.25)$$

where

$$\sum z_k a_k = - \, \omega X \qquad\qquad\qquad (7.26)$$

$$a_k \equiv \sum_i a_{ik}$$

$$\bar{\gamma} \equiv \frac{1}{d} \int_0^d \gamma(x) \; dx$$

The evaluation of the remaining term in Eq. (7.21) is the most difficult part. Schlögl offered the following solution, which results from transformation of Eq. (4.37a):

$$\frac{A_k}{\Sigma A_k} = \frac{(1/z_k - q_1)\ (1/z_k - q_2) \cdots (1/z_k - q_{N-1})}{\prod_{j \neq k} (1/z_k - 1/z_j)} \qquad (7.27)$$

The values of the parameters $q_1 \cdots q_{N-1}$ are determined by the implicit equation (see also Section 4.2.e):

$$\frac{F}{RT} \Delta\phi = q \ \ln \left[\sum_k \frac{z_k a_k(0)}{q - 1/z_k} \ \Big/ \ \sum_k \frac{z_k a_k(d)}{q - 1/z_k} \right] \qquad (7.28)$$

The system of Eqs. (7.21) and (7.24) - (7.28) allows to rigorously assess the fluxes of all ions for a given value of $\Delta\phi$ and for given boundary conditions on both sides of the membrane. To evaluate the surface values of ion activities and electrical potential, one has to make use of Eqs. (7.12), (7.13), and (7.26). As the general calculation procedure turns out to be rather laborious, we restrict the following discussion to four special cases.

For ion-exchange membranes containing only one class of diffusing counterions, it holds that $A_k/\Sigma A_k = 1$, $z_k a_k(x) = -\omega X$, and $\gamma(x) \cong \bar{\gamma}$. Hence the relations of Schlögl's theory reduce immediately to Goldman's flux equation (7.4). This result is somewhat surprising because the classical Goldman equation is usually applied to site-free membranes, whereas here the fixed charge concentration of the membrane is

$$c = |\omega| X/\bar{\gamma}$$

A further simplification is obtained for experimental situations where the applied voltage $V = \phi' - \phi''$ is kept sufficiently high, i. e.

$$V \approx - \Delta\phi \gg RT/F$$

In this case the Goldman equation simply reduces to

$$J_m = z_m D_m c_m(0) \frac{F}{RT} \frac{V}{d} , \qquad (7.29a)$$

respectively, for anion-exchange membranes:

$$J_x = z_x D_x c_x(d) \frac{F}{RT} \frac{V}{d} \qquad (7.29b)$$

Finally, insertion of Eqs. (7.12), (7.16), and (7.26) leads to

$$J_m = D_m \frac{k_m a_m'}{\Sigma \, k_m a_m'} c \frac{F}{RT} \frac{V}{d} \qquad (7.30a)$$

or

$$J_x = - D_x \frac{k_x a_x''}{\Sigma \, k_x a_x''} c \frac{F}{RT} \frac{V}{d} \qquad (7.30b)$$

These results are formally identical to those obtained for symmetrical bathing solutions (see Section 7.3). Equations (7.30a,b) clearly show, however, that it is primarily the activities on the side of the ion uptake into the membrane that are decisive for the magnitude of the ion fluxes.

In site-free ion-permeable membranes, both cations and anions are mobile. Because of $\omega X = 0$ the electroneutrality condition reads:

$$\Sigma |z_m| a_m(x) = \Sigma |z_x| a_x(x) = a(x)$$

The parameter q is therefore determined, for two ion classes, as follows:

$$q = - \frac{F\Delta\phi/RT}{\ln[a(d)/a(0)]}$$

For such systems Schlögl's theory yields the transport equation first proposed by Behn [11] (for ideal solutions of 1:1 electrolytes) and later extended by Linderholm [12]. The limiting case obtained for high voltages is given by

$$J_m = D_m \frac{k_m a_m'}{\Sigma \ k_m a_m'} \ \bar{c} \ \frac{F}{RT} \ \frac{V}{d} \tag{7.31a}$$

and

$$J_x = - D_x \frac{k_x a_x''}{\Sigma \ k_x a_x''} \ \bar{c} \ \frac{F}{RT} \ \frac{V}{d} \tag{7.31b}$$

Evidently, these flux relationships are of the same form as in the preceding example. The ion activities in Eqs. (7.31a, b) again refer to the external solution on that side where either cations or anions enter the membrane. However, the mean ion concentration in the membrane is here defined as

$$\bar{c} = \frac{a(d)-a(0)}{\bar{\gamma} \ln[a(d)/a(0)]}$$

A comparison of the present results with the analysis given by Linderholm [12], who used a Debye-Hückel approach to strictly account for activity coefficients in the diffusion layer, reveals that the overall coefficient $\bar{\gamma}$ does not depend on the shape of the membrane-internal concentration profiles. This implies that the concentration parameter $\bar{c}$ is only a function of the boundary concentrations, but is independent of the applied voltage.

The treatment offered by Schlögl [4, 13] was primarily devoted to the classical ion-exchangers (fixed-site membranes) which are easily permeable for different sorts of counterions and, to a minor degree, also for coions. This corresponds to a situation intermediate to the former limiting cases realized for ideally permselective membranes and for site-free extraction systems, respectively. In the present context, we consider in more detail the transport behavior of cation-exchange membranes at relatively high current densities. Of course, analogous results may be obtained for anion-exchangers. For systems with counterions of the same charge z_m and coions of the same charge z_x, Eqs. (7.28) and (7.26) yield the root:

$$q \cong - \frac{\Sigma \; z_m^2 a_m(d) \; + \; \Sigma \; z_x^2 a_x(d)}{|z_m z_x \omega| X}$$

Hence the Schlögl formalism reduces to the following relations:

$$J_m = D_m \frac{k_m a_m'}{\Sigma \, k_m a_m'} \frac{\Sigma \, |z_m| a_m(d)}{\bar{\gamma}} \frac{F}{RT} \frac{V}{d} \qquad (7.32a)$$

$$J_x = - D_x \frac{k_x a_x''}{\Sigma \, k_x a_x''} \frac{\Sigma \, |z_x| a_x(d)}{\bar{\gamma}} \frac{F}{RT} \frac{V}{d} \qquad (7.32b)$$

with $\Sigma |z_m| a_m(d) = |\omega| X + \Sigma |z_x| a_x(d)$. A further simplification is obtained for the electrodialytic transport of a single salt:

$$J_m = z_m D_m \frac{a_m(d)}{\bar{\gamma}} \frac{F}{RT} \frac{V}{d} \qquad (7.33a)$$

$$J_x = z_x D_x \frac{a_x(d)}{\bar{\gamma}} \frac{F}{RT} \frac{V}{d} \qquad (7.33b)$$

Again the flux equations agree with those derived previously, except for the terms describing the mean equivalent concentrations of cations and anions in the interior of the membrane. We can recognize that here the profiles of total ion concentrations tend to flatten out, reflecting the situation at the interface ($x = d$) where the coions drift into the membrane [4, 13]. In the end, it is the distribution of cations and anions across this interface that determines whether the membrane approximates permselective behavior or not. The membrane boundary at $x = 0$, on the other hand, regulates the individual uptake of counterions and therefore gives rise to selectivity among different counterions.

As the last example, we focus on cation-exchange membranes exposed to different classes of counterions, I and J. For simplicity, we restrict considerations to the electrical transference of cations at high voltages, assuming ideal perm-selectivity of the membrane (i. e., exclusion of permeating anions). The corresponding solution for q reads, in analogy to the former case:

$$q \cong \frac{\Sigma \; z_i^2 a_i(0) + \Sigma \; z_j^2 a_j(0)}{|z_i z_j| \omega | X}$$

Insertion into Eqs. (7.21) and (7.24) - (7.27) leads to the flux equations

$$J_i = z_i D_i \; \frac{a_i(0)}{\bar{\gamma}} \; \frac{F}{RT} \; \frac{V}{d} \qquad\qquad (7.34a)$$

$$J_j = z_j D_j \; \frac{a_j(0)}{\bar{\gamma}} \; \frac{F}{RT} \; \frac{V}{d} \qquad\qquad (7.34b)$$

These results agree perfectly with the expectations from the simple model presented in Section 7.3. The implications of the classical Nernst-Planck descriptions of ion transport are summarized below.

7.5. ELECTRICAL PROPERTIES AND ION-TRANSPORT SELECTIVITY OF BULK MEMBRANES

The fundamental relationships derived in the preceding sections offer a very detailed picture of the ion transport behavior and the electrical characteristics of bulk membranes. They represent a basis for the understanding of ion-selective membranes and may thus help to design potentiometric or electrodialytic experiments in which ion selectivity can be investigated or exploited. Since the results for the zero-current membrane potential have already been summarized in Chapter 6, we will here discuss the behavior of thick ion-permeable membranes at relatively high current densities. It has been shown that electrodiffusion in symmetrical membrane cells can be straightforwardly characterized by the flux equation (7.19). Accordingly, the electrical properties of such systems are found to reflect the membrane-internal concentrations and mobilities of permeating species.

The electric conductance of the membrane (given in $\Omega^{-1}cm^{-2}$) is obtained as

$$G = \frac{F \Sigma z_i J_i}{V} = \frac{F^2}{RTd} \Sigma z_i^2 D_i c_i \tag{7.35}$$

In the Goldman-Hodgkin-Katz treatment of thin membranes, the membrane concentrations c_i were directly related to the outside activities a_i, witness Eq. (7.5):

$$G \cong \frac{F^2}{RTd} \Sigma z_i^2 D_i k_i a_i = \frac{F^2}{RT} \Sigma z_i^2 P_i a_i \tag{7.36}$$

Obviously, the conductance is then mainly dictated by those kinds of ions which are preferably extracted and transported by the membrane. Thus conductance measurements on microscopic membranes can bear information on the ion selectivity of such systems (see Chapter 8). In contrast, the above approach is untenable for thick electroneutral membranes. Inspection of Eq. (7.30) shows, for example, that the conductance of ion-exchange membranes represents a measure of the fixed-site density rather than of the permeability selectivity. Straightforward information on the ion selectivity can be obtained, however, if the simultaneous transference of different ions is studied.

The electric transference number is defined as

$$t_i = \frac{z_i J_i}{\Sigma \ z_i J_i} = \frac{z_i^2 D_i c_i}{\Sigma \ z_i^2 D_i c_i} \tag{7.37}$$

$$\Sigma \ t_i = 1$$

For two ions of the same charge, I^z and J^z, the following ratio of transference numbers is obtained:

$$\frac{t_i}{t_j} = \frac{D_i c_i}{D_j c_j} \quad ; \quad t_i + t_j = 1 \tag{7.38}$$

Recalling Eq. (7.5) or (7.12), we can readily write [9, 10]:

$$\frac{t_i}{t_j} = \frac{a_i}{K_{ij} a_j} \tag{7.39}$$

with $K_{ij} = D_j k_j / D_i k_i$. Hence the transference of ions I^z and J^z is governed by the same selectivity, K_{ij}, as is exhibited in potentiometric measurements on identical bulk membranes. Such correlations between selectivity data were indeed established experimentally for sodium-selective membranes (Figure 7.1) and for enantiomer-selective membranes (Figure 7.2). These results, being in excellent agreement with theory, clearly indicate that ion-selective systems may be exploited for electrolytic separations of ions. Similar separation

8*

103

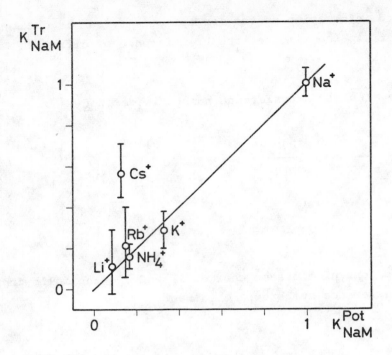

<u>Figure 7.1.</u> Transport selectivities K_{NaM}^{Tr} (transference number ratios) and potentiometric selectivities K_{NaM}^{Pot} of a Na^+-carrier PVC membrane [14].

effects are achieved when ion pumping is propelled by a chemical potential gradient instead of a voltage (see Figure 7.3).

The relations presented in Sections 7.3 and 7.4 can also be applied to describe the transport selectivity of thick

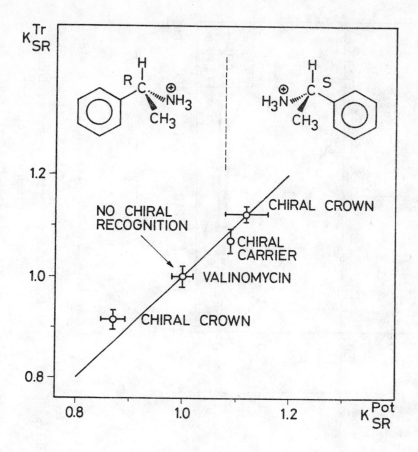

Figure 7.2. Transport selectivity and potentiometric selectivity between (R)- and (S)-phenylethylammonium ions for PVC membranes based on different enantiomer-selective ionophores [14, 15].

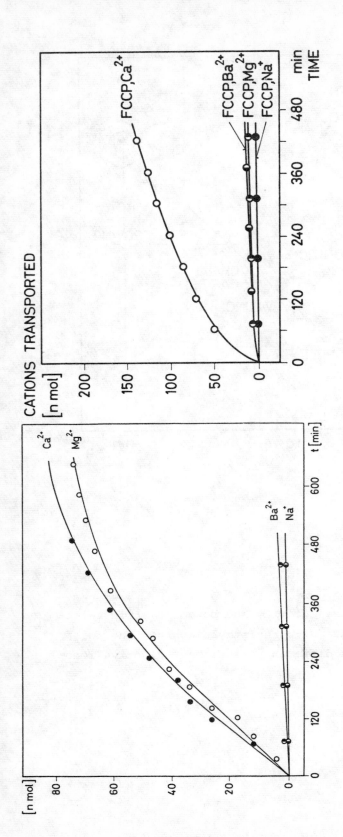

Figure 7.3. Transport of different cations across membranes selective for divalent ions. Cation transport was induced by applying a pH-gradient across the membranes (countertransport of protons) [16]. Left: PVC membrane based on the charged ligand 1,3-bis(p-chlorophenyl)-1,3-propanedion. Right: PVC membrane containing the neutral carrier 8 (p. 115) and the proton carrier FCCP. The observed transport rates agree with the potentiometric selectivity sequence $Ca^{2+} \gtrsim Mg^{2+} >> Na^+$, Ba^{2+} (left) and $Ca^{2+} >> Ba^{2+}$, Mg^{2+}, Na^+ (right), respectively.

membranes towards ions of different charge. For two cations I^{2+} and J^+, we get in analogy to Eq. (7.39) [9, 10, 17]:

$$\frac{t_i}{t_j} = \frac{D_i}{D_j} \left[\sqrt{\frac{a_i}{\frac{1}{4} K_{ij} a_j^2} + 1} - 1 \right] \quad ; \quad t_i + t_j = 1 \qquad (7.40)$$

where the selectivity factor, $K_{ij} = k_j^2/2Xk_i$, again conforms to the monovalent/divalent ion selectivity defined for potentiometric measuring cells. Figure 7.4 shows results of the simultaneous transport of Ca^{2+} and Na^+ across permselective liquid membranes. It is manifest that one given membrane may exhibit specificity for divalent cations in relatively diluted aqueous solutions, and increasing selectivity for monovalent cations in highly concentrated solutions. The documented agreement between theory and experiment - electrodialysis as well as potentiometry - is excellent and corroborates the basic membrane model presented here.

Another explicit description obtains for the transference of cations M^+ and anions X^- across ion-exchangers (fixed-site membranes). In this case, combination of the former Eqs. (7.12), (7.26), and (7.37) leads to the results:

$$\frac{t_m}{t_x} = \frac{D_m}{D_x} \frac{\sqrt{1 + Q^2} + 1}{\sqrt{1 + Q^2} - 1} \qquad \text{(for cation-exchangers)} \qquad (7.41a)$$

$$\frac{t_m}{t_x} = \frac{D_m}{D_x} \frac{\sqrt{1 + Q^2} - 1}{\sqrt{1 + Q^2} + 1} \qquad \text{(for anion-exchangers)} \qquad (7.41b)$$

with $t_m + t_x = 1$. The parameter Q was defined previously in Chapter 6, Eq. (6.19), and was shown to constitute a measure of permselectivity. Indeed, ideal permselectivity is found

107

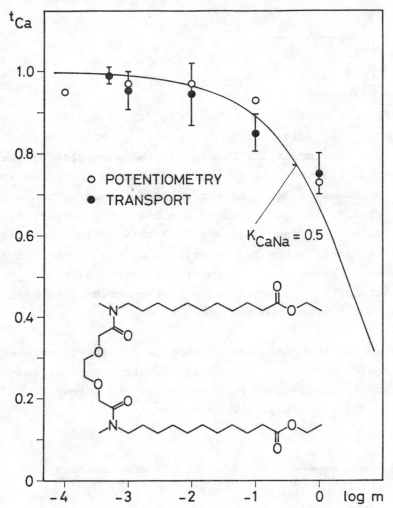

TRANSFERENCE NUMBER

○ POTENTIOMETRY
● TRANSPORT

$K_{CaNa} = 0.5$

Figure 7.4. Transference number for Ca^{2+}, as obtained in the electrodialysis of Ca^{2+} and Na^+ (both at concentration m) through a Ca^{2+}-carrier membrane [9, 17]. Full circles: experimental values. Solid line: theoretical curve according to Eq. (7.40) with $D_{Ca}=D_{Na}$. Open circles: values calculated from potentiometric selectivity data.

when the condition $Q \ll 1$ is fulfilled which, according to Eq. (7.41), is equivalent to a transference number of $t_m=1$ (for cation-exchangers) or $t_x=1$ (for anion-exchangers). In this case, all current is carried through the membrane by the counterions. In the other extreme, $Q \gg 1$, a complete loss of permselectivity occurs. Then the membrane becomes nearly equally permeable for cations and anions, i. e. $t_m/t_x = D_m/D_x$.

The above examples clearly demonstrate that there exists a causal relationship between the ion transport behavior and the potentiometric characteristics of membranes. Generally, the electrodialytic transference numbers can be formulated in terms of potentiometric selectivity parameters. Conversely, the emf-response of any kind of membrane electrode must be expressible in terms of ionic transference numbers. This is easily exemplified for cells containing the primary ion I^{z_i} and one interfering species J^{z_j}. In this case, the membrane-internal diffusion potential is obtained from the generalized Teorell equation (4.31) as follows:

$$E_D = \frac{D_i - D_j}{z_i D_i - z_j D_j} \frac{RT}{F} \ln \frac{z_i^2 D_i a_i(0) + z_j^2 D_j a_j(0)}{z_i^2 D_i a_i(d) + z_j^2 D_j a_j(d)} \qquad (7.42)$$

Using the definition given in Eq. (7.37), we can substitute transference numbers into the numerator and the denominator of the logarithmic term:

$$E_D = \frac{D_i}{z_i D_i - z_j D_j} \frac{RT}{F} \ln \frac{a_i(0)/t_i'}{a_i(d)/t_i''} - \frac{D_j}{z_i D_i - z_j D_j} \frac{RT}{F} \ln \frac{a_j(0)/t_j'}{a_j(d)/t_j''} \qquad (7.43)$$

109

Elimination of the boundary activities according to Eq.(7.12) then leads to a very fundamental relationship for the zero-current membrane potential:

$$E_M = \frac{D_i}{z_i D_i - z_j D_j} \frac{RT}{F} \ln \frac{a_i'/t_i'}{a_i''/t_i''} - \frac{D_j}{z_i D_i - z_j D_j} \frac{RT}{F} \ln \frac{a_j'/t_j'}{a_j''/t_j''} \qquad (7.44)$$

An analogous expression holds for the emf of the membrane electrode cell [17]. The universal description offered by Eq. (7.44) incorporates virtually all of the more specific solutions presented in Chapter 6; these results may be verified by substituting Eqs. (7.39) - (7.41) into Eq. (7.44). Evidently, the selectivity-determining parameters entering in Eq. (7.44) are the diffusion coefficients and the transference numbers of the permeating ions in the membrane phase. The latter quantities, by definition, refer to and follow from electrodialysis experiments using the same membrane (as for the ion-selective electrode application) in contact with bathing solutions identical to solution (') or ("). Finally, these experimental transference numbers allow the straightforward construction of the expected emf-response curve of the corresponding membrane electrode system. The results given in Figure 7.5 confirm most impressively the fundamental relationship between potentiometric and ion transport characteristics of ion-selective membranes.

In conclusion, we have learned that ion transport experiments offer an independent method for assessing the ion selectivity of diffusion-type membranes. As a rule, membrane electrodes will give a specific emf-response to that ionic species the extraction and permeation of which is most favored.

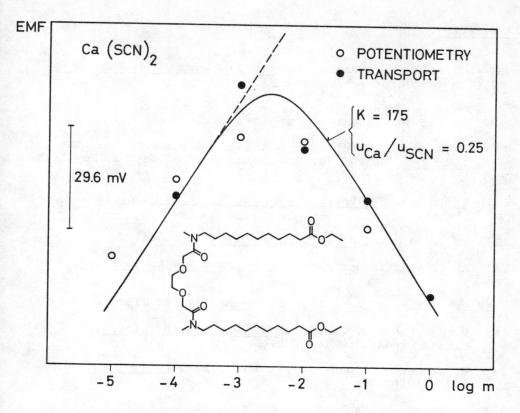

Figure 7.5. EMF-response of a Ca^{2+}-carrier membrane electrode to solutions of $Ca(SCN)_2$ (primary ion Ca^{2+}, interfering ion SCN^-; salt concentration m).

Open circles: experimental emf values. Full circles: values calculated from Eq. (7.44) using experimental transference numbers. Solid line: theoretical curve based on Eq. (7.44), fitting the individual results obtained from transport experiments (for details, see [9, 17]).

REFERENCES

[1] N. Lakshminarayanaiah, Transport Phenomena in Membranes,
     Academic Press, New York, 1969.

[2] P. Läuger and B. Neumcke, in Membranes, Vol. 2 (G. Eisen-
     man, ed.), M. Dekker, New York, 1973, p. 1.

[3] S. M. Ciani, G. Eisenman, R. Laprade,and G. Szabo, in
     Membranes, Vol. 2 (G. Eisenman, ed.), M. Dekker, New York,
     1973, p. 61.

[4] R. Schlögl, Z. Phys. Chem. (Frankfurt am Main) 1, 305
     (1954).

[5] D. E. Goldman, J. Gen. Physiol. 27, 37 (1943).

[6] A. L. Hodgkin and B. Katz, J. Physiol. (London) 108, 37
     (1949).

[7] H. H. Ussing, Acta Physiol. Scand. 19, 43 (1949).

[8] A. Kotyk and K. Janáček, Cell Membrane Transport,
     Plenum Press, New York, 1975.

[9] W. E. Morf, P. Wuhrmann, and W. Simon, Anal. Chem. 48,
     1031 (1976).

[10] W. Simon, W. E. Morf, E. Pretsch, and P. Wuhrmann, in
     Calcium Transport in Contraction and Secretion (E. Cara-
     foli, ed.), North-Holland, Amsterdam, 1975, p. 15.

[11] U. Behn, Ann. Phys. 62, 54 (1897).

[12] H. Lindérholm, Acta Physiol. Scand. Suppl. 27, 97 (1952).

[13] R. Schlögl, Ber. Bunsenges. Phys. Chem. 82, 225 (1978).

[14] W. E. Morf and W. Simon, in Ion-Selective Electrodes
     (E. Pungor, ed.), Akadémiai Kiadó, Budapest, 1977, p. 25.

[15] W. Simon, W. E. Morf, and A. P. Thoma, in Ion-Selective
     Electrodes (E. Pungor, ed.), Akadémiai Kiadó, Budapest,
     1977, p. 13.

[16] D. Erne, W. E. Morf, S. Arvanitis, Z. Cimerman, D. Ammann,
     and W. Simon, Helv. Chim. Acta,62, 994 (1979); W. E. Morf,
     S. Arvanitis, and W. Simon, Chimia 33, 452 (1979).

[17] W. E Morf and W. Simon, Conference on Ion-Selective
     Electrodes - Budapest, 1977 (E. Pungor, ed.), Akadémiai
     Kiadó, Budapest, 1978, p. 149.

CHAPTER 8

FREE AND CARRIER-MEDIATED ION TRANSPORT ACROSS BILAYER MEMBRANES

During the nineteen sixties, two new classes of artificial organic membranes were developed that show permeability for lipophilic ions. First, the black or bilayer lipid membranes were introduced by Mueller and coworkers [1] as primitive models of biological membranes, the membrane thickness being on the order of only 10 nm. Second, the preparation of solvent polymeric membranes was pioneered by Shatkay [2, 3] and later refined by Moody and Thomas [4]. Such bulk membranes became important as the working principle of ion-selective electrodes, and a wide range of accessible ion selectivities was obtained by the incorporation of different ion-complexing agents [4 - 9]. In contrast to the bilayer membranes, where electroneutrality must not be maintained in the membrane interior, the macroscopic counterparts provide ionic sites for the compensation of the charge of the permeating ions [10, 11].

A new area of ion transport studies was opened with the discovery of carrier molecules or ionophores that facilitate the permeation of hydrophilic ions in organic membranes [12 - 15]. Some typical examples for electrically neutral, natural or synthetic carriers are given in Figure 8.1. Such lipophilic compounds are capable of selectively complexing certain alkali or alkaline-earth metal ions, thereby solubilizing these cations in the organic phase, and transporting them across the membrane by carrier translocation.

The theoretical description of carrier-mediated electrical properties of lipid bilayers was set forth mainly by the groups of Eisenman and Läuger (for a review, see [16 - 19]). In the classical approach by Läuger and Stark [20], the membrane

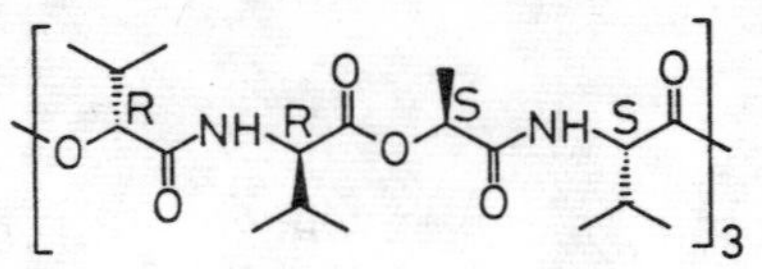

VALINOMYCIN 1

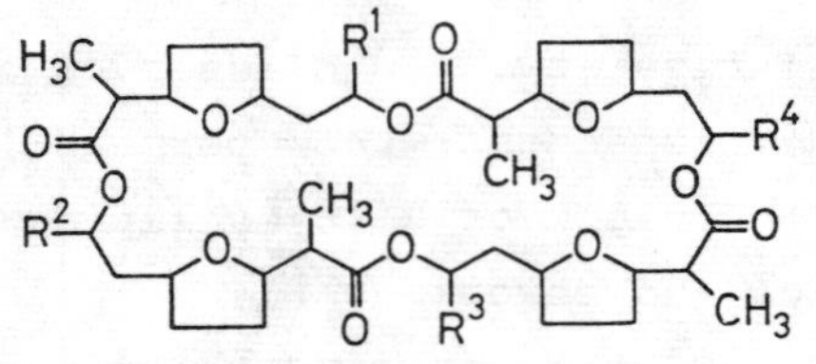

$R^1 = R^2 = R^3 = R^4 = CH_3$ — NONACTIN 2

$R^1 = R^2 = R^3 = CH_3$, $R^4 = C_2H_5$ — MONACTIN 3

$R^1 = R^3 = CH_3$, $R^2 = R^4 = C_2H_5$ — DINACTIN 4

$R^1 = CH_3$, $R^2 = R^3 = R^4 = C_2H_5$ — TRINACTIN 5

$R^1 = R^2 = R^3 = R^4 = C_2H_5$ — TETRANACTIN 6

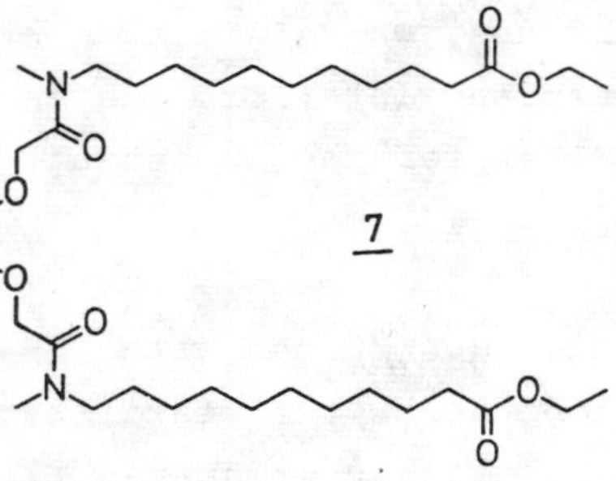

7

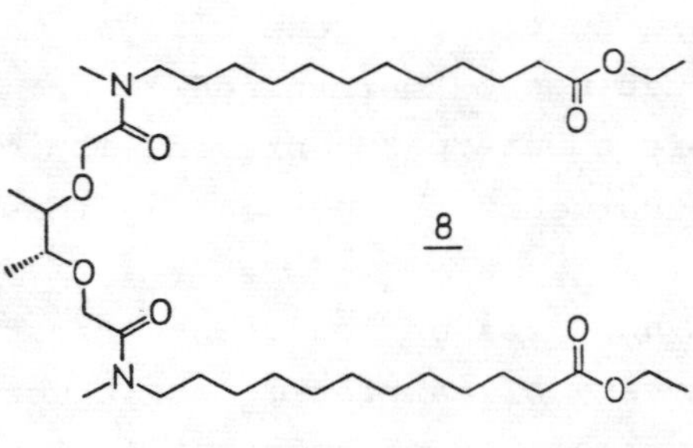

8

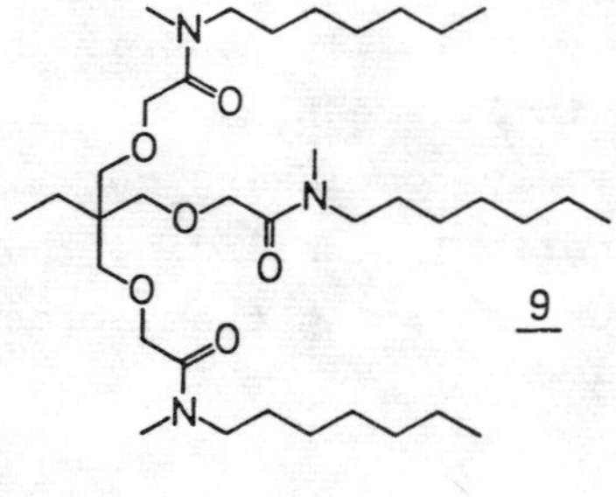

9

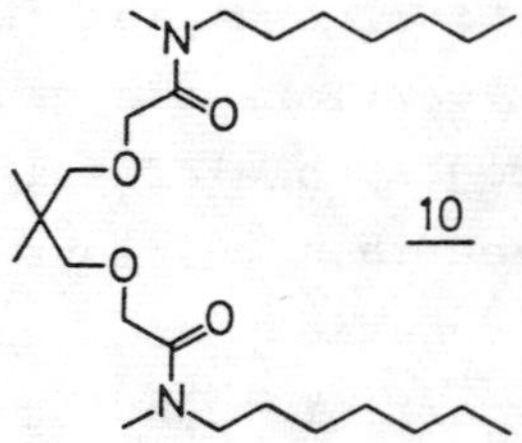

10

Figure 8.1. Structures of some electrically neutral, natural or synthetic ionophores for cations. 1: valinomycin, 2-6: macrotetrolide antibiotics, 7, 8: synthetic carriers for Ca^{2+} [6, 9], 9: carrier for Na^+ [9], 10: carrier for Li^+ [6, 9].

interior was considered as a single sharp energy barrier of the Eyring type. Since this model was not capable of rationalizing quantitatively the carrier-induced ion transport behavior of bilayer membranes, Hladky [21, 22] and more recently Ciani [23, 24] developed an extended theory which allows for a voltage dependence of the interfacial reactions (complexation and decomplexation), as well as for a trapezoidal shape of the internal free energy barrier for translocation of carrier complexes. This rather intriguing description was extremely successful in the interpretation of the conductances and zero-current transmembrane potentials observed on bilayers [24, 25] but, unfortunately, appears to be at variance with the earlier accepted [17] Läuger-Stark formalism when it is applied to the same limiting case (see, e. g., equations 3 and 5 in Reference 24). A different model was finally evolved by Morf et al. [26, 27] to account for the ion transport behavior of bulk membranes based on neutral carriers.

In this chapter a generalized theory is developed which incorporates all the strategic results of the earlier treatments mentioned above. Two formally different but equivalent models for the shape of the membrane-internal free energy profile are described which both cover the wide range between microscopic (bilayer lipid membranes) and macroscopic membranes (solvent polymeric membranes). These are the trapezoid-barrier model outlined by Hladky and Ciani and, alternatively, a multi-barrier concept which leads to a different interpretation of the membrane parameters involved.

8.1. DESCRIPTION OF THE ION FLUX ACROSS THE MEMBRANE INTERIOR

a) Trapezoid-barrier model

It has been shown in Chapters 4 and 7 that the steady-state flux of a species I between two positions (1) and (2) within an idealized membrane can be described by the Nernst-Planck equation, written here in the general form[*]

$$J_i = D_i \frac{c_i' \, e^{f_i'} - c_i'' \, e^{f_i''}}{\int\limits_{(1)}^{(2)} e^{f_i(x)} \, dx} \tag{8.1}$$

$$f_i(x) = \frac{\mu_i^o(x)}{RT} + \frac{z_i F \, \phi(x)}{RT} \tag{8.2}$$

where:

- J_i: flux density
- D_i: diffusion coefficient
- c_i', c_i'': concentration at position (1) resp. (2)
- z_i: ionic charge
- $f_i(x)$: reduced free energy function
- f_i', f_i'': value of $f_i(x)$ at position (1) resp. (2)
- $\mu_i^o(x)$: chemical standard potential
- $\phi(x)$: electrical potential
- R: gas constant
- T: absolute temperature
- F: Faraday constant

of species I in the membrane

For bulk membrane phases, the chemical standard potential of ions is practically invariant throughout the membrane (independent of x) and hence gives no contribution in Eq. (8.1). The presumed large effects of image forces and other, more specific interactions in the case of very thin membranes [19],

---

[*] In contrast to the former Eq. (7.2), the present description allows for local variations of the chemical standard potentials μ_i^o. For simplicity, Eqs. (8.1) and (8.2) are given in the ideal-solution form.

however, lead here to a pronounced inconstancy of the potential energy function μ_i^o/RT, respectively f_i. This means that the interior of bilayer membranes acts as an activation barrier for the passage of ions.

To cover both cases, namely a flat standard potential function or a high barrier located at $x=d/2$ (d is the membrane thickness), the free energy profile of an ion in the central portion of the membrane may be approximated by a trapezoid [22, 23]. This has the mathematical consequence:

$$\frac{1}{RT} \frac{d\mu_i^o}{dx} = 0 \qquad \text{[flat top of the trapezoid]} \qquad (8.3)$$

$$\frac{1}{RT} \frac{d\mu_i^o}{dx} = \frac{\omega_i}{\frac{1}{2}a-P_1a} \quad \text{respectively} \quad - \frac{\omega_i}{\frac{1}{2}a-P_1a} \qquad (8.4)$$

[remaining portions of the trapezoid]

where a is the basis of the trapezoid between positions (1) and (2), P_1a is the half width of the top of the trapezoid, and $\omega_i = \Delta\mu_i^o/RT$ is its height (see Figure 8.2). In addition we will use the constant-field approximation. It has been shown in Chapter 7 that this is an acceptable assumption for bilayer membranes [19, 28] as well as for bulk membranes under symmetrical conditions [26]. Hence:

$$\frac{z_iF}{RT} \frac{d\phi}{dx} = \frac{z_iF}{RT} \frac{\phi(2)-\phi(1)}{a} = \frac{z_i\phi_m}{a} \qquad (8.5)$$

where ϕ_m is the reduced membrane-internal electrical potential difference. If the height (ω_i) and the width ($2P_1a$) of the free energy barrier are sufficiently large, we can replace

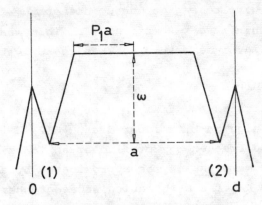

TRAPEZOID-BARRIER MODEL

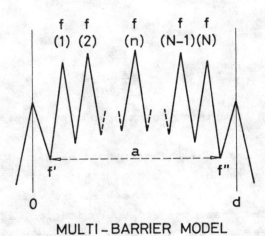

MULTI-BARRIER MODEL

Figure 8.2. Schematic diagram of different activation barriers (free energy profiles at zero voltage) considered for the translocation of ions across the interior of organic membranes (see text).

118

the integral in Eq. (8.1) by the portion of the integral along
the flat top of the barrier, i. e. between the positions
$x = d/2 - P_1a$ and $x = d/2 + P_1a$:

$$\int_{(1)}^{(2)} e^{f_i(x)} \, dx \cong \frac{a}{z_i \Phi_m} \left[e^{f_i(d/2+P_1a)} - e^{f_i(d/2-P_1a)} \right]$$

$$= 2P_1a \; e^{f_i(d/2)} \; \frac{\sinh(P_1 z_i \Phi_m)}{P_1 z_i \Phi_m} \tag{8.6}$$

The form of Eq. (8.6) is kept more general than the analogous
expressions used by Ciani who discusses two cases separately
(see equations 19 and 21 in Reference 23). Insertion into
Eq. (8.1) finally leads to

$$J_i = \frac{\bar{k}_i}{F_i(\Phi)} \left[c_i' \, e^{-z_i \Phi_m/2} - c_i'' \, e^{z_i \Phi_m/2} \right] \tag{8.7}$$

with

$$F_i(\Phi) = \frac{\sinh(P_1 z_i \Phi_m)}{P_1 z_i \Phi_m} \tag{8.8}$$

$$\bar{k}_i = D_i/2P_1a \; e^{\omega_i} \tag{8.9}$$

$F_i(\Phi)$ is a function of the applied voltage drop Φ, and $\bar{k}_i$
represents a rate constant for the translocation of ions
across the interior of the membrane which obviously depends
on the parameters of the activation barrier.

9*   119

Two limiting cases are easily derived from Eqs. (8.7) and (8.8). For a rectangular shape of the free energy barrier, we have $P_1 = 0.5$:

$$F_i(\Phi) = \frac{\sinh(0.5\ z_i \Phi_m)}{0.5\ z_i \Phi_m} \tag{8.10}$$

and immediately obtain a result that is formally identical to the well-known Goldman equation (7.4):

$$J_i = -z_i \Phi_m \bar{k}_i \frac{c_i'' e^{z_i \Phi_m/2} - c_i' e^{-z_i \Phi_m/2}}{e^{z_i \Phi_m/2} - e^{-z_i \Phi_m/2}} \tag{8.11}$$

In contrast, a sharp activation barrier is characterized by $P_1 \to 0$ or

$$F_i(\Phi) = 1 \tag{8.12}$$

and leads to an Eyring-type description of the ionic flux across the membrane interior, this having also been the basis of the treatment given by Läuger and Stark [20]:

$$J_i = \bar{k}_i c_i' e^{-z_i \Phi_m/2} - \bar{k}_i c_i'' e^{z_i \Phi_m/2} \tag{8.13}$$

As will be shown below, the same fundamental results, namely Eqs. (8.7), (8.11) and (8.13), may be deduced from a different conceptual framework.

b) Multi-barrier model

An alternative approach to the ion transport through membranes originates from the pioneering work by Eyring et al.
[29, 30]. In the absolute rate theory of membrane permeation
developed by these authors, the membrane is generally treated
as a series of N sharp activation barriers (see Figure 8.2).
Accepting this model for the membrane interior between positions (1) and (2), which represent local energy minima near
the interfaces, we get the flux equation in the following form
(see also Läuger and Neumcke [19]):

$$J_i = \frac{kT}{h} \, l_i \, \frac{c_i' \, e^{f_i'} - c_i'' \, e^{f_i''}}{\sum\limits_{n=1}^{N} e^{f_i(n)}} \tag{8.14}$$

where: $f_i(n)$: value of the free energy
function at the top of the
n-th barrier

l_i : distance between the interfacial activation barrier
and the neighboring first
barrier of the membrane
interior

for species I

k: Boltzmann's constant

h: Planck's constant

The other parameters of Eq. (8.14) have already been introduced in Eq. (8.1). Evidently, these two expressions constitute equivalent descriptions - in spite of the formal differences.

For simplicity, we shall assume that each of the N barriers
has the same height ω_i and the same basis a/N, and that again

the approximation of a constant electrical field can be applied. Then we may write:

$$f_i(n) \cong \frac{f_i' + f_i''}{2} + \omega_i - \frac{N+1-2n}{2N} z_i \Phi_m \qquad (8.15)$$

Finally, a result of the same form (8.7) can be derived

$$J_i = \frac{\bar{k}_i}{F_i(\Phi)} \left[c_i' e^{-z_i \Phi_m/2} - c_i'' e^{z_i \Phi_m/2} \right] \qquad (8.7)$$

if the following relations hold:

$$F_i(\Phi) = \frac{1}{N} \sum_{n=1}^{N} \exp\left[-\frac{N+1-2n}{2N} z_i \Phi_m \right] \qquad (8.16)$$

$$\bar{k}_i = \frac{kT}{h} \frac{l_i}{N} e^{-\omega_i} \qquad (8.17)$$

It goes without saying that the limiting cases according to Eq. (8.11) and Eq. (8.13) correspond to the situation $N \rightarrow \infty$ and $N = 1$, respectively:

$$[F_i(\Phi)]_{N \rightarrow \infty} = \int_0^1 \exp\left[-\frac{1-2y}{2} z_i \Phi_m \right] \cdot dy = \frac{\sinh(0.5 z_i \Phi_m)}{0.5 z_i \Phi_m} \qquad (8.18)$$

$$[F_i(\Phi)]_{N=1} = 1 \qquad (8.19)$$

The first case is certainly fulfilled in bulk liquid membranes whereas the second one may be realized in bilayer lipid membranes.

8.2. DESCRIPTION OF THE ION FLUX ACROSS THE INTERFACES

The transition of ions across the interfacial barriers between membrane and aqueous solutions is commonly characterized by a description of the Eyring type [21, 22, 26, 29, 30]. This approach turns out to be analogous to the classical formalism introduced by Butler [31], Erdey-Gruz, and Volmer [32] to describe kinetics of charge transfer reactions at metal-solution interfaces (see also Chapter 3). If the possibility of a formation of $1:n_i$ complexes between the species I and carriers S in the membrane is taken into account, the general form of the flux equation reads:

$$J_i = \vec{k}_i a_i' (c_s')^{n_i} \exp[-\alpha z_i (\Phi_o' + \Delta \Phi')]$$

$$- \overleftarrow{k}_i c_i' \exp[(1-\alpha) z_i (\Phi_o' + \Delta \Phi')] \tag{8.20a}$$

respectively for the other interfacial barrier:

$$J_i = \overleftarrow{k}_i c_i'' \exp[(1-\alpha) z_i (\Phi_o'' + \Delta \Phi'')]$$

$$- \vec{k}_i a_i'' (c_s'')^{n_i} \exp[-\alpha z_i (\Phi_o'' + \Delta \Phi'')] \tag{8.20b}$$

where:　　$\vec{k}_i$: rate constant of ion transfer from the aqueous
solution into the membrane (rate of the complexa-
tion reaction)

$\overleftarrow{k}_i$: rate constant of ion transfer out of the membrane
(rate of the decomplexation reaction)

a_i', a_i'': activity of the uncomplexed species I in the
aqueous solutions contacting the membrane

c_i', c_i'': concentration of the $1:n_i$ carrier complexes of
species I inside the interfacial barriers

c_s', c_s'': concentration of the free carriers in the membrane

Φ_o', Φ_o'': reduced surface potential difference existing
across the interfacial barriers at voltages $\Phi \approx 0$
(measured in units RT/F)

$\Delta\Phi'$, $\Delta\Phi''$: part of the applied voltage Φ dropping across the
interfacial barriers (overpotential)

$\alpha \approx 0.5$: transfer factor (part of the surface potential
difference operating on the aqueous side of the
interfacial barrier)

In most of the earlier theories devoted to bilayer membranes,
the potential contributions Φ_o' and Φ_o'' were neglected or tacitly
assumed to be included in the rate constants of the interfacial
reactions. These surface potentials are a major factor determining the ion selectivity of the membrane, however, and should
not be overlooked in the case of bulk membranes (e. g. solvent
polymeric membranes).

At equilibrium, $J_i = 0$ holds and the two terms on the right
hand side of Eq. (8.20a) or (8.20b) become equal. This leads
to the following relationships for the equilibrium distribution
of species I between the aqueous solutions and the first free
energy minima inside the membrane (for $\Delta\Phi' = \Delta\Phi'' = 0$):

$$c'_{i,o} = (\vec{k}_i/\overleftarrow{k}_i) \; a'_i c_s^{n_i} \; e^{-z_i \phi'_o} \tag{8.21a}$$

$$c''_{i,o} = (\vec{k}_i/\overleftarrow{k}_i) \; a''_i c_s^{n_i} \; e^{-z_i \phi''_o} \tag{8.21b}$$

where $c'_{i,o}$ and $c''_{i,o}$ are the equilibrium values of c'_i and c''_i, respectively, and c_s is the equilibrium concentration of free carriers within the membrane. This quantity will be described in the following.

8.3. CONSEQUENCES OF A CLOSED-CIRCUIT FLUX OF CARRIERS

The mechanism of carrier-mediated ion transport through membranes involves a translocation of carrier complexes in one direction and a subsequent back-diffusion of free carriers in the opposite direction. In the steady-state, the total flux of carrier molecules approximates zero. Since the total carrier flux is composed of contributions by each sort of complex $(n_i J_i)$ as well as by the free carrier (J_s), we thus obtain the relationship

$$\sum_i n_i J_i + J_s = 0 \tag{8.22}$$

The mass fluxes entering into Eq. (8.22) can be rigorously formulated on the basis of Eq. (8.1), respectively (8.14). The corresponding expression for the electrically neutral species S assumes the simple form

$$J_s = \bar{k}_s (c'_s - c''_s) \tag{8.23}$$

125

Another steady-state condition follows from the assumption of conservation of free carriers within the membrane. This reasonable approximation was basically imposed in all the aforementioned theoretical treatments of carrier-modified membranes and relates the carrier concentrations c_s' and c_s'' to the value c_s at zero fluxes:

$$c_s' + c_s'' = 2c_s \qquad (8.24)$$

A detailed description of the equilibrium concentration of free carriers in the membrane, c_s, in terms of outside concentrations and rate constants of partitioning reactions was offered by Ciani [23][*]. There is plenty of experimental evidence on lipid bilayers indicating that c_s is normally proportional to the concentration (activity) of carriers available in the aqueous solutions.

Combination of Eqs. (8.22) – (8.24) finally leads to

$$c_s' = c_s \left(1 - \frac{\Sigma n_i J_i}{2\bar{k}_s c_s}\right) \qquad (8.25a)$$

$$c_s'' = c_s \left(1 + \frac{\Sigma n_i J_i}{2\bar{k}_s c_s}\right) \qquad (8.25b)$$

These results illustrate that a positive flux of cation/carrier complexes through the membrane interior (as induced, e. g., by an applied voltage) clearly leads to a certain

---

[*] In this article, the concentrations $c_i [\text{mol cm}^{-3}]$ were replaced by so-called surface concentrations $N_i = c_i \cdot d/2 \ [\text{mol cm}^{-2}]$ throughout.

126

accumulation of free carriers at the side of the unloading of ions, respectively to a depletion at the side of the loading of ions [10, 26]. These deviations from a symmetrical distribution of free carriers are negligible, however, as long as $\Sigma n_i J_i << \bar{k}_s c_s$.

8.4. DERIVATION OF THE GENERAL RESULT

The aim of the present section is to combine the former Eqs. (8.7) and (8.20a,b) into a closed general formula for the ion flux J_i. To this end, a series of additional assumptions and definitions concerning the potential contributions Φ_m, $\Delta\Phi'$, and $\Delta\Phi''$ have to be imposed.

First, we introduce a dimensionless function Φ that is related to the actual transmembrane potential V as follows:

$$\Phi = \frac{F}{RT} V + \Phi'_o - \Phi''_o \qquad (8.26)$$

Recalling that V is the sum of membrane-internal and interfacial potential contributions[*]:

$$-\frac{F}{RT} V = \Phi_m + (\Phi'_o + \Delta\Phi') - (\Phi''_o + \Delta\Phi'') \qquad (8.27)$$

---

[*] The sign of the transmembrane potential V conforms to the convention that is usually accepted for cell potentials in electrolysis cells. For potentiometric cell arrangements, the use of the electromotive force $E = -V_o$ (at zero-current) is preferable.

we immediately get

$$\Phi_m + \Delta\Phi' - \Delta\Phi'' = -\Phi \qquad (8.28)$$

Second, we postulate that the same overpotential is produced across the two, oppositely oriented interfacial barriers,
i. e.

$$\Delta\Phi' = -\Delta\Phi'' \qquad (8.29)$$

The validity of this assumption has been substantiated for
both microscopic [17, 21 - 24] and macroscopic membranes (see
equation 8a in Reference 26).

For an adequate translation of the potential functions
entering into Eqs. (8.7), (8.8), (8.16), and (8.20) in terms
of Φ, we finally introduce two pivotal membrane parameters as
follows:
a) The portion of the membrane-internal potential difference
   that operates along <u>half of the flat top of a trapezoidal
   barrier</u> (Figure 8.2) corresponds to a fraction P_2 of the
   total applied voltage.
b) The voltage drop occurring across the <u>region between the
   center of the membrane and the top of an interfacial barrier</u>
   (Figure 8.2) amounts to a fraction P_3 of the total applied
   voltage.

In the bilayer models worked out by Hladky [22] and Ciani [23]
the total applied voltage (transmembrane potential) was assumed
to fall with a constant gradient across the physical range of
the membrane, i. e. between the surfaces at x=0 and x=d. There-

128

fore, the symbols P_2 and P_3 introduced by Eisenman et al. [24] are purely geometrical parameters, characterizing the half-width of the plateau of the "diffusion barrier" and the position of the "reaction plane", respectively (both in units of membrane thickness). Generally, the formal definitions of P_2 and P_3 are the following:

$$-P_2 \Phi = P_1 \Phi_m \qquad [0 < P_2 \leqslant P_3] \qquad (8.30)$$

$$-P_3 \Phi = \frac{1}{2} \Phi_m + (1-\alpha) \, \Delta\Phi' \qquad [P_2 \leqslant P_3 \leqslant 0.5] \qquad (8.31)$$

From Eqs. (8.28) - (8.31) we may obtain an interrelation between the membrane parameters P_1, P_2, and P_3:

$$P_2/P_1 = \frac{2P_3 - (1-\alpha)}{\alpha} \approx 4P_3 - 1 \qquad (8.32)$$

Since the ratio P_2/P_1 precisely relates the membrane-internal to the total voltage drop (see Eq. (8.30)), one can finally rewrite all the potential functions in terms of Φ, P_2 (trapezoid-barrier model) or N (multi-barrier model), P_3, and α.

Combination of the flux equations (8.7) and (8.20) is now accomplished straightforwardly although the elimination of the unknown concentration and potential terms does bear some algebraic problems. The general result for J_i reads:

$$J_i \left\{ F_i(\Phi) + \frac{\bar{k}_i}{\overleftarrow{k}_i} \left[\frac{\exp(P_3 z_i \Phi)}{\exp((1-\alpha) z_i \Phi_o')} + \frac{\exp(-P_3 z_i \Phi)}{\exp((1-\alpha) z_i \Phi_o'')} \right] \right\}$$

$$= \bar{k}_i c_{i,o}' \left[1 - \frac{\Sigma n_i J_i}{2\bar{k}_s c_s} \right]^{n_i} \exp(z_i \Phi / 2)$$

$$- \bar{k}_i c_{i,o}'' \left[1 + \frac{\Sigma n_i J_i}{2\bar{k}_s c_s} \right]^{n_i} \exp(-z_i \Phi / 2) \qquad (8.33)$$

The equilibrium concentrations $c_{i,o}'$ and $c_{i,o}''$ are given by Eqs. (8.21a) and (8.21b), respectively, and the barrier function $F_i(\Phi)$ is obtained as follows:

$$F_i(\Phi) = \frac{\sinh(P_2 z_i \Phi)}{P_2 z_i \Phi} \qquad \text{[trapezoid-barrier]} \qquad (8.34)$$

$$F_i(\Phi) = \frac{1}{N} \sum_{n=1}^{N} \exp\left[\frac{N+1-2n}{2N} \frac{2P_3 - (1-\alpha)}{\alpha} z_i \Phi \right] \quad \text{[multi-barrier]} \quad (8.35)$$

$$\approx \frac{1}{N} \sinh\left[\frac{1}{2}(4P_3 - 1) z_i \Phi \right] \Big/ \sinh\left[\frac{1}{2N}(4P_3 - 1) z_i \Phi \right]$$

where $\quad \Phi = \frac{F}{RT} V + \Phi_o' - \Phi_o''$.

The implicit formalism of these fundamental relationships appears to be rather cumbersome. On the other hand, it should be pointed out that they offer a universal description for the transport of carrier-bound ions ($z_i \neq 0$, $n_i > 0$), free ions ($z_i \neq 0$, $n_i = 0$), or even uncharged species ($z_i = 0$, $n_i = 0$ or $n_i > 0$) in bilayer or bulk membranes ! Among the

variety of membrane phenomena that may be rationalized directly on the basis of Eqs. (8.33) - (8.35), current-voltage or
conductance-voltage characteristics, zero-current membrane potentials, as well as ion-selectivity criteria are of special
interest. Besides, an interpretation of current saturation
(e. g., for $\Sigma n_i J_i \rightarrow 2\bar{k}_s c_s$ [23, 26]) and rectification phenomena
(for $c'_{i,o} \neq c''_{i,o}$ [22, 28, 33] and/or $\Phi'_o \neq \Phi''_o$ [19]) is easily
accomplished.

The following discussion of the present theory is restricted
to the limiting cases that correspond to the treatments by
Läuger and Stark [20], Ciani, Eisenman, and Krasne [23, 24],
and Morf et al. [26].

8.5. BILAYER MODEL BY LÄUGER AND STARK

In the classical treatment by Läuger and Stark [20] the
membrane interior between the local energy minima near the
interfaces was considered as a single sharp activation barrier.
This corresponds to the situation $F_i(\Phi) = 1$ (see Eqs. (8.12)
and (8.19)). The membrane was apparently assumed to consist
of two chemically identical layers of lipids, so that
$\Phi'_o = \Phi''_o = \Phi_o$. In addition, the model is restricted to one sort
of cations I^+ of the activity $a'_i = a''_i = a_i$ that form exclusively 1:1 complexes with neutral carriers S. For such a simplified
system, Eq. (8.33) reduces to the following explicit description of the steady-state current density j[*)]:

[*)] The terms $\bar{k}_i c_i$, $\bar{k}_i c_i/w_i$, and $\bar{k}_s c_s$ in Eq. (8.36) correspond
to $k_{MS} N_{MS}$, $k_D N_{MS}$, and $k_S N_S$ in Läuger's terminology [20, 34].
N_{MS} and N_S are the surface concentrations of complexes and free
carriers in the membrane at equilibrium, k_{MS} and k_S are the rate
constants for translocation of the respective species across
the membrane interior, and k_D is the rate constant of the
decomplexation reaction. Whereas the dimension of $\bar{k}_i$, $\overleftarrow{k}_i$, and $\bar{k}_s$
is cm sec^{-1}, the rate constants k_{MS}, k_D, and k_S are given in
sec^{-1} [34].

$$\frac{j}{z_i F} = J_i = \frac{2\,\sinh(z_i \Phi/2)}{\dfrac{1}{\bar{k}_i c_i} + \dfrac{2\,w_i}{\bar{k}_i c_i}\cosh(P_3 z_i \Phi) + \dfrac{1}{\bar{k}_s c_s}\cosh(z_i \Phi/2)} \tag{8.36}$$

with:

$$c_i = c'_{i,o} = c''_{i,o} = (\vec{k}_i/\overset{\leftarrow}{k}_i)\, a_i c_s{}^{n_i}\, e^{-z_i \Phi_o} \quad \begin{array}{l}[\text{here } n_i=1 \\ \text{and } z_i=1]\end{array} \tag{8.37}$$

and

$$w_i = \bar{k}_i/\overset{\leftarrow}{k}_i \; \exp[(1-\alpha)\, z_i \Phi_o] \tag{8.38}$$

In the paper by Läuger and Stark, the value of P_3 was 0.5. The zero-current conductance of symmetrical membrane arrangements turns out to be formally independent of the parameters P_3 and P_2 or N, however, although the latter are included implicitly in the terms $\bar{k}_i$ and w_i. From Eq. (8.33) or (8.36), we indeed obtain the following general relationship for the zero-current membrane conductance G_o in the presence of only one sort of ions ($a'_i = a''_i = a_i$; $\Phi'_o = \Phi''_o = \Phi_o$):

$$G_o = \lim_{\Phi \to 0} \frac{j}{(RT/F)\,\Phi} = \frac{(z_i F)^2}{RT} \; \frac{1}{\dfrac{1}{\bar{k}_i c_i} + \dfrac{2\,w_i}{\bar{k}_i c_i} + \dfrac{n_i^2}{\bar{k}_s c_s}} \tag{8.39}$$

The series of reciprocal terms appearing in the denominator of this expression is related to the resistance to ion migration. The contributions to the total membrane resistance are

132

evidently due to the membrane-internal ion translocation, the interfacial reactions, as well as the back diffusion of free carriers (terms from left to right in Eq. (8.39), respectively in Eq. (8.36)).

If the pure ion translocation is the rate-limiting step, we immediately get a "hyperbolic" [24] shape of the current-voltage characteristic:

$$\frac{j}{z_i F} = 2 \bar{k}_i c_i \sinh(z_i \phi /2) \quad [\bar{k}_i c_i << \bar{k}_s c_s \text{ and } w_i \approx 0] \qquad (8.40)$$

In this limit, the zero-current membrane conductance is simply given as

$$G_o = \frac{(z_i F)^2}{RT} \bar{k}_i c_i = \frac{(z_i F)^2}{RT} \bar{k}_i (\vec{k}_i / \overleftarrow{k}_i) \, a_i c_s^{n_i} \, e^{-z_i \phi_o} \qquad (8.41)$$

In agreement with experimental evidence from lipid bilayers, Eq. (8.41) leads to the following conclusions:

The zero-current membrane conductance is
1) proportional to the outside activity a_i of permeating ions [17 - 20, 34],
2) proportional to the n_i-th power of the carrier concentration c_s [17, 18, 20, 34],
3) heavily influenced by the surface potential ϕ_o of the bilayer, which is mainly given by the nature of the lipid [18, 35].

Since the validity of Eqs. (8.39) and (8.41) is rather general, as mentioned earlier, the rules 1) - 3) are also true when different assumptions concerning the parameters P_2 or N are made (see Section 8.6).

In contrast to Eq. (8.40), a current-voltage curve of the "saturation" type is predicted if the back-flow of free carriers in the membrane or the decomplexation reaction at the interface (for $P_3 = 0.5$) is the rate-limiting process:

$$\frac{j}{z_i F} = \frac{2}{\dfrac{2\,w_i}{\bar{k}_i c_i} + \dfrac{1}{\bar{k}_s c_s}} \tanh(z_i \Phi/2) \quad [\bar{k}_i c_i \gg \bar{k}_s c_s \text{ or } w_i \gg 1] \qquad (8.42)$$

Although the cases of "hyperbolic" and "saturation" type current-voltage characteristics are well known for the carrier-mediated transport of cations in bilayers (see next section), neither Eq. (8.40) nor Eq. (8.42) (or Eq. (8.36)) are capable of a quantitative interpretation of the corresponding experimental results. This lack of theory led Eisenman's group [23, 24, 33] and others [9, 21, 22] to the creation of a more sophisticated membrane model. However, a current-voltage behavior according to Eq. (8.40) was obtained earlier for the transport of several lipophilic anions ($z_i = -1$, $n_i = 0$) across dioleoyl lecithin bilayers. The correlation presented in Figure 8.3 is surprising and suggests that the internal free energy profile for anion translocation in such lipid membranes is approximated merely by a single sharp activation barrier. In contrast, a broad barrier or a couple of free energy peaks must be operative in the translocation of cation-carrier complexes across the same membranes.

8.6. BILAYER MODEL BY CIANI, EISENMAN, AND KRASNE

The principal merit of the work by Ciani, Eisenman, and

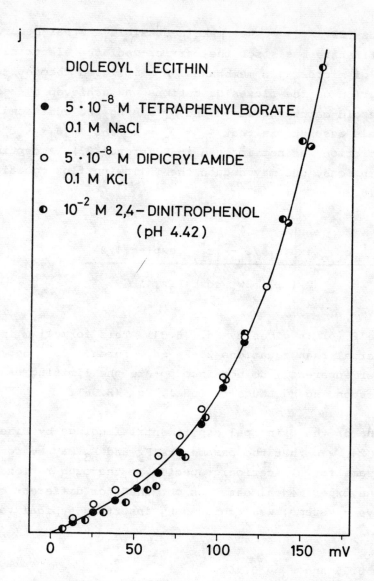

Figure 8.3. Current-voltage characteristics for dioleoyl lecithin bilayer membranes in the presence of different lipophilic
anions (25°C). The experimental points are taken from conductance or current data reported by Läuger et al. [19, 36]; the
scales of current densities j were transformed to yield a
common point at 128.5 mV (Φ=5). The solid line was calculated
according to Eq. (8.40).

Krasne [23 - 25] was to make accessible a straightforward quantitative basis for the carrier-mediated electrical properties of microscopic membranes. A simplified theoretical description of the fluxes of cations was achieved by using the trapezoid-barrier concept and by making the additional reasonable assumptions that $\Phi_o' = \Phi_o'' = \Phi_o$ and $\Sigma n_i J_i << \bar{k}_s c_s$ (the last restriction was not imposed in Reference 23). Accepting these assumptions, one may deduce the following flux equation from Eq. (8.33):

$$J_i = \frac{\bar{k}_i c_{i,o}' \exp(z_i \Phi/2) - \bar{k}_i c_{i,o}'' \exp(-z_i \Phi/2)}{F_i(\Phi) + 2w_i \cosh(P_3 z_i \Phi)} \qquad (8.43)$$

where $F_i(\Phi)$ is given by Eq. (8.34). This formalism is more universal than equations 23-26 in Reference 23[*] because the latter apparently do not incorporate the limiting case $P_2 \to 0$ corresponding to Läuger's model, Eq. (8.36).

One of the principal experimental findings by Eisenman et al. [24] was that the parameters P_2 and P_3 may be considered the same for all cationic species. A thorough rationalization of the experimental data, as obtained for different carrier-bilayer-systems, was achieved by inserting typical values of

$$P_2 \cong 0.35 \text{ and } P_3 \cong 0.46 \qquad (8.44)$$

---

[*] The terms $\bar{k}_i c_i$ correspond to $\tilde{A}_{is}^* N_{is}^*/P_2$ in the terminology of Eisenman's group [23 - 25], and w_i is transformed into $\tilde{w}_i/P_2$.

into Eq. (8.43), respectively Eq. (8.34). A deeper understanding of these remarkable findings follows from the multi-barrier model presented in this work. By treating the interior of the bilayer as a couple of barriers (here $N = 2$) of the Eyring type, we deduce from Eq. (8.35):

$$F_i(\Phi) = \cosh \left[\frac{2P_3 - 1 + \alpha}{4\alpha} \, z_i \Phi \right]$$

$$\cong \cosh[(4P_3 - 1) z_i \Phi / 4], \text{ for } \alpha \cong 0.5 \tag{8.45}$$

This result turns out to be practically identical to Eisenman's term (8.34)

$$F_i(\Phi) = \frac{\sinh(P_2 z_i \Phi)}{P_2 z_i \Phi}$$

in the usual voltage range 0 to 150 mV (for $z_i = 1$) if it holds that

$$P_2 \cong 0.42 \, (4P_3 - 1) \tag{8.46}$$

Such a relationship between P_2 and P_3 is in fact fulfilled for the carrier-mediated cation transport in lipid bilayers, as may be seen from the experimental values (8.44). Finally, the corresponding flux equation for a symmetrical membrane arrangement ($c'_{i,o} = c''_{i,o} = c_i$) reads:

137

$$J_i = \frac{2\bar{k}_i c_i \sinh(z_i \Phi/2)}{\cosh(P_3 z_i \Phi - z_i \Phi/4) + 2w_i \cosh(P_3 z_i \Phi)} \tag{8.47}$$

This expression may be replaced by the following approximation for $P_3 \cong 0.5$ and $w_i \cong 0$ ("equilibrium domain"):

$$J_i = 4\bar{k}_i c_i \sinh(z_i \Phi/4) \tag{8.48}$$

which is a result intermediate to the former case (8.40) and the ohmic behavior usually fulfilled for macroscopic membranes (see next section).

A current-voltage behavior according to (8.47), respectively (8.48), is illustrated in Figure 8.4 for the transport of K^+ across phosphatidylserine membranes in the presence of the carrier antibiotics valinomycin 1 and monactin 3 (see Figure 8.1), and in Figure 8.5 for the transport of Ca^{2+} across lecithin bilayers modified by the synthetic ionophore 8. Whereas the current-voltage curves for K^+-monactin and Ca^{2+} show the normal "hyperbolic" shape according to Eq. (8.48), the characteristic for K^+-valinomycin is of the "saturation" type, which indicates that here reaction kinetics at the membrane/solution interfaces become rate-limiting ("kinetic domain" with $w_i > 0$). Further results were discussed by Eisenman et al. [24, 25].

A readily accessible estimate of the cation selectivity of carrier-bilayer-systems is obtained from the study of zero-current membrane conductances or zero-current membrane potentials [17, 18, 23 - 25, 34]. A straightforward description of these electrical properties is based on flux equations of the type (8.43). Recalling Eqs. (8.21a,b), we can replace the interfacial complex-concentration values $c'_{i,o}$ and $c''_{i,o}$ in

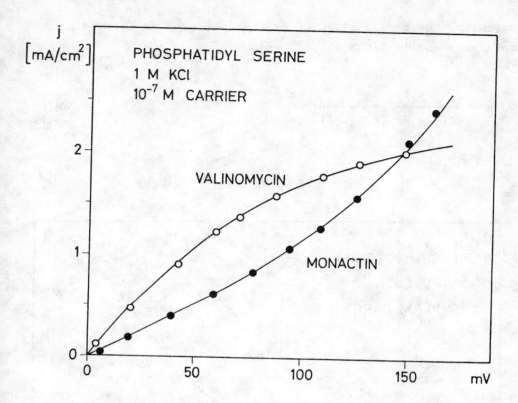

Figure 8.4. Current-voltage characteristics for phosphatidyl serine bilayer membranes in the presence of valinomycin (<u>1</u>) or monactin (<u>3</u>). The aqueous solution contained 1M KCl (25°C). The experimental points are taken from fig. 10 in [34]. The solid lines are theoretical curves according to Eq. (8.47) (valinomycin; $P_3 = 0.5$ and $w_i = 0.46$) respectively Eq. (8.48) (monactin; $P_3 = 0.5$ and $w_i = 0$).

(8.43) by the cation activities a_i' and a_i'' of the outside solutions, respectively. We then may write:

$$J_i = P_i \, [a_i' \exp(z_i \Phi/2) - a_i'' \exp(-z_i \Phi/2)] \qquad (8.49)$$

Herewith, the permeability P_i of the species I is defined most naturally and is given as follows:

$$P_i = \frac{\bar{k}_i (\vec{k}_i / \overleftarrow{k}_i) c_s^{n_i} e^{-z_i \Phi_o}}{F_i(\Phi) + 2w_i \cosh(P_3 z_i \Phi)} \qquad (8.50)$$

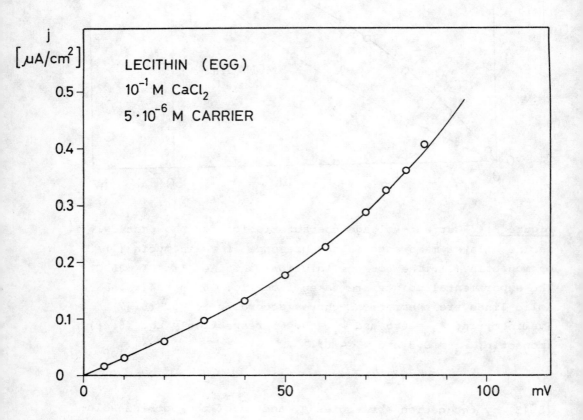

Figure 8.5. Current-voltage characteristics for lecithin bi-layer membranes in the presence of the synthetic Ca^{2+}-carrier 8 (Fig. 8.1). The aqueous solution contained 0.1M $CaCl_2$ and 5μM ligand (25°C) [37]. The theoretical curve was drawn according to Eq. (8.48).

140

The magnitude of this strategic parameter obviously depends on the transmembrane potential, $V = (RT/F)\Phi$, except when it holds that $F_i(\Phi) = 1$ and $w_i = 0$. The permeability ratio P_j/P_i between two ions of the same charge, I^{z+} and J^{z+}, represents directly a measure for the ion selectivity of the carrier-membrane system studied. This selectivity parameter can be determined from zero-current membrane potentials V_o measured in mixed aqueous solutions of the two ions [17, 23, 24]:

$$V_o = \frac{RT}{zF} \ln \frac{a_i'' + \left(\frac{P_j}{P_i}\right)_{V_o} a_j''}{a_i' + \left(\frac{P_j}{P_i}\right)_{V_o} a_j'} \qquad (8.51)$$

or from zero-current membrane conductances $G_o(i)$ and $G_o(j)$, as obtained from separate measurements using solutions of the species I^{z+} or J^{z+} of the same activity $a' = a''$ [17, 23, 24]:

$$\frac{G_o(j)}{G_o(i)} = \left(\frac{P_j}{P_i}\right)_{V \to 0} \qquad (8.52)$$

The fundamental relationships (8.51) and (8.52) follow immediately from flux equations of the type (8.49) after imposing the condition of zero current. The symbols $(P_j/P_i)_{V_o}$ and $(P_j/P_i)_{V \to 0}$ have been used to indicate that the permeability ratios often remain a function of the transmembrane potential V.

Since the potential functions entering into Eq. (8.50) are identical for different cations of the same charge, the ion-selectivity parameter P_j/P_i becomes potential-independent as long as $w_i \cong 0$ and $w_j \cong 0$. This limit was called "equilibrium domain" [17, 23, 24] and is characterized by

141

$$\frac{P_j}{P_i} = \left(\frac{P_j}{P_i}\right)_{Eq} = \frac{\bar{k}_j (\vec{k}_j / \overleftarrow{k}_j) c_s^{n_j}}{\bar{k}_i (\vec{k}_i / \overleftarrow{k}_i) c_s^{n_i}} \qquad [w_i, w_j \cong 0] \qquad (8.53)$$

It was clearly demonstrated by Eisenman et al. [18, 24, 25] that the equilibrium ion-selectivity of a given carrier (with $n_i = n_j = 1$ throughout) is virtually the same for different lipid compositions of the bilayer membrane. Moreover, it does not depend on the method of determination, i. e. potentiometry or conductance measurements. The experimental studies by Eisenman's group showed, however, that it is possible to encounter situations for which the rates of the interfacial reactions become comparable to the rates of transport of cationic complexes across the membrane interior. This limit was called "kinetic domain" and is characterized by values of the kinetic parameters of $w_i > 0$ and/or $w_j > 0$ (see Eq. (8.38)). Here, the selectivity of ion transport generally depends on the transmembrane potential. Only for rather hypothetical systems where the rates of cation translocation are much higher than the effective rates of decomplexation (i. e. w_i, $w_j \gg 1$), the permeability ratios would again assume constant values:

$$\frac{P_j}{P_i} = \left(\frac{P_j}{P_i}\right)_{Eq} \cdot \frac{w_i}{w_j} = \frac{\vec{k}_j c_s^{n_j}}{\vec{k}_i c_s^{n_i}} \qquad [w_i, w_j \gg 1] \qquad (8.54)$$

The exciting work presented by Eisenman et al. [18, 24, 25] revealed that Eq. (8.53) offers an adequate description for most carrier-bilayer-systems. Distinct effects of interfacial kinetics were observed, however, for bilayers formed from glyceryl dioleate and related lipids. From the experimental data reported [24], we get a surprising correlation between the ratios of kinetic parameters, w_j / w_i, and the corresponding equilibrium selectivities, $(P_j / P_i)_{Eq}$, for alkali metal ions and different carriers (see Figure 8.6):

$$\frac{w_j}{w_i} \approx \left(\frac{P_j}{P_i}\right)_{Eq} \qquad \text{[alkali ions, } n_i = n_j \text{]} \qquad (8.55)$$

or, according to Eq. (8.54):

$$\frac{\vec{k}_j}{\vec{k}_i} \approx 1 \qquad \text{[alkali ions, } n_i = n_j \text{]} \qquad (8.56)$$

This means nothing less than that the rate of the complexation reaction at the membrane-solution interface is nearly the same for all alkali ions (for a given carrier-bilayer-system). Thus, if substantial kinetic limitations at the interfaces come into play, a serious loss in the ion selectivity will occur, as is evident from Eqs. (8.54) - (8.56). Such a kinetic effect was predicted earlier [26] but is here corroborated by experimental evidence. We may conclude that the complexation and decomplexation steps in the carrier-mediated ion transport through membranes should be relatively fast processes. This is a prerequisite for both a high rate of and a high selectivity in ion permeation, such as obtains in many biological membrane systems.

8.7. COMPARISON WITH THE BULK MEMBRANE MODEL BY MORF, WUHRMANN, AND SIMON

The concept of bulk membranes differs significantly from bilayers since here electroneutrality must be maintained in the membrane interior. Whereas permselectivity of bilayers for cations (relative to anions) may easily be interpreted by

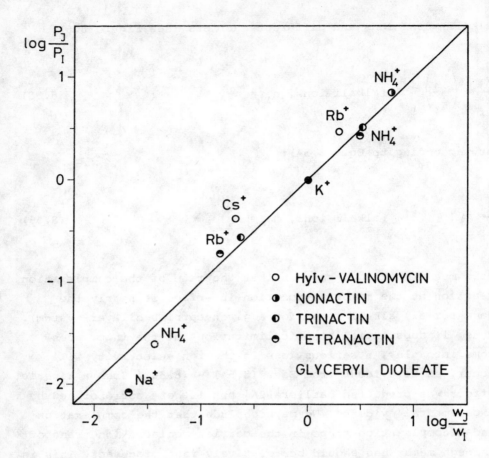

Figure 8.6. Correlation between the ratios of equilibrium
permeabilities $(P_J/P_I)_{Eq}$ and the ratios of kinetic parameters
(w_J/w_I) for different ion-carrier combinations in glyceryl
dioleate bilayers. The experimental values are taken from [24]
$(I^{z+}: K^+)$. The carrier hydroxyisovalerate-valinomycin is a
homolog of 1, containing exclusively isopropyl substituents
(see Fig. 8.1).

a preferential uptake of these species - solubilized by lipophilic ligand shells - into the membrane, such a space-charge
model becomes untenable in the case of macroscopic membranes

[38, 39]. Nevertheless, all the analytically relevant bulk membrane types based on neutral carriers are capable of producing permselectivity for cations, which fact is observable by an exclusive transport of cations [8 - 11, 26, 39]. The underlying mechanism will be elucidated in Chapter 12.

A previous theoretical approach to the ion transport behavior of thick, electroneutral carrier membranes [26, 27] assumed constancy of the concentration c of anionic sites existing within the membrane phase (see also Section 7.3). Hence:

$$\Sigma z_i c'_{i,o} = \Sigma z_i c''_{i,o} = c = \text{const} \qquad (8.57)$$

For simplicity, the translocational rate constants $\bar{k}_i$ were assumed to be the same for all counter-ions, i. e. $\bar{k}_i = \bar{k}$. This is a reasonable assumption for "isosteric" cation-carrier complexes and, in addition, justifies the use of the constant-field approximation which was applied in the derivation of Eq. (8.7). Since the rate constant for crossing the membrane interior, $\bar{k}$, assumes very low values in the case of thick membranes, the "equilibrium domain" of ion transport is usually not exceeded. We therefore shall restrict the present discussion to purely diffusion-controlled systems where it holds that $w_i = 0$ and $\Sigma n_i J_i << \bar{k}_s c_s$. Such bulk membrane systems can be adequately characterized by values of the parameters P_1-P_3 of 0.5. Finally, the general flux equation (8.33) may be applied in the following reduced form:

$$J_i = \frac{\bar{k}}{F_i(\Phi)} \left[c'_{i,o} \exp(z_i \Phi/2) - c''_{i,o} \exp(-z_i \Phi/2) \right]$$

$$= z_i \Phi \bar{k} \frac{c'_{i,o} \exp(z_i \Phi/2) - c''_{i,o} \exp(-z_i \Phi/2)}{\exp(z_i \Phi/2) - \exp(-z_i \Phi/2)} \tag{8.58}$$

This expression turns out to be analogous to the Goldman equation (8.11).

We first consider the results obtained for the electrical properties of symmetrical membrane arrangements. For $a'_i = a''_i = a_i$ and $\Phi'_o = \Phi''_o = \Phi_o$, Eq. (8.58) further reduces to

$$J_i = \bar{k} \, z_i c_i \Phi = \bar{k} \, z_i c_i \frac{F}{RT} V \tag{8.59}$$

Equation (8.59) conforms to the former expression (7.19); it is valid, in the framework of the present model, for thick cation-permselective membranes under symmetrical conditions. For systems with only one sort of cations we immediately get

$$\frac{j}{F} = z_i J_i = z_i \bar{k} c \frac{F}{RT} V \tag{8.60}$$

Equation (8.60) predicts ohmic behavior of the membrane[*], the conductance being roughly independent of the nature

---

[*] Saturation of the current occurs only at voltages on the order of several volts and is due to finite back-diffusion of free carriers in the membrane [26].

of the cation. This agrees with experimental observations on 1.5 mm thick valinomycin-heptane liquid membranes for which a conductance ratio of $K^+: Na^+ \approx 2$ was reported [40]. Such behavior is in striking contrast to the findings for lipid bilayer membranes where the zero-current conductance was a direct measure of their ion selectivity. Nevertheless, a pronounced selectivity in the cation transport of thick carrier-based membranes can be observed if different ions are permeating simultaneously and if their transference number is studied (Section 7.5).

A straightforward description of the zero-current membrane potential of carrier-based bulk membranes becomes also available from Eqs. (8.58) and (8.21a,b). The solution for two cations of the same charge z is the well known Nicolsky equation:

$$V_o = \frac{RT}{zF} \ln \frac{a_i'' + K_{ij}^{Pot} a_j''}{a_i' + K_{ij}^{Pot} a_j'} \tag{8.61}$$

This result is formally identical to Eisenman's equation (8.51) but was derived here for the case of electroneutral membranes. The selectivity coefficient K_{ij}^{Pot} is given as follows:

$$K_{ij}^{Pot} = \frac{(\vec{k}_j / \overleftarrow{k}_j) \ c_s^{n_j}}{(\vec{k}_i / \overleftarrow{k}_i) \ c_s^{n_i}} \tag{8.62}$$

. Evidently, this selectivity parameter of bulk membranes is equivalent to the equilibrium permeability ratio of bilayers, as defined in Eq. (8.53):

$$K_{ij}^{Pot} \stackrel{\wedge}{=} \left(\frac{P_j}{P_i} \right)_{Eq} \tag{8.63}$$

147

For membranes modified by the carrier antibiotic valinomycin or the macrotetrolides, which form predominantly 1:1 complexes with alkali metal ions, the cation selectivity was shown to be dictated mainly by the stability constants of the complexes formed [5 - 9, 11, 15 - 18, 26, 41]. This may offer an explanation for the experimentally corroborated fact that one given carrier usually induces the same selectivity pattern, independent of the membrane type and the method used (see Figure 8.7 and [5 - 11, 18, 24 - 26, 39]). A review on electrically neutral carriers and their behavior in ion-selective electrodes is given in Chapter 12.

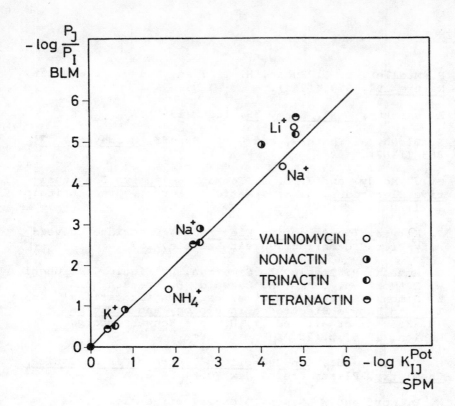

Figure 8.7. Comparison of the selectivities of neutral-carrier-modified solvent polymeric membranes (SPM) and bilayer membranes (BLM) [42]. The permeability ratios P_J/P_I (at "equilibrium" as far as available) fulfilled for the glyceryl dioleate BLM's are taken from figs. 10 and 11 in [24]. Values on the SPM's were obtained using 0.1 M solutions of the aqueous chlorides and membranes of the composition: 33.1 wt.-% polyvinyl chloride, 66.2 wt.-% dioctyl adipate, 0.7 wt.-% carrier. For the macro-tetrolides I^{z+}: NH_4^+; for valinomycin I^{z+}: K^+.

REFERENCES

[1] P. Mueller, D. O. Rudin, H. Ti Tien, and W. C. Wescott,
     Nature 194, 979 (1962).

[2] A. Shatkay, Anal. Chem. 39, 1056 (1967).

[3] R. Bloch, A. Shatkay, and H. A. Saroff, Biophys. J. 7,
     865 (1967).

[4] G. J. Moody and J. D. R. Thomas, Selective Ion Sensi-
     tive Electrodes, Merrow, Watford (Herts., Great Britain),
     1971.

[5] J. Koryta, Ion-Selective Electrodes, Cambridge Univer-
     sity Press, Cambridge, Great Britain, 1975.

[6] D. Ammann, R. Bissig, Z. Cimerman, U. Fiedler, M. Güggi,
     W. E. Morf, M. Oehme, H. Osswald, E. Pretsch, and
     W. Simon, 'Synthetic neutral carriers for cations', in
     Ion and Enzyme Electrodes in Biology and Medicine
     (M. Kessler et al., eds.), Urban & Schwarzenberg,
     Munich, 1976, p. 22.

[7] H. Freiser, ed., Ion-Selective Electrodes in Analytical
     Chemistry, Plenum Press, New York, 1978.

[8] W. E. Morf and W. Simon, 'Ion-selective electrodes
     based on neutral carriers', chapter 3 of ref. [7].

[9] W. E. Morf, D. Ammann, R. Bissig, E. Pretsch, and W. Si-
     mon, 'Cation selectivity of neutral macrocyclic and
     nonmacrocyclic complexing agents in membranes', in
     Progress in Macrocyclic Chemistry, Vol. 1 (R. M. Izatt
     and J. J. Christensen, eds.), Wiley-Interscience, New
     York, 1979.

[10] A. P. Thoma, A. Viviani-Nauer, S. Arvanitis, W. E. Morf,
     and W. Simon, Anal. Chem. 49, 1567 (1977).

[11] W. E. Morf and W. Simon, Proceedings of the Internatio-
     nal Conference on Ion-Selective Electrodes - Budapest,
     1977 (E. Pungor, ed.), Akadémiai Kiadó, Budapest, 1978,
     p. 149.

[12] C. Moore and B. C. Pressman, Biochem. Biophys. Res.
     Commun. 15, 562 (1964).

[13] Z. Štefanac and W. Simon, Chimia (Switzerland) 20, 436
     (1966); Microchem. J. 12, 125 (1967).

[14] P. Mueller and D. O. Rudin, Biochem. Biophys. Res. Commun. 26, 398 (1967).

[15] W. Simon, W. E. Morf, and P. Ch. Meier, Struct. Bonding 16, 113 (1973).

[16] G. Eisenman, ed., Membranes, Vol. 2, Dekker, New York, 1973.

[17] S. M. Ciani, G. Eisenman, R. Laprade, and G. Szabo, 'Theoretical analysis of carrier-mediated electrical properties of bilayer membranes', chapter 2 of ref. [16].

[18] G. Szabo, G. Eisenman, S. M. Ciani, R. Laprade, and S. Krasne, 'Experimentally observed effects of carriers on the electrical properties of membranes', chapter 3 of ref. [16].

[19] P. Läuger and B. Neumcke, 'Theoretical analysis of ion conductance in lipid bilayer membranes', chapter 1 of ref. [16].

[20] P. Läuger and G. Stark, Biochim. Biophys. Acta 211, 458 (1970).

[21] D. A. Haydon and S. B. Hladky, Q. Rev. Biophys. 5, 187 (1972).

[22] S. B. Hladky, Biochim. Biophys. Acta 352, 71 (1974).

[23] S. Ciani, J. Membrane Biol. 30, 45 (1976).

[24] G. Eisenman, S. Krasne and S. Ciani, Ann. N. Y. Acad. Sci. 264, 34 (1975).

[25] S. Krasne and G. Eisenman, J. Membrane Biol. 30, 1 (1976)

[26] W. E. Morf, P. Wuhrmann, and W. Simon, Anal. Chem. 48, 1031 (1976).

[27] W. Simon, W. E. Morf, E. Pretsch, and P. Wuhrmann, in Calcium Transport in Contraction and Secretion (E. Carafoli, ed.), North-Holland Publishing Company, Amsterdam, 1975.

[28] D. E. Goldman, J. Gen. Physiol. 27, 37 (1943).

[29] F. H. Johnson, H. Eyring, and M. J. Polissar, The Kinetic Basis of Molecular Biology, Wiley, New York, 1954.

[30] R. B. Parlin and H. Eyring, in Ion Transport across Membranes (H. T. Clarke, ed.), Academic Press, New York, 1954.

[31]   J. A. V. Butler, Trans. Faraday Soc. 19, 729 (1924).

[32]   T. Erdey-Gruz and M. Volmer, Z. Phys. Chem. (Leipzig) 150, 203 (1930).

[33]   J. E. Hall, C. A. Mead, and G. Szabo, J. Membrane Biol. 11, 75 (1973).

[34]   P. Läuger, Science 178, 24 (1972).

[35]   D. A. Haydon, Ann. N. Y. Acad. Sci. 264, 2 (1975).

[36]   P. Läuger, 'Ion transport across lipid bilayer membranes', in Physical Principles of Biological Membranes (F. Snell, ed.), Gordon & Breach, New York, 1970.

[37]   P. Vuilleumier, Dissertation ETH, Juris, Zürich, 1978.

[38]   J. H. Boles and R. P. Buck, Anal. Chem. 45, 2057 (1973).

[39]   W. E. Morf and W. Simon, 'Transport properties of neutral carrier membranes', in Ion-Selective Electrodes (E. Pungor, ed.), Akadémiai Kiadó, Budapest, 1977.

[40]   A. A. Lev, V. V. Malev, and V. V. Osipov, 'Electrochemical properties of thick membranes with macrocyclic antibiotics', chapter 7 of ref. [16].

[41]   E. Eyal and G. A. Rechnitz, Anal. Chem. 43, 1090 (1971); 44, 370 (1972).

[42]   W. Simon, Sixteenth Solvay Conference on Chemistry, Brussels, November 22-26, 1976.

Chapter 9

Summary of Fundamental Relationships

To facilitate the use of the theoretical relations worked out in Part A, the key results and the underlying assumptions are summarized below. These digests, especially the ones catalogued in Sections 9.b - e, offer the basis for a thorough discussion of the different types of ion-selective membrane electrodes (see Part B).

a) Definitions_and_notations

The response of potentiometric sensors was generally formulated as

$$E = E_0 + E_J + E_M \qquad (1.2/9.1)$$

where E is the emf (cell potential at zero current), E_0 a reference-electrode potential term, E_J the liquid-junction potential, and E_M the membrane potential. The last term, which is normally the decisive one, was subdivided into a membrane-internal contribution E_D and an interfacial contribution E_B:

$$E_M = E_B + E_D \qquad (2.1/9.2)$$

These potential contributions were evaluated in terms of ionic activities a_i, charges z_i, distribution coefficients k_i, and mobilities u_i referring to the membrane phase. In the following equations, the left boundary of a membrane (or a liquid junction) is denoted by the coordinate $x = 0$, and the right boundary by $x = d$. The notation (') is used for the contacting solution on the left side of the membrane, and (") for the right side.

b) Descriptions_of_the_phase-boundary_potential_(Donnan_term)

Based on assumptions I - IV specified in Chapter 2, the boundary potential contribution E_B was described by the equation

$$E_B = \frac{RT}{z_i F} \ln \frac{k_i a_i'}{a_i(0)} - \frac{RT}{z_i F} \ln \frac{k_i a_i''}{a_i(d)} = \frac{RT}{z_i F} \ln \frac{a_i'\, a_i(d)}{a_i''\, a_i(0)} \qquad (3.12/9.3)$$

The subscript i signifies any species that be distributed across the membrane/solution interfaces. The classical Donnan potential corresponds to the special case where $a_i(0) = a_i(d)$.

A very useful formulation was obtained by summarizing ions of the same valency class:

$$E_B = \frac{RT}{z_i F} \ln \frac{\Sigma\, w_i k_i a_i'}{\Sigma\, w_i a_i(0)} - \frac{RT}{z_i F} \ln \frac{\Sigma\, w_i k_i a_i''}{\Sigma\, w_i a_i(d)} \qquad (3.15/9.4)$$

where w_i is any additional weighting factor.

The following solution was given for permselective membranes and two classes of counterions, I^{2z} and J^z:

$$E_B = \frac{RT}{zF} \ln \frac{\sqrt{8X\,\Sigma\, k_i a_i' + (\Sigma\, k_j a_j')^2} + \Sigma\, k_j a_j'}{\sqrt{8X\,\Sigma\, k_i a_i'' + (\Sigma\, k_j a_j'')^2} + \Sigma\, k_j a_j''} \qquad (3.19/9.5)$$

where X is the activity of dissociated ion-exchange sites within the membrane.

c) Solutions for the diffusion potential (liquid-junction potential)

Explicit or at least definite solutions for the membrane-internal diffusion potential E_D were derived from the flux equations by using assumptions I and IV - IX (see Chapter 2). The first three relations are also applicable to aqueous diffusion layers and thus allow ready computation of liquid-junction potentials, $E_J \cong E_D$.

The most frequently used expression for E_D (and especially for E_J) is the Henderson approximation, Eq. (9.6), which is based on the additional assumption of linear activity profiles for all ions within the diffusion layer:

$$E_D \cong \frac{\Sigma|z_m|u_m \Delta a_m - \Sigma|z_x|u_x \Delta a_x}{\Sigma\, z_m^2 u_m \Delta a_m + \Sigma\, z_x^2 u_x \Delta a_x} \cdot \frac{RT}{F} \ln \frac{\Sigma\, z_m^2 u_m a_m(0) + \Sigma\, z_x^2 u_x a_x(0)}{\Sigma\, z_m^2 u_m a_m(d) + \Sigma\, z_x^2 u_x a_x(d)} \qquad (4.41/9.6)$$

with $\Delta a_i = a_i(d) - a_i(0)$. While in practice the Henderson equation is commonly written in terms of ion concentrations, here the use of ion activities is advised.

A more rigorous description is offered by the extended Planck relation which was derived in a new form:

$$E_D = \frac{\bar{u}_m}{|z_m|\bar{u}_m + |z_x|\bar{u}_x} \frac{RT}{F} \ln \frac{\Sigma\, a_m(0)}{\Sigma\, a_m(d)} - \frac{\bar{u}_x}{|z_m|\bar{u}_m + |z_x|\bar{u}_x} \frac{RT}{F} \ln \frac{\Sigma\, a_x(0)}{\Sigma\, a_x(d)} \qquad (4.21/9.7)$$

Equation (9.7) applies to electroneutral membranes or aqueous diffusion layers containing only one class of cations M^{z_m} and one class of anions X^{z_x}. At steady-state, the mean mobilities $\bar{u}_i$, characteristic of each ion class, are determined by

154

$$\bar{u}_i = \frac{\Sigma\, u_i a_i(d) \cdot \exp(z_i FE_D/RT) - \Sigma\, u_i a_i(0)}{\Sigma\, a_i(d) \cdot \exp(z_i FE_D/RT) - \Sigma\, a_i(0)}$$
(4.22/9.8)

For liquid junctions formed by two equimolar solutions of 1:1 electrolytes, the Planck solution was found to reduce to <u>Goldman's equation</u>:

$$E_D = \frac{RT}{F} \ln \frac{\Sigma\, u_m c_m(0) + \Sigma\, u_x c_x(d)}{\Sigma\, u_m c_m(d) + \Sigma\, u_x c_x(0)}$$
(4.25/9.9)

In the literature, this formula is much more frequently used to describe the internal potential of very thin (space-charge) membranes which meet the assumption of a constant electric field.

The Planck relation and the Goldman equation can be considered as special cases of the <u>generalized Teorell equation</u>. The latter allows for any two classes of diffusing ions, I^{z_i} and J^{z_j}, as well as for additional fixed sites in the membrane. The corresponding result for E_D reads:

$$E_D = \frac{\bar{u}_i - \bar{u}_j}{z_i \bar{u}_i - z_j \bar{u}_j} \frac{RT}{F} \ln \frac{\Sigma\, z_i^2 \bar{u}_i a_i(0) + \Sigma\, z_j^2 \bar{u}_j a_j(0)}{\Sigma\, z_i^2 \bar{u}_i a_i(d) + \Sigma\, z_j^2 \bar{u}_j a_j(d)}$$
(4.31/9.10)

A more convenient solution was obtained for the diffusion potential of permselective membranes when all permeating ions are of the same charge z_i:

$$E_D = \frac{RT}{z_i F} \ln \frac{\Sigma\, u_i a_i(0)}{\Sigma\, u_i a_i(d)}$$
(4.15/9.11)

This expression, which followed directly from assumptions I and IV - IXb, is basic to the <u>Horovitz-Eisenman equation</u>. A simplified version was established for ideally ion-specific membranes:

$$E_D = \frac{RT}{z_i F} \ln \frac{a_i(0)}{a_i(d)}$$
(6.1/9.12)

The trivial solution

$$E_D = 0$$
(9.13)

was found either for ideally homogeneous membranes (Eq. (4.28)) or for ion-exchange membranes when identical mobilities were assumed for all permeating species (Eq. (4.33)).

d) Results_for_the_membrane_potential_and_for_the_emf_of_ion-selective_electrode_cells

Descriptions of the membrane potential E_M were generally obtained by combining appropriate expressions for E_B and E_D (see Eqs. (9.3) - (9.13)). The simplest result applies to ideally homogeneous (solid-state) membranes and other types of ion-specific membranes:

$$E_M = \frac{RT}{z_i F} \ln \frac{a_i'}{a_i''} \qquad\qquad (3.14, 6.3/9.14)$$

The emf-response of such idealized membrane electrodes ($a_i'' = const$) evidently follows the Nernst equation:

$$E = E_i^o + \frac{RT}{z_i F} \ln a_i' \qquad\qquad (6.4, 6.5/9.15)$$

E_i^o represents a reference or standard potential; the subscript i denotes the primary ion for which an equilibrium is established between the membrane and the contacting solution of activity a_i'.

If different ions of the same charge z_i have to be considered as the permeating species in a permselective membrane (solid or liquid ion-exchanger), the result for E_M reads:

$$E_M = \frac{RT}{z_i F} \ln \frac{\Sigma\, u_i k_i a_i'}{\Sigma\, u_i k_i a_i''} \qquad\qquad (6.6/9.16)$$

and the emf-response function is described by the classical Nicolsky equation (Horovitz equation, Eisenman equation):

$$E = E_i^o + \frac{RT}{z_i F} \ln \left[a_i' + \sum_{j \neq i} K_{ij}^{Pot} a_j' \right] \qquad\qquad (6.7/9.17)$$

The potentiometric selectivity for an interfering ion (subscript j) relative to the primary ion is characterized by the coefficient

$$K_{ij}^{Pot} = \frac{u_j k_j}{u_i k_i} \qquad\qquad (6.8/9.18)$$

An explicit solution for E_M was also elaborated for the case of permselective membranes with two classes of diffusing counterions, I^{2z} and J^z (see Eq. (6.15)

156

where $z = +1$ was inserted). For two species having the same diffusion mobilities, $u_i = u_j$, the following emf-relationship was derived:

$$E = E_i^o + \frac{RT}{zF} \ln \left[\sqrt{a_i' + \frac{1}{4} K_{ij}^M a_j'^2} + \sqrt{\frac{1}{4} K_{ij}^M a_j'^2} \right] \qquad (6.10/9.19)$$

where the monovalent/divalent ion selectivity is given by

$$K_{ij}^M = \frac{k_j^2}{2 X k_i} \qquad (6.11/9.20)$$

Equation (9.19) is at variance with semiempirical extensions of the Nicolsky equation; the latter were found to be rigorously valid only for the more hypothetical situations. Thus, one version of Nicolsky-type equation was theoretically confirmed for $u_i \gg u_j$:

$$E = E_i^o + \frac{RT}{2zF} \ln \left[a_i' + K_{ij}^{Pot} a_j'^2 \right] \qquad (6.12/9.21)$$

$$K_{ij}^{Pot} = \frac{u_j}{2u_i} K_{ij}^M \qquad (6.16/9.22)$$

whereas the alternative formulation was obtained in the limit $u_j \gg u_i$:

$$E = E_j^o + \frac{RT}{zF} \ln \left[a_j' + K_{ji}^{Pot} a_i'^{1/2} \right] \qquad (6.13/9.23)$$

$$K_{ji}^{Pot} = \frac{2u_i}{u_j} (K_{ij}^M)^{-1/2} \qquad (6.17/9.24)$$

Deviations from permselectivity of fixed-site membranes were described by the Teorell-Meyer-Sievers equation, Eq. (6.18). When this relation is applied to cation-exchange membranes ($\omega = -1$) that contain counterions M^+ and coions X^-, the emf-response function becomes

$$E = E_m^o + \frac{RT}{F} \ln a_m' - \frac{RT}{F} \ln \frac{\sqrt{1 + Q'^2} + 1}{2} + q \frac{RT}{F} \ln \frac{\sqrt{1 + Q'^2} + q}{1 + q} \qquad (9.25)$$

with $q = \frac{u_m - u_x}{u_m + u_x}$ and $Q'^2 = \frac{4 k_m a_m' k_x a_x'}{x^2}$.

Ideal permselectivity occurs for values of $Q' \approx 0$ and results in a Nernstian behavior of the membrane electrode system. Failure of coion exclusion, on the other hand, is indicated by a mixed emf-response of the type

$$E = E_{mx}^{o} + \frac{u_m - u_x}{u_m + u_x} \frac{RT}{F} \ln a_{\pm}' \qquad [Q' \gg 1] \qquad (6.21/9.26)$$

A near-zero slope of the function E vs. log $a_{\pm}$ is characteristic of such systems.

A more general description of the membrane potential was obtained from <u>extension of the Planck theory</u>. The following relationship was derived for liquid (diffusion-type) membranes that are permeable to both cations M^{z_m} and anions X^{z_x}:

$$E_M = (1 - \tau_x) \frac{RT}{z_m F} \ln \frac{\Sigma \, k_m a_m'}{\Sigma \, k_m a_m''} + \tau_x \frac{RT}{z_x F} \ln \frac{\Sigma \, k_x a_x'}{\Sigma \, k_x a_x''} \qquad (6.22/9.27)$$

The integral anionic transference number τ_x was defined as a function of the mean ionic mobilities $\bar{u}_i$:

$$\tau_x = \frac{|z_x| \bar{u}_x}{|z_m| \bar{u}_m + |z_x| \bar{u}_x} \qquad (6.23/9.28)$$

where

$$\bar{u}_i = \frac{\Sigma \, u_i k_i a_i'' \cdot \exp(z_i F E_M/RT) - \Sigma \, u_i k_i a_i'}{\Sigma \, k_i a_i'' \cdot \exp(z_i F E_M/RT) - \Sigma \, k_i a_i'} \qquad (6.24/9.29)$$

An explicit formulation results when no significant differences exist between the individual mobilities of cations and anions, respectively, that is for $u_m \approx \bar{u}_m$ and $u_x \approx \bar{u}_x$.

e) <u>Relationships between the potentiometric and the ion-transport characteristics of membranes</u>

As a rule, the potentiometric ion selectivity of thick electroneutral membranes was found to be strictly a consequence of the transference of the ions involved. A universal formula was presented that describes the zero-current potential of any kind of membrane in the presence of two permeating species (cations or anions):

$$E_M = \frac{u_i}{z_i u_i - z_j u_j} \frac{RT}{F} \ln \frac{a_i'/t_i'}{a_i''/t_i''} - \frac{u_j}{z_i u_i - z_j u_j} \frac{RT}{F} \ln \frac{a_j'/t_j'}{a_j''/t_j''} \qquad (7.44/9.30)$$

Here the ionic transference numbers t' and t'' refer to electrodialysis experiments using the same membrane in contact with bathing solutions identical to solution (') and ("), respectively. Equation (9.30) incorporates virtually all of the relations compiled in Section 9.d. For example, the classical Nicolsky equation (9.17) can be rewritten as (see Eq. (7.39))

$$E = E_i^o + \frac{RT}{z_i F} \ln \frac{a_i'}{t_i'} \qquad (9.31)$$

It becomes evident that a transference number of the primary ions of unity is required for rendering a membrane ion-specific.

The response to monovalent ions of extremely thin space-charge membranes (bilayers) that meet the assumption of a constant electric field is commonly characterized by the Goldman-Hodgkin-Katz equation:

$$E_M \cong \frac{RT}{F} \ln \frac{\Sigma\, P_m a_m' + \Sigma\, P_x a_x''}{\Sigma\, P_m a_m'' + \Sigma\, P_x a_x'} \qquad (7.11/9.32)$$

The selectivity-determining parameters P_i, called the ion permeabilities, are again derivable from ion-transport studies, such as conductance measurements on bilayers. The potentiometric selectivity between two ions of the same charge is simply given by the corresponding permeability ratio, i. e.,

$$E = E_i^o + \frac{RT}{z_i F} \ln \left[a_i' + \frac{P_j}{P_i} a_j' \right] \qquad (8.51,\ 8.63/9.33)$$

These relations suggest that ion-transport experiments offer an independent method for assessing the ion selectivity of membranes.

f) Analysis of the ion-transport behavior of membranes (current-voltage characteristics)

General treatments of the ion transport across membranes were presented in Chapters 7 and 8. The steady-state ion fluxes in macroscopic membranes were rigorously formulated on the basis of the Schlögl theory, Eqs. (7.21) - (7.28). For practical purposes, a simplified version of the theory was introduced which permits the electrical behavior of symmetrical membrane cells to be rationalized:

$$J_i = z_i u_i c_i F \frac{V}{d} = z_i D_i c_i \frac{F}{RT} \frac{V}{d} \qquad (7.19/9.34)$$

159

J_i is the mass flux and D_i the diffusion coefficient of the species I^{z_i}, V is the applied voltage, and d the membrane thickness. The ion concentrations in the membrane, c_i, depend on the boundary conditions as well as on the concentration level of fixed or stationary ionic sites. Equation (9.34) applies in the normal case where the translocation of the permeating ions across the bulk of the membrane is rate-limiting; it predicts ohmic behavior of the membrane.

Contrasting electrical behavior was reported for lipid bilayer membranes. A unified approach to the theory of such systems was offered (see Eqs. (8.33) – (8.35)) which incorporates the Läuger-Stark model, the Ciani-Eisenman-Krasne theory, as well as other special treatments. The transport of lipid-soluble anions across certain bilayers in symmetrical electrodialysis cells was found to obey the relationship

$$J_i \cong 2 \bar{k}_i c_i \sinh(z_i FV/2\ RT) \qquad\qquad (8.40/9.35)$$

where $\bar{k}_i$ is the rate constant for the translocation of ions. Equation (9.35) implies that here the membrane interior can be approximated by a single sharp activation barrier. In contrast, a couple of free-energy peaks or a broad barrier were suggested to be operative in the carrier-mediated translocation of cations across bilayers. The corresponding result for the "equilibrium domain" of cation transport was

$$J_i \cong 4 \bar{k}_i c_i \sinh(z_i FV/4\ RT) \qquad\qquad (8.48/9.36)$$

The last two expressions predict a "hyperbolic" shape of the current-voltage characteristic, the zero-current conductance of the membrane being given by

$$G_o(i) = \frac{(z_i F)^2}{RT} \bar{k}_i c_i = \frac{(z_i F)^2}{RT} P_i a_i \qquad\qquad (8.41,\ 8.49/9.37)$$

where P_i is the permeability and a_i the external activity of the species I^{z_i}.

In the "kinetic domain" of ion transport, the rate $(\bar{k}_i/w_i)$ of ion transfer out of the membrane was found to be limiting. The following approximation was obtained:

$$J_i \cong (\bar{k}_i/w_i)\ c_i \tanh(z_i FV/2\ RT) \qquad\qquad (8.42,\ 8.47/9.38)$$

160

A current-voltage curve of the "saturation" type results in this case. Saturation of the current is also expected to occur for carrier membranes when the back-flow of free carriers is the rate-limiting process:

$$J_i \cong (2/n_i)\ \bar{k}_s c_s\ \tanh(z_i FV/2n_i RT) \qquad\qquad (8.33,\ 8.42/9.39)$$

Here $\bar{k}_s$ is the translocational rate constant and c_s the mean concentration of free carriers in the membrane, and n_i is the stoichiometry number of the ion/carrier complexes.

Part B

Ion-Selective Electrodes

CHAPTER 10

SOLID-STATE MEMBRANE ELECTRODES

That solid-state membranes could potentially serve as the active principle in ion-selective electrodes was recognized as early as 1921. Pioneering work in this respect was done by Trümpler [1], and later by Kolthoff and Sanders [2] who proposed membrane materials in the form of disks cast from silver salt melts. Tendeloo [3] reported on a calcium fluoride electrode. Some of these historical attempts at designing useful solid-state membrane electrodes were not met by success, analytically speaking, though, primarily because of a lack of experimental and theoretical know-how in connection with such materials, and secondly, because reasonably sensitive electronic measuring equipment became available only in the sixties.

One of the first workable electrodes based on crystalline material was realized in 1961 by Pungor and Hollós-Rokosinyi [4] who used silver iodide precipitate embedded in an inert matrix as iodide-sensitive membrane. Such heterogeneous solid-state membranes were later brought to perfection and were applied in a variety of modifications based on different materials (for a review, see Pungor and Tóth [5-7]). A real break-through in the field of ion-selective electrodes, however, came with the development by Frant and Ross in 1966 [8] of a homogeneous ion-exchange membrane composed of a single crystal. Their lanthanum trifluoride electrode for fluoride determination became one of the most important ion sensors. Indeed, the specificity of this electrode remains outstanding since the only significant interfering species is hydroxyl ion, the observable selectivity $K_{F,OH}^{Pot}$ being on the order of 0.1 [9-11].

10.1. <u>CHARACTERIZATION OF MEMBRANE MATERIALS</u>

The requirements solid-state- and other ion-exchange materials would have to conform to for the construction of practical devices were summarized by Buck [12]. Suitable materials are nearly insoluble in ionogenic solvents, although they may absorb these solvents to a certain degree. The ion-exchange process at the interface between the membrane phase and the contacting solution must be rapid and reversible for one or the other ion. Such rapidity is measured by the exchange current density or flux, which must be large compared with the current passed by the measuring circuit. Then, the ion-exchange processes at the membrane surfaces maintain local thermodynamic equilibrium (see Chapter 3), and eventual consecutive chemical reactions exert thermodynamically predictable effects on the ion-exchange processes. An additional and basic requirement to these materials is that they be essentially ionic conductors with resistances small compared with the input impedance of the measuring device. In solid electrolytes such as silver halides or lanthanum fluoride, the ionic charge transport can proceed only by a defect mechanism which generally leads to considerable resistances (10^6 - $10^8 \Omega$, for convenient membrane dimensions). Some reduction of the membrane resistance may be achieved by the addition of certain chemical impurities ("dopants" such as europium in lanthanum fluoride) that generate extrinsic defects in the membrane material. However, this problem is reduced in severity with the availability of modern measuring circuits (input impedance on the order of 10^{12} - $10^{14} \Omega$). An electronic component of the solid membrane conductivity does not in principle have a deleterius effect if the above requirements are met. In fact, the presence of some electronic conductivity is beneficial in that a metal/salt contact can be used in place of the inner reference solution by establishing a reversible electron exchange rather than a reversible ion exchange at one interface. A drawback of high electronic conductivity is that parasitic, concurrent redox processes are encouraged and there may be some slowing of the rate of ion exchange as well (partly cited from Buck [12]).

The number of solid-state materials possessing the necessary properties at room temperature is limited. Among those used or considered for selective electrode applications are:
- rare earth- and alkaline earth metal fluorides;
- halides of silver, lead, mercury and thallium (I);
- sulfides and other chalcogenides of silver, copper, lead, mercury, cadmium and zinc;
- silver thiocyanate, cyanide, azide, chromate and phosphate;
- mixtures of different halides or chalcogenides with silver sulfide.

Reports on the use of these materials as components in ion-selective electrodes were summarized in earlier reviews [5 - 24]. The forms selected for electrode application include single crystals [8, 10, 25, 26], disks cast from melts [2, 27], sintered materials [28-30], pressed polycrystalline pellets [10, 15, 31-33], as well as heterogeneous combinations of precipitates held in hydrophobic polymer binders [4-7, 34].

For a sound analytical application of solid-state membrane electrodes, a thorough knowledge of different parameters, especially selectivities and detection limits, is of utmost importance. The early theoretical treatments available in this field were of limited scope: the anion selectivity of silver halide [5, 10, 13-15, 35-39] and LaF_3 [10, 36] membrane electrodes, and the detection limits of AgCl [36, 40, 41] and LaF_3 [41] solid-state electrodes. A generally applicable model, which went beyond the pioneering contributions by Jaenicke [35] and by Buck [36], was offered only in 1974 by Morf, Kahr, and Simon [42]. This theory describes the selectivity of different solid-state membrane electrodes towards cations, anions, and other species forming complexes with membrane components, as well as the detection limits of the systems in question. In the following, we shall discuss the framework of this theory but also include some more recent approaches and suggestions. Because of the special analytical significance of membranes prepared from silver compounds, the basic equations are given and tested for these systems (see also [42]). However, the derived principles are valid in a more general sense and may help to characterize different types of solid-state membrane electrodes.

10.2. BASIC THEORETICAL ASPECTS OF SOLID-STATE MEMBRANE ELECTRODES

The general considerations set forth in Part A have shown that solid-state membranes usually conform to a simplified membrane model. Thus, in the application of the universal theoretical formalism to homogeneous solid-state membranes, contributions to the emf of the cell due to diffusion potentials within the membrane may be neglected [36, 42, 43]. The membrane potential will finally be given by the Donnan term (9.3) alone:

$$E_M = \frac{RT}{z_i F} \ln \frac{a_i' \; a_i(d)}{a_i'' \; a_i(0)} \tag{10.1}$$

Here, $a_i(0)$ and $a_i(d)$ are the activities of the primary ion I^{z_i} on the membrane surfaces contacting the sample and inner reference solution, respectively. If the membrane can be treated as an ideally homogeneous phase, the activities $a_i(0) = a_i(d)$ cancel. The ionic activities a_i' and a_i'' refer to the boundary zones of sample and reference solution. These boundaries are in direct contact and in thermodynamic equilibrium with the membrane phase, a fact which must not necessarily hold for the bulk of sample (activity a_i) and inner

solution. For membrane electrodes having a given internal reference system[*], the value of a_i'' is constant throughout. Hence,
Eq. (10.1) can be reduced and transformed into the following
basic relationship that describes the emf of the cell as

$$E = E_i^o + \frac{RT}{z_i F} \ln a_i'$$
(10.2)

The term E_i^o may be considered the standard potential of the
membrane electrode assembly. However, it also includes the
liquid-junction potential arising from the outer reference-
electrode system (see Chapter 5) and may therefore be subject
to corresponding variations.

Equation (10.2) shows that solid-state membrane electrodes
are, in principle, capable of responding to those kind of ions
that are components of the active membrane material. As long as
the activity of these primary ions remains constant throughout
the sample solution, i. e. $a_i' \cong a_i$ where a_i refers to the bulk
of sample solution, the ion sensor approximates a so-called
Nernstian behavior according to Eq. (10.2a):

$$E \cong E_i^o + \frac{RT}{z_i F} \ln a_i$$
(10.2a)

This practical form of the electrode response function is less

---

[*] A profound theoretical discussion of the so-called "all-solid-
state" membrane electrodes, which involve a direct membrane/
metal contact in place of the inner reference solution, was
given by Buck [44].

universal than Eq. (10.2), however. In fact, Eq. (10.2) offers
the basis for an easy understanding of nearly all emf-response
phenomena observed for solid-state ion-selective membrane
electrodes (see below). The basic requirement of rapid ion-exchange at the membrane-solution interface, as needed in the
derivation of Eq. (10.1) resp. (10.2), seems to be fulfilled
for the aforementioned electrode systems.

In the following, the implications of theory shall be discussed extensively for membranes prepared from silver compounds
Ag_zX. Here the general result for the electrode response obviously reads:

$$E = E^o_{Ag} + \frac{RT}{F} \ln a'_{Ag} \tag{10.3}$$

If the following convenient membrane cell is applied for the
emf measurements:

Hg; Hg_2Cl_2; KCl(satd.)|salt bridge|sample solution

|Ag_zX membrane | inner solution   ; AgCl; Ag   (25^oC)
                     (e.g. 0.1 M $AgNO_3$)

the reference potential term is given by $E^o_{Ag} \approx 555$ mV [15, 45].
An alternative description of the emf is obtained by using the
solubility product L_{Ag_zX}:

$$L_{Ag_zX} = a'^z_{Ag} \, a'_X \tag{10.4}$$

Equation (10.3) can then be rewritten in the form

$$E = E^o_X - \frac{RT}{zF} \ln a'_X \tag{10.5}$$

170

where

$$E^O_X = E^O_{Ag} + \frac{RT}{zF} \ln L_{Ag_zX} \qquad (10.6)$$

Equation (10.5) could also be deduced directly from Eq. (10.2). The use of this description of the emf-response of solid-state membrane electrodes is preferable for test solutions containing as anions only X^{z-}. It should be recapitulated that the activity a'_X detected by the sensor refers again to the boundary film of sample contacting the membrane, and not to the bulk solution. This important point deserves further discussion.

10.3. POTENTIAL RESPONSE AND DETECTION LIMIT OF SILVER COMPOUND MEMBRANES IN UNBUFFERED SOLUTIONS OF THE PRIMARY IONS

The intrinsic lower detection limit, as dictated by the membrane material, is related to the minimal activity $a'_{Ag,min}$ which is established by dissolution processes at the membrane/solution interface. For a successful rationalization of all experimental observations [15, 42] at least two such processes must be taken into account, namely

1) dissolution governed by the solubility product of the salt forming the active component of the membrane, and
2) leaching out of silver ions originating from coprecipitated soluble salts or reversibly adsorbed components, or produced by air oxidation of the membrane material.

In the original work by Morf et al. [42], the leached cations were assumed to be due to a distribution of silver ions between interstitial sites at the membrane surface (defects) and the contacting solution film. This hypothesis was later criticized by Buck [19, 21] who instead stressed and explained the fatal role of impurities, introduced during the procedure of membrane preparation, as well as of ions adsorbed at the membrane surface ("chemisorption" [7]). On the other hand, he clearly advocated the validity of the final theoretical results which, of course, remain unaffected by the more academic question of the mechanism of leaching. In the event, the theory discussed here [42] (see also [24]) was the first capable of a quantitative interpretation of various response phenomena, including those that could not be accounted for on the basis of the classical solubility-equilibrium approaches.

The leached cation activity α at the membrane surface is found to be roughly constant for a given set of experimental parameters but may be changed by a different preparation technique or conditioning of the membrane [15, 42]. In addition, it is believed to depend on the stirring rate of the sample [19, 21]. Assuming a constant value of α for a given silver compound as membrane material, the following relationship is acceptable to describe the deviations in activities between the boundary (a_i') and the bulk (a_i) of unbuffered sample solutions [42]:

$$a_{Ag}' - a_{Ag} = z(a_X' - a_X) + \alpha \qquad (10.7)$$

The left-hand side of this activity balance equation considers the total amount of silver ions released at the membrane surface, whereas the terms on the right-hand side specify the contributions due to dissolution (1) and leaching processes (2). By combining Eqs. (10.7) and (10.4), one obtains the following expression which holds for sample solutions of a silver salt that contains no anions interfering with the membrane material ($a_X = 0$, e. g. for a AgNO$_3$ solution):

$$a_{Ag}'^{z+1} - a_{Ag}'^{z}(a_{Ag}+\alpha) - z\ L_{Ag_zX} = 0 \qquad (10.8)$$

The corresponding result for sample solutions containing only the anion X^{z-} native to the membrane as well as cations not interfering with the membrane material (i. e. for $a_{Ag} = 0$) assumes the analogous form:

$$a_X'^{\frac{z+1}{z}} - a_X'^{\frac{1}{z}}(a_X - \frac{\alpha}{z}) - \frac{L_{Ag_zX}^{\frac{1}{z}}}{z} = 0 \qquad (10.9)$$

The last two expressions offer a mathematical solution for the activities a_{Ag}', respectively a_X', as maintained in unbuffered samples near the ion-sensing membrane surface and as detected by the electrode assembly. The parameters entering into Eqs. (10.8) and (10.9) are the solubility product of the silver compound forming the membrane and the leached cation activity. Depending on the magnitude of these parameters, two cases may be distinguished which will be explored in the following.

a) Membrane_material_AgX_(z=1)_with_$L_{AgX} \geq\geq \alpha^2$

This situation is applicable, for instance, to a AgCl-membrane electrode. Equations (10.3) and (10.8) here lead to a function for the silver ion response of the electrode which is in agreement with the equivalent one derived by Buck [36], Baucke [40], and Havas [41] and is similar to an equation used by Pungor and Tóth [37]:

$$E = E_{Ag}^{o} + \frac{RT}{F}\ \ln \frac{a_{Ag} + \sqrt{a_{Ag}^2 + 4L_{AgX}}}{2} \qquad (10.10)$$

173

Table 10.1. Thermodynamic values characterizing the reactions of silver compound membranes [a]

Compound	Solubility product (25°C) $-\log L_{Ag_zX}$ [46]	Complex	Stability constant (25°C) $\log \beta_n$ [47]
Ag_2S	48.54	$Ag(CN)_2^-$	18.75
AgI	16.08	$Ag(S_2O_3)_2^{3-}$	13.2
AgBr	12.30	$Ag(SC(NH_2)_2)_2^+$	12[b]
AgSCN	11.92	$Ag(SO_3)_2^{3-}$	8.45
AgCl	9.75	$Ag(NH_3)_2^+$	7.2
AgOH ($1/2\ Ag_2O + 1/2\ H_2O$)	7.68	HgI^+	12.87
		HgI_2	23.82
		HgI_3^-	27.60
		HgI_4^{2-}	29.83

a See also Sections 10.3 - 10.5.

b Value estimated from emf data [15].

According to Eq. (10.10), a so-called Nernstian response (linear region of the practical response function E vs. log a_{Ag}, see Eq. (10.2a)) is exhibited only at sample activities $a_{Ag} \gg \sqrt{L_{AgX}}$, as shown in Figure 10.1. The detection limit of the sensor is evidently dictated by the solubility of the membrane material since it holds that

$$a'_{Ag,min} = \sqrt{L_{AgX}} \qquad (10.11)$$

An analogous result is obtained for the anion response of membrane materials such as AgCl:

$$E = E_X^o - \frac{RT}{F} \ln \frac{a_X + \sqrt{a_X^2 + 4L_{AgX}}}{2} \qquad (10.12)$$

Figure 10.1 confirms the symmetric response to cations and anions as demanded by Eqs. (10.10) and (10.12).

It should be noted that the mechanism of anion response of silver compound membranes may be understood as a buffering of free silver cations near the membrane surface, which ions are primarily sensed by the electrode. According to the solubility product of AgCl (see Table 10.1), a chloride activity of 0.1 M would then correspond to a free silver activity of around 10^{-9} M. This buffered value of a'_{Ag} is much lower than the detection limit predicted by Eq. (10.11) for unbuffered systems, $a'_{Ag,min} = 1.3 \cdot 10^{-5}$ M. This may exemplify the principal differences between activity measurements in unbuffered and in buffered samples. When buffer systems are used to establish the activity level of free ions via superimposed solubility or complexation equilibria, the linear electrode-response function usually extends over the whole conceivable range of activities

175

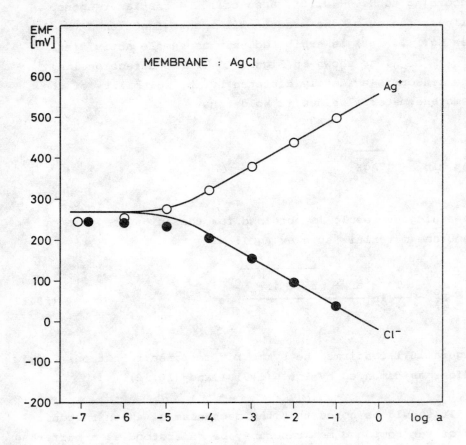

<u>Figure 10.1.</u> EMF response of a silver chloride membrane electrode to Ag^+ and Cl^- ($25^\circ C$) [42].
Solid lines: calculated from Eq. (10.10), resp. (10.12).
Experimental points: Ag^+ (o); Cl^- (•).

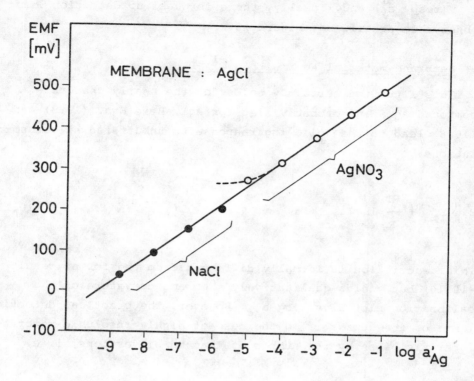

Figure 10.2. Wide-range response of a silver chloride membrane electrode to silver ion activities.
Open circles: $AgNO_3$ solutions, 10^{-1} - 10^{-5} M (see Figure 10.1).
Full circles: silver ion buffers (NaCl solutions, 10^{-4} - 10^{-1}M) establishing free silver ion activities in the range $1.8 \cdot 10^{-6}$ - $1.8 \cdot 10^{-9}$ M.
Solid line: theoretical response according to Eq. (10.3).

a'_{Ag} (resp. a'_X) and, ideally, there is quasi no detection limit observable at all (see Figure 10.2).

b) <u>Membrane material Ag_zX with $\alpha^{z+1} >> L_{Ag_zX}$</u>

This situation certainly holds for the nearly insoluble compound Ag_2S and probably also for AgI. Here Eqs. (10.3) and (10.8) lead to the following response to unbuffered silver solutions:

$$E = E^o_{Ag} + \frac{RT}{F} \ln \ (a_{Ag} + \alpha) \tag{10.13}$$

This description is formally different from the former result (10.10) but yields qualitatively the same curvature of the calibration plot E vs. log a_{Ag}. However, the practical detection limit of the sensor is in the case of highly insoluble membrane materials given by the number of leached silver ions, i. e.

$$a'_{Ag,min} = \alpha \tag{10.14}$$

Thus, a Nernstian behavior according to Eq. (10.2a) is observed only for $a_{Ag} >> \alpha$ (unbuffered samples). The experimental calibration plots shown in Figures 10.3 and 10.4 indicate that, in fact, the practical detection limits of AgI- and Ag_2S-membrane electrodes may be located at considerably higher activities than might be guessed from pure solubility considerations. The corresponding values $a'_{Ag,min}$ estimated from the solubility products alone would be $9.1 \cdot 10^{-9}$ M for AgI and $8.3 \cdot 10^{-17}$ M for Ag_2S (see Table 10.1 and Eq. (10.8) with $\alpha=0$) whereas the α-values fitting the experimental situation in Figures 10.3 and 10.4 are 10^{-6} M for AgI and $3.2 \cdot 10^{-6}$ M for Ag_2S [42]. The use of buffered test solutions, on the other

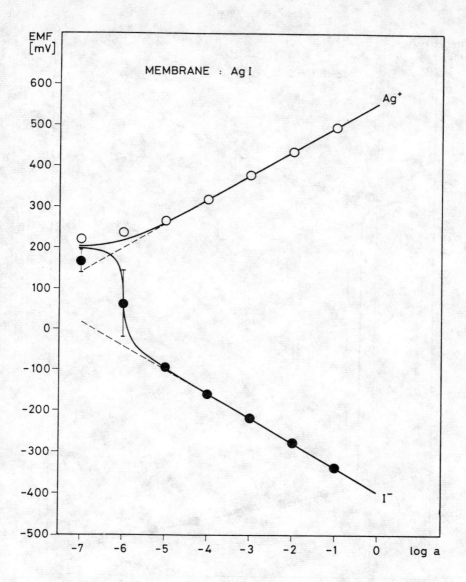

Figure 10.3. EMF response of a silver iodide membrane electrode to Ag^+ and I^- (25^oC) [42].

Solid lines: calculated from Eq. (10.13), resp. (10.15), with $\alpha = 10^{-6}$ M.

Dashed lines: calculated with $\alpha = 10^{-8}$ M.

Experimental points: Ag^+ (o); I^- (●).

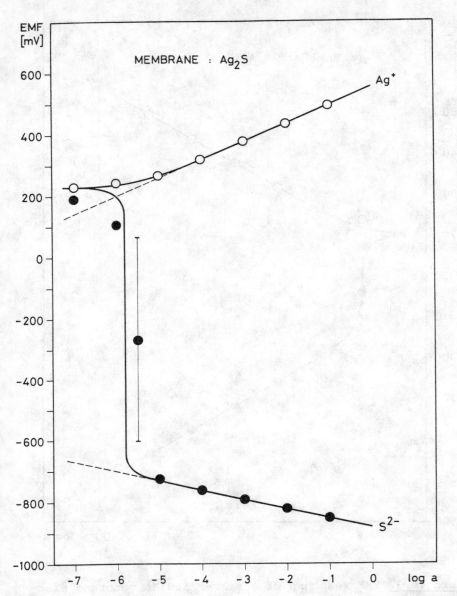

Figure 10.4. EMF response of a silver sulfide membrane electrode to Ag^+ and S^{2-} (25°C) [42].

Solid lines: calculated from Eq. (10.13), resp. (10.15), with
$\alpha = 10^{-5.5}$ M.

Dashed lines: calculated with $\alpha = 10^{-8}$ M.

Experimental points: Ag^+ (o); S^{2-} (•).

hand, again permits apparent activity reading far below these values to be reached. An Ag_2S-membrane electrode, for example, will be capable of responding to solutions having free silver activities down to 10^{-25} M in the presence of complexed silver ion [25, 48] (see Figure 10.5), or free sulfide ion activities down to 10^{-17} M or below in the presence of sulfide complexes [48]. In such cases, however, the electrode does definitely not respond to these free ion forms, as might be concluded from earlier statements [48], since the theoretically inferred least silver ion activity of around 10^{-25} M (for a 0.1 M sulfide solution, see Figure 10.5) is found to correspond to only 1 free ion per 10 litres of sample ! Although the effects documented in Figures 10.3 and 10.4 may finally be interpreted as "spurious experimental artifacts that disappear when buffered test solutions are used" [19, 21], they seem to be of considerable importance for conventional electrode practice and should be kept in mind.

In contrast to the findings for AgCl, the anion response of membrane materials such as AgI and Ag_2S becomes more complex [42]:

$$\text{for } a_X > \frac{\alpha}{z} : \quad E = E_X^o - \frac{RT}{zF} \ln \left(a_X - \frac{\alpha}{z} \right) \qquad (10.15a)$$

$$\text{for } a_X < \frac{\alpha}{z} : \quad E = E_{Ag}^o + \frac{RT}{F} \ln \left(\alpha - z a_X \right) \qquad (10.15b)$$

According to Eqs. (10.15a) and (10.15b), a sudden increase in the emf with decreasing activity a_X must occur at $a_X \approx \alpha/z$. This means, of course, that the emf is less stable in this region when using unbuffered samples, and this imposes a practical detection limit (see Figures 10.3 and 10.4). These predicted and experimentally found response functions are at variance

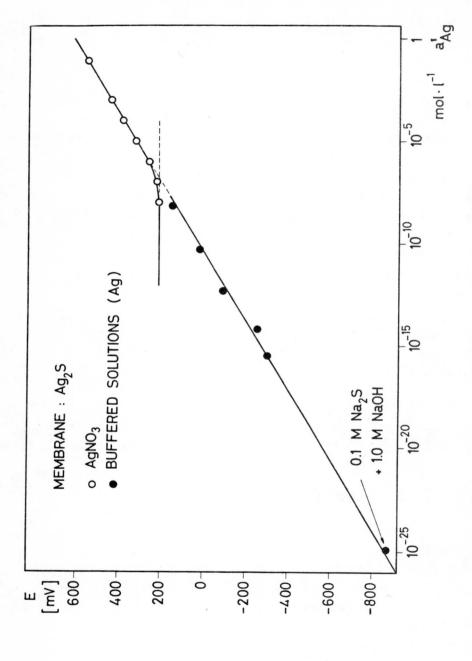

Figure 10.5. Wide-range response of a silver sulfide membrane electrode to silver ion activities [48] (for similar plots, see [25, 49]).

with usual expectations. Commonly, deviations from Nernstian behavior at the detection limit of the sensor occur in the direction towards a constant potential level, whereas here a highly super-Nernstian region is obtained. Yet, the anion response curves given in Figures 10.3 and 10.4 may be conveniently interpreted as a titration of leached/adsorbed silver ions with the respective anions. It was mentioned before (see also [19, 21]) that the leached cation activity α depends on the actual experimental parameters. Thus, a selected preparation technique for AgI membranes and their continuous use in iodide solutions will lead to some reduction of α, and the corresponding emf-response curves may be mimicked by the type found for AgCl membranes (dashed lines in Figure 10.3). However, the same idealized behavior can never be fully attained for Ag_2S-membrane electrodes in unbuffered sulfide solutions. The unusual (super-Nernstian) response characteristics of this electrode system at the detection limit, as predicted by theory and documented in Figure 10.4, was in the meantime confirmed by several workers [50-52] although the observed effects were found to be less pronounced, i. e. the detection limit was shifted down to activity levels of 10^{-7} to 10^{-8} M.

10.3. POTENTIAL RESPONSE OF SILVER HALIDE MEMBRANES TO DIFFERENT CATIONS

Silver halides and especially silver sulfide are widely used as membrane materials in electrodes selective for Ag^+. An excellent review on the characteristics of the commercially available sensors is given in Koryta's book on ion-selective electrodes [17], for example.

One of the few cations which reacts with silver compound membrane materials is Hg^{2+}. The overall reaction with silver halides AgX may be described as follows:

$$Hg^{2+} + nAgX \rightleftharpoons HgX_n^{2-n} + nAg^+ \tag{10.16}$$

As a reasonable steady-state approximation, the following relation holds:

$$a'_{Hg} + \sum_n a'_{HgX_n} = a_{Hg} \tag{10.17}$$

In similarity to Eq. (10.7), the activity a'_{Ag} of the silver ions formed according to reaction (10.16) becomes

$$a'_{Ag} = \sum_n n a'_{HgX_n} + \alpha \tag{10.18}$$

In Eq. (10.18) it is assumed that the sample solution initially contains no silver ions ($a_{Ag} = 0$). With the expressions for the complex formation and membrane solubility at the phase boundary:

$$a'_{HgX_n} = \beta_n a'_{Hg} a'^n_X \tag{10.19}$$

$$a'_{Ag} a'_X = L_{AgX} \tag{10.20}$$

one obtains an implicit function and, therefore, numerical values for a'_{Ag}:

$$a'_{Ag} = \frac{\sum_n n\beta_n (L_{AgX}/a'_{Ag})^n}{1 + \sum_n \beta_n (L_{AgX}/a'_{Ag})^n} a_{Hg} + \alpha \tag{10.21}$$

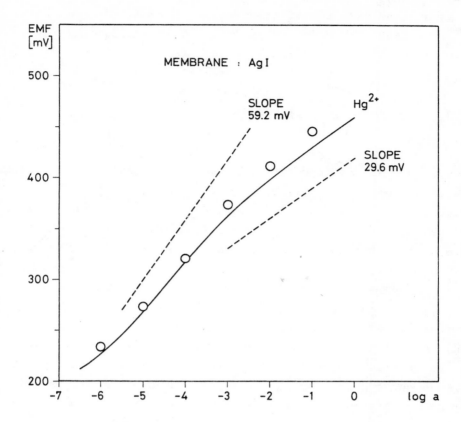

Figure 10.6. EMF response of a silver iodide membrane elec-
trode to Hg^{2+} (25°C) [42].
Calculated: solid line.
Experimental: circles.

Since the weighting factor of a_{Hg} in Eq. (10.21) is identical
to the average number $\bar{n}$ of halide ions coordinated to the
Hg^{2+} ions at the phase boundary (mean degree of complex formation), Eqs. (10.3) and (10.21) may be combined:

$$E = E^o_{Ag} + \frac{RT}{F} \ln (\bar{n}a_{Hg} + \alpha) \qquad (10.22)$$

In Figure 10.6, the computed response (Eqs. (10.21) and (10.22)
with $\alpha=10^{-6}$ M; Table 10.1) of an AgI membrane electrode to
variable activities of Hg^{2+} is compared to experimental values.
Obviously, there is both experimental and theoretical evidence that the response of such sensors is neither Nernstian
nor linear. This behavior follows from the pronounced acti-
vity-dependence of the selectivity-determining parameter $\bar{n}$.
In the lower activity range of Hg^{2+} (10^{-4} to 10^{-5} M), the
response curve may very crudely be approximated by a straight
line; the calculated slope, however, is only about 48 mV. At
high activities (above 10^{-1} M) the slope of the calculated
response curve approximates an asymptotic value of 29.6 mV.
This means, of course, that very careful calibrations are
required for a competent application of Hg^{2+} sensors based
on AgI membranes. For practical purposes, indirect determination procedures (titration of Hg^{2+} with I^-) are therefore
recommended [53].

10.5. SELECTIVITY OF SILVER HALIDE MEMBRANES TOWARDS DIFFERENT
       ANIONS

Membranes consisting of pure silver halides AgX or of AgX
mixed with Ag_2S [10, 15] may be used for the determination of
anions Y^{z-} that form sparingly soluble salts Ag_zY with Ag^+.
Such a direct determination of species Y^{z-} is, in principle,

possible only if there exists an equilibrium between the boundary of the sample solution and the component Ag_zY deposited on the membrane. It is therefore necessary that the surface of the membrane electrode be covered - at least partly - by a new phase containing Ag_zY. Three types of such coatings will be discussed below. Although additional potentials may develop between the primary AgX phase and the new boundary phase, such contributions are neglected in the following treatment.

a) Single phase Ag_zY covering completely the AgX phase

Such a phase is formed only when contacting an AgX-membrane electrode with a sample solution containing anions Y^{z-} of sufficiently high activity a_Y (e. g. a 0.1 M solution). The silver ion activity a'_{Ag} established near the membrane surface will then be determined exclusively by the solubility product of the component Ag_zY:

$$a'^z_{Ag} \, a_Y = L_{Ag_zY} \tag{10.23}$$

Equation (10.3) immediately leads to the following emf-response function:

$$E = E^o_Y - \frac{RT}{zF} \ln a_Y \tag{10.24}$$

where, in analogy to Eq. (10.6),

$$E^o_Y = E^o_{Ag} + \frac{RT}{zF} \ln L_{Ag_zY} \tag{10.25}$$

Since the selectivity realized for the species Y^{z-} relative
to the primary anion X^- may be characterized by the term
K_{XY}^{Pot} as follows:

$$E = E_X^o - \frac{RT}{F} \ln \left[K_{XY}^{Pot} \, a_Y^{1/z} \right] \tag{10.26}$$

we obtain the fundamental relationship

$$K_{XY}^{Pot} = \exp \left[- \frac{F}{RT} (E_Y^o - E_X^o) \right] = \frac{L_{AgX}}{L_{Ag_zY}^{1/z}} \tag{10.27}$$

It becomes evident that the potentiometric selectivity coefficients of AgX-membrane electrodes ideally constitute a measure
describing the following ion-exchange equilibria:

$$z \, AgX + Y^{z-} \rightleftharpoons Ag_zY + z \, X^- \tag{10.28}$$

The agreement between these theoretical selectivities and
experimental results is perfect (see Figure 10.7 and References 5-7, 15, 37) as long as the above requirements are
met.

b) Mixed_phase_AgX-AgY

For sample solutions containing both X^- and Y^- at activities a_X and a_Y well above the detection limits, it is conceivable that a mixed phase or mixed adsorption isotherm is
formed on the membrane surface the composition of which
$[a_X(0), a_Y(0)]$ is in equilibrium with that of the contacting
aqueous solution. Recalling equilibrium (10.28), this activity relationship is formulated as

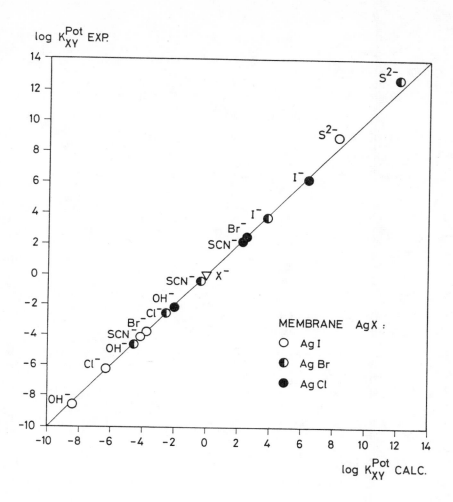

Figure 10.7. Comparison of the experimental and the calculated
anion selectivity of different silver halide membrane electrodes (25°C) [42]. Calculations are based on Eq. (10.27) and
Table 10.1. Experimental values were obtained from emf-measurements on 0.1 M separate solutions.

$$\frac{a_Y(0)}{a_X(0)} = K_{XY} \frac{a_Y'}{a_X'} \cong K_{XY} \frac{a_Y}{a_X} \tag{10.29}$$

where the ion-exchange constant may be assumed to be identical to the ratio of solubility products, as given in Eq. (10.27) [36, 43]. By following the suggestion of Rothmund and Kornfeld [54], the ratio of the activities of two counterions on the surface of a solid ion-exchanger is generally given by

$$\frac{a_Y(0)}{a_X(0)} = \left(\frac{N_Y}{N_X}\right)^n \quad ; \quad N_X + N_Y = 1 \tag{10.30}$$

where N are the corresponding mole fractions of counterions (here the fractions of sites Ag^+ in the mixed phase occupied by the subscripted anions). This has been referred to as n-type behavior [55] (see also Chapter 13) and has been demonstrated to conform to the regular-solution theory of binary mixtures*) [56]. In addition to Eq. (10.30), we may write:

$$\frac{a_X(d)}{a_X(0)} = \left(\frac{1}{N_X}\right)^n \tag{10.31}$$

These expressions allow us to replace the unknown membrane-activity terms entering into the generalized potential relationship, Eq. (10.1) for i=X, by outside activities. The

---

*) It holds as an approximation that $n = 1 - W_{X-Y}/2RT$, where W_{X-Y} is the excess interaction energy of neighboring ions X^- and Y^- [56]. Thus, repulsion between the two counterions is indicated by values $n < 1$ (and not $n > 1$ as claimed earlier [56]).

final result for the emf-response assumes the form:

$$E = E_X^O - \frac{nRT}{F} \ln \left[(a_X)^{1/n} + (K_{XY}a_Y)^{1/n} \right] \tag{10.32}$$

This general result is given here for the first time, but four limiting cases have been characterized earlier (see also Figure 10.8):

1) For $n \to 1$: Ideal mixed phase or mixed adsorption isotherm AgX–AgY [36, 42, 43].
2) For $n \to 0$: Strongly hindered formation of a mixed phase, i. e. rapid conversion from AgX to AgY, or vice versa, at the point where $a_X = K_{XY} a_Y$ [10, 36, 42].
3) For $a_Y \to 0$: Pure phase AgX, see Section 10.3.
4) For $a_X \to 0$: Pure phase AgY, see paragraph a) of this section.

Since Eq. (10.32) is found to be formally different from the commonly approved Nicolsky equation:

$$E = E_X^O - \frac{RT}{F} \ln \left[a_X + K_{XY}^{Pot}(app.) \ a_Y \right] \tag{10.33}$$

except for n=1, the apparent selectivity coefficients herewith obtained for mixed systems are sensitive to experimental conditions.

c) Two-phase patchwork AgX+AgY

When AgX-membrane electrodes are exposed to solutions of anions Y^- of intermediate activities (e. g. 10^{-3} M), it is often observed that the apparent selectivity between species Y^- and X^-, as determined from Eq. (10.33), is far less than

191

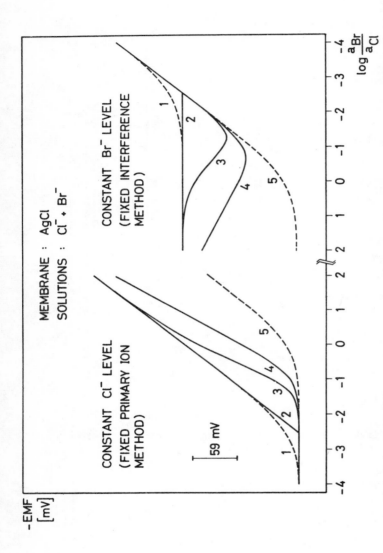

Figure 10.8. Calculated EMF-response of a silver chloride membrane electrode to mixed aqueous solutions of chloride and bromide (left: constant primary ion; right: constant interfering ion). Different models were used to describe the electrode surface at steady-state:

1: ideal mixed phase AgCl-AgBr [36, 42, 43], Eq. (10.32) with n=1; 2: strongly hindered mixed phase [10, 36, 42], Eq. (10.32) or Eqs. (10.40) and (10.41) with n→0; 3: reversible two-phase patchwork AgCl+AgBr, Eqs. (10.40) and (10.41) with n=1; 4: the same as for 3 but with n=2; 5: irreversible two-phase patchwork [58], Eq. (10.38) or Eqs. (10.40) and (10.41) with n→∞.

the theoretical value expected from (10.27). This phenomenon
was ascribed to an incomplete covering of the membrane by the
deposited phase AgY, which hypothesis has been underscored by
both experimental [57] and theoretical evidence [58] on solid-
state membrane electrodes. The reaction of AgCl with aqueous
iodide solution, for example, was shown to lead to the growth
of discrete AgI crystals on the membrane surface [57]. By this
reaction of the type (10.28), I^- ions are consumed and, simultaneously, Cl^- ions are liberated into the boundary zone of
the sample solution. The membrane electrode finally senses
the generated Cl^- rather than the originally present I^-. This
behavior is reflected, typically, by an apparent selectivity
coefficient K_{ClI}^{Pot}(app.) on the order of only 10^2 [57, 59], instead of the theoretical value of around 10^6 exhibited for
more ideal situations (Figure 10.7).

   A formal interpretation of the apparent selectivity coefficients of solid-state membrane electrodes was offered first
by Hulanicki and Lewenstam [58]. These authors considered the
case where two solid phases coexist in thermodynamic equilibrium with ions in solution. Accepting the equilibrium condition for the AgX and AgY phase, which both coexist at the
same value of a_{Ag}' and E, respectively, we obtain:

$$E = E_X^o - \frac{RT}{F} \ln a_X'$$

$$= E_X^o - \frac{RT}{F} \ln [K_{XY} a_Y'] \qquad (10.34)$$

where $K_{XY} = L_{AgX}/L_{AgY}$. Accordingly, there exists a well-defined
relationship between the activities a_X' and a_Y' that are maintained near the membrane surface via the solubility equilibria.
Such a relationship is usually not fulfilled for the bulk of
sample solution (activities a_X and a_Y), however. This implies

that activity gradients build up between the boundary and the bulk of sample solution which give rise to diffusion of ions. If the Nernst approximation [60] is used to describe the ionic fluxes across the diffusion layer (thickness δ') at zero-current steady-state, one may write:

$$J_X = - \frac{D_X'}{\delta'} (a_X' - a_X) \tag{10.35}$$

$$J_Y = - \frac{D_Y'}{\delta'} (a_Y' - a_Y) \tag{10.36}$$

$$J_X + J_Y = 0 \tag{10.37}$$

Combination of Eqs. (10.34) - (10.37) then leads to the following result for the activity a_X' sensed by the membrane electrode system:

$$a_X' = \frac{D_X' K_{XY}}{D_X' K_{XY} + D_Y'} a_X + \frac{D_Y' K_{XY}}{D_X' K_{XY} + D_Y'} a_Y \tag{10.38}$$

This expression, suggested by Hulanicki and Lewenstam, includes some interesting features. For experimental situations where the two-phase patchwork on the membrane surface remains intact even in pure solutions of the anions Y^-, the potentiometric selectivity will be obtained as

$$K_{XY}^{Pot}(app.) = \frac{D_Y' L_{AgX}}{D_X' L_{AgX} + D_Y' L_{AgY}} \tag{10.39}$$

194

Accordingly, the intrinsic selectivity value dictated by the solubility products is approximated only for anions Y^- discriminated by the AgX electrode, whereas in the other extreme an upper limit of around 1 is predicted for the apparent selectivity coefficient:

$$K_{XY}^{Pot}(app.) = \frac{L_{AgX}}{L_{AgY}} \quad , \text{ for } L_{AgX} \ll L_{AgY} \qquad (10.39a)$$

$$K_{XY}^{Pot}(app.) = \frac{D_Y'}{D_X'} \quad , \text{ for } L_{AgX} \gg L_{AgY} \qquad (10.39b)$$

Such discrepancies in the observed selectivity coefficients, as indicated by Eqs. (10.27) and (10.39b), were often reported in the literature (see Table 10.2, Figure 10.8, and Reference 58). Although the anion selectivity sequence is commonly found the same for all silver compound membranes, i. e.

$$S^{2-} \gg I^- > Br^- \geqslant SCN^- > Cl^- > OH^-,$$

the extent of preference for a given species may highly depend on the experimental conditions (membrane material, ionic composition and concentration of the test solution). This fact is of considerable practical importance because it becomes feasible, by select methods of electrode application, to create an optimum in the ion selectivity behavior of a sensor.

The principal drawback of the diffusion model by Hulanicki and Lewenstam [58] is that both phases AgX and AgY are claimed to coexist in comparable quantities on the membrane surface, irrespective of the composition of the sample solution. Thus, for membranes equilibrated with pure solutions of X^-

(Eq. (10.38) with $a_Y = 0$), it leads to the absurd situation that the predicted selectivity for the primary ion depends on parameters of a hypothetical interfering ion. This dilemma may be overcome if Eqs. (10.35) and (10.36) of the former treatment are replaced by the following ones:

$$J_X = - \frac{D_X'}{\delta'} (a_X' - a_X) \, x_X \tag{10.35a}$$

$$J_Y = - \frac{D_Y'}{\delta'} (a_Y' - a_Y) \, x_Y \tag{10.36a}$$

where x_X and $x_Y = 1 - x_X$ are the fractions of the membrane surface covered by the phase AgX and AgY, respectively. These corrected expressions for the mean flux densities take into account that diffusion of a species is restricted to that surface area of the membrane that is occupied by the corresponding silver salt. By using these equations instead of the former ones, we arrive at the following generalized result:

$$a_X' = \frac{D_X' K_{XY} x_X}{D_X' K_{XY} x_X + D_Y' x_Y} \, a_X + \frac{D_Y' K_{XY} x_Y}{D_X' K_{XY} x_X + D_Y' x_Y} \, a_Y \tag{10.40}$$

This modified description includes both the normal limiting cases realized for concentrated solutions of one sort of anions (for $a_X = 0$ and $x_X = 0$, or $a_Y = 0$ and $x_Y = 0$) and the intermediate case discussed by Hulanicki and Lewenstam (for $x_X = x_Y = 0.5$). On the other hand, the quantities x_X and x_Y entering into Eq. (10.40) must be considered as additional variables that cannot be expressed strictly in terms of outside ionic activities. To facilitate the illustration of the encountered selectivity phenomena, notwithstanding, the following empirical approximation was introduced which is

Table 10.2. Calculated and experimental selectivity coefficients of silver halide membrane electrodes

Electrode system AgX/X^-, Y^-	K_{XY}^{Pot} according to Eq.(10.27)[a]	K_{XY}^{Pot} (app.) according to Eq.(10.39)[a]	K_{XY}^{Pot} (meas.) Values determined by Hulanicki and Lewenstam [58]	K_{XY}^{Pot} (meas.) Literature data compiled by Hulanicki and Lewenstam [58]
$AgCl/Cl^-$, I^-	$2.1 \cdot 10^6$	1.0	86.5^{b} $1.8 \cdot 10^6$ $(10^{-1}_M)^{c}$	—
$AgCl/Cl^-$, Br^-	$3.5 \cdot 10^2$	1.0	2.1 (10^{-3}_M) $3.3 \cdot 10^2$ (10^{-1}_M)	$1.2 - 3.3 \cdot 10^2$
$AgBr/Br^-$, SCN^-	0.42	0.28	0.34 (10^{-3}_M) 0.37 (10^{-1}_M)	$0.20 - 0.65$
$AgBr/Br^-$, Cl^-	$2.8 \cdot 10^{-3}$	$2.8 \cdot 10^{-3}$	$5.6 \cdot 10^{-3}$ (10^{-3}_M) $2.9 \cdot 10^{-3}$ (10^{-1}_M)	$1.8 \cdot 10^{-3} - 1.0 \cdot 10^{-2}$
AgI/I^-, Cl^-	$4.7 \cdot 10^{-7}$	$4.7 \cdot 10^{-7}$	$6.6 \cdot 10^{-6}$ ^{b} $5.6 \cdot 10^{-7}$ $(10^{-1}_M)^{c}$ $3.7 \cdot 10^{-7}$ ^{d}	—

a Solubility products according to Table 10.1; $D'_I/D'_{Cl}=1.01$, $D'_{Br}/D'_{Cl}=1.03$, $D'_{SCN}/D'_{Br}=0.84$.
b Values given in Ref. 59. c Values given in Figure 10.7. d Value given in Refs.5, 7 and 37.

analogous to the formalism applied for mixed phases:

$$\frac{x_Y}{x_X} \approx \left(\frac{K_{XY}a_Y}{a_X}\right)^{1/n} \qquad (10.41)$$

Figure 10.8 shows some calculated emf-response curves for mixed aqueous solutions. It clearly demonstrates that the formation of a two-phase patchwork on the membrane surface may lead to unusual response- and selectivity phenomena; only the case $n \to 0$ is found to agree with the corresponding results obtained for a mixed-phase covering of the electrode. The practical selectivity of such systems will finally be determined as

$$K_{XY}^{Pot}(app.) = \left[\left(\frac{D_Y'}{D_X'}\right)^n \frac{L_{AgX}}{L_{AgY}}\right]^{\frac{1}{n+1}} \qquad (10.42)$$

where n is an experimental parameter. For the combination $AgCl/Cl^-$, Br^- characterized in Figure 10.8 and Table 10.2, the corresponding "experimental" selectivity coefficients are obtained as 19 (n=1), 7 (n=2), or even only 1 ($n \to \infty$), instead of the "theoretical" value of 355 (n=0) deduced from solubility data alone.

10.6. POTENTIAL RESPONSE AND SELECTIVITY OF SILVER HALIDE MEMBRANES TOWARDS DIFFERENT LIGANDS

In contrast to the previous section, negatively charged ligands as well as neutral species forming discrete complexes with Ag^+ are considered here. Since such ligands react with the membrane material according to:

$$nL^{\nu-} + AgX \rightleftharpoons AgL_n^{1-n\nu} + X^- \tag{10.43}$$

forming X^-, a detection of these ligands becomes possible [16]. For a sample solution containing the ligand $L^{\nu-}$ and possibly anions X^- at an activity a_L and a_X, respectively, the approximation holds:

$$a_L' + \sum_n na_{AgL_n}' = a_L \tag{10.44}$$

A somewhat more rigorous treatment may be found in the appendix of Ref. 42. For sufficiently high activities a_L or a_L', the complexed silver ions dominate relative to free Ag^+ at the phase boundary and therefore Eq. (10.7) is applicable in the form:

$$\sum_n a_{AgL_n}' = a_X' - a_X + \alpha \tag{10.45}$$

In addition, the solubility product, Eq. (10.20), and the description of the complex formation:

$$a_{AgL_n}' = \beta_n a_{Ag}' a_L'^n \tag{10.46}$$

are used in the following treatment. Since Ag^+ generally forms linear 1:2 complexes (n=2 in Eqs. (10.43) - (10.46)) the following quadratic equation in a_{Ag}' may be derived:

$$4\beta_2 \left[a'_{Ag} \left(\frac{1}{2} a_L + a_X \right) - L_{AgX} \right]^2 + a'_{Ag} (a_X - \alpha) - L_{AgX} = 0 \qquad (10.47)$$

This equation may be further simplified for the cases discussed below.

a) Sample solution containing ligand $L^{\nu-}$ only

For $a_X = 0$, Eqs. (10.5), (10.20), and (10.47) combine to:

$$E = E_X^o - \frac{RT}{F} \ln \frac{2\kappa a_L^2}{4\kappa a_L + \alpha + \sqrt{4\kappa a_L^2 + \alpha^2}} \qquad (10.48)$$

$$\kappa = \beta_2 L_{AgX} \qquad (10.49)$$

The parameter κ stands for the relative complex formation constant, defined by reaction (10.43), and is a major factor determining the selectivity exhibited by the ligand/membrane combination. In Figure 10.9, the relative emf of a given sensor $(E-E_X^o)$ is plotted as a function of κ for different activities a_L (for Ag_2S membranes, see later). The surprisingly good agreement corroborates the basic assumptions made. Three major parts are discerned in Figures 10.9a-c.

Region 1 (right hand side in Figures 10.9a-c): $4\kappa \gg 1$. Equation (10.48) simplifies to

$$E = E_X^o - \frac{RT}{F} \ln \left(\frac{1}{2} a_L \right) \qquad (10.50)$$

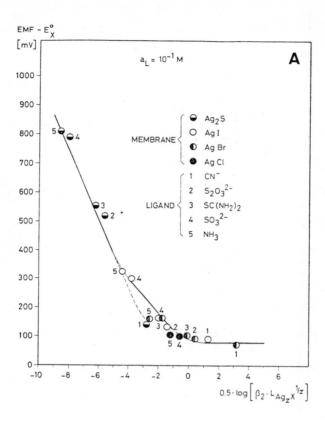

Figure 10.9. Comparison of the calculated (solid and dashed lines) and the experimental (circles) EMF response of different ligand/membrane combinations (25°C) [42]. The ligand activities a_L are 10^{-1}, 10^{-2}, and 10^{-3} M for A, B, and C, respectively. Dashed lines denote silver sulfide membranes.

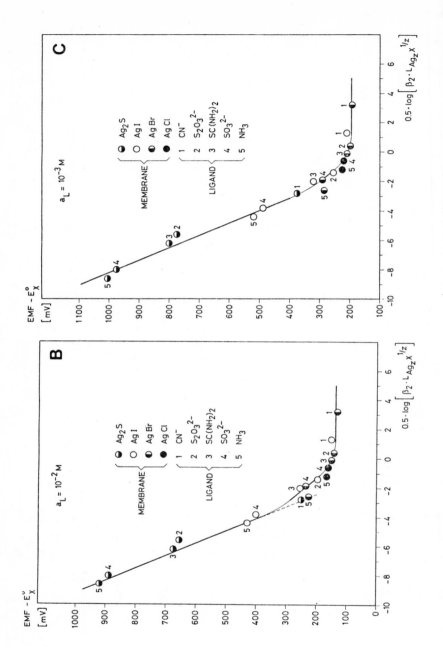

The term $1/2\ a_L$ is due to the fact that the complexes $AgL_2^{1-2\nu}$ are predominant at the phase boundary and therefore a formation of $a_X' = 1/2\ a_L$ (reaction (10.43)) has to be expected. For such values of κ, the emf of the corresponding electrochemical cells is independent of κ (horizontal lines in Figures 10.9a -c) and the linear range of the response of the electrode system to the ligand activity in the sample solution (e. g., a_{CN^-}) shows a slope of -59.16 mV ($25^{\circ}C$).

Region 2 (center part in Figures 10.9a-c): $1 \gg 4\kappa \gg (\alpha/a_L)^2$. Equation (10.48) is reduced to

$$E = E_X^o - \frac{RT}{F} \ln (\sqrt{\kappa} a_L)\ .$$
(10.51)

In this case, the free ligands $L^{\nu-}$ dominate at the phase boundary; the activity of the anions formed by reaction (10.43) is, however, larger than the concentration α of leached/adsorbed silver ions. In similarity to region 1, the emf-response has a slope of -59.16 mV ($25^{\circ}C$). In contrast to the previous case, the emf now depends both on the ligand activity as well as the complex formation behavior of the ligand. The center portions of the plots in Figure 10.9, therefore, have a slope of -59.16 mV, as demanded by Eq. (10.51).

Region 3 (left hand side in Figures 10.9a-c): $4\kappa \ll (\alpha/a_L)^2$. Equation (10.48) may be rewritten in the form:

$$E = E_X^o + \frac{RT}{F} \ln \alpha - 2 \frac{RT}{F} \ln (\sqrt{\kappa} a_L)$$
(10.52)

In this case, the ligand does not really dissolve the membrane material but may still react with the fraction of leached or reversibly adsorbed silver ions. Such a behavior has to be expected, in general, for extremely insoluble silver compounds and ligands forming relatively weak complexes with Ag^+. The slope of the electrode response function becomes -118.32 mV (25°C). Similarly, the emf depends linearly on $\log \sqrt{\kappa}$ with a slope of -118.32 mV (Figure 10.9). For $\kappa \ll 1$ this means:

$$\text{slope } s = \left(\frac{\partial E}{\partial \log a_L}\right)_\kappa = \left(\frac{\partial E}{\partial \log \sqrt{\kappa}}\right)_{a_L} \qquad (10.53)$$

When using Ag_2S membranes, the ligands generally used no longer dissolve this material to an appreciable extent. A detailed study has shown that Eq. (10.52) is applicable to this case where:

$$\kappa = \beta_2 L_{Ag_2S}^{1/2} \qquad (10.54)$$

Hence a straightforward explanation is obtained for the uncommon slope s of -103 mV reported for the emf-response of an Ag_2S membrane to cyanide ions [25]. The theoretical slope according to Eqs. (10.52) and (10.53) takes on the value -118.32 mV (25°C). The corresponding functions are presented by dashed lines in Figure 10.9. Since α has been assumed to be the same for Ag_2S and $AgX-Ag_2S$, which were used to obtain the experimental results [15], the dashed lines coincide with the solid lines for decreasing κ.

Figures 10.9 may be used to assess the practical applicability of different sensor materials in the determination of the activity of a given ligand. If a point representing a given combination of sensor material and ligand falls into one and the same region at different levels of a_L, this system is theoretically suitable and exhibits a linear emf-response function.

The lower detection limit for ligands $L^{\nu-}$ is given by expressions which are similar to those for anions (see Section 10.3). This is reasonable since membrane-active species, such as CN^-, exert a thermodynamically predictable effect on the activity of the primary anions, which is sensed by the electrode. A formal description of the emf in the range of low activities could be obtained by including the activity of free silver ions, a'_{Ag}, in Eq. (10.45).

b) Sample_solution_containing_both_ligands_$L^{\nu-}$_and_anions_X^-

For activities $a_X \gg \alpha$, Eq. (10.48) has to be replaced by a different emf-relationship [42]. The following limiting case is obtained for $\kappa \gg 1$:

$$E = E_X^o - \frac{RT}{F} \ln \left[a_X + \frac{1}{2} a_L \right]$$

(10.55)

and for $\kappa \ll 1$:

$$E = E_X^o - \frac{RT}{F} \ln \left[a_X + \kappa \frac{a_L}{a_X} a_L \right]$$

(10.56)

Equation (10.55) shows in comparison with (10.50) that the selectivity of an AgX membrane is the same for separate solutions containing either $L^{\nu-}$ or X^- and mixed solutions containing both species. This has experimentally been demonstrated to hold for $L^{\nu-} = CN^-$ by Pungor and Tóth [37] and Bound, Fleet, von Storp, and Evans [39]. A refined theoretical model, taking into account the diffusion of CN^- and X^- across the aqueous boundary layer (in analogy to Section

205

10.5.c , leads to the following selectivity constant [16, 35, 39, 42]:

$$K^{Pot}_{X,CN} = \frac{D'_{CN}}{2\ D'_X}$$ (10.57)

A value of 0.61 is calculated for the cyanide/iodide selectivity which compares favorably with the approximative value of 0.5 derived above as well as with experimental results [5, 37, 39, 61]. An even more complicated theory was called upon to quote a value of 0.59 [58].

In contrast to ligands being considered as strong complex formers ($\kappa \gg 1$), those forming weak complexes ($\kappa \ll 1$) behave differently. A comparison of Eq. (10.56) with (10.51) and (10.52) indicates that the selectivity now highly depends on the composition of the sample solution. The use of the term "selectivity constant" is therefore not appropriate. The interference in the determination of anions by ligands forming weak complexes with Ag^+ may easily be estimated using Eq. (10.56).

REFERENCES

[1]   G. Trümpler, Z. Phys. Chem. 99, 9 (1921).

[2]   I. M. Kolthoff and H. L. Sanders, J. Am. Chem. Soc. 59, 416 (1937).

[3]   H. J. C. Tendeloo, J. Biol. Chem. 113, 333 (1936).

[4]   E. Pungor and E. Hollós-Rokosinyi, Acta Chim. Hung. 27, 63 (1961).

[5]   E. Pungor and K. Tóth, Pure Appl. Chem. 34, 105 (1973).

[6]   E. Pungor and K. Tóth, Pure Appl. Chem. 36, 441 (1973).

[7]   E. Pungor and K. Tóth, in Ion-Selective Electrodes in Analytical Chemistry (H. Freiser, ed.), Plenum Press, New York, 1978.

[8]   M. S. Frant and J. W. Ross, Jr., Science 154, 1553 (1966).

[9]   R. A. Durst, ed., Ion-Selective Electrodes, National Bureau of Standards Spec. Publ. 314, Washington, D. C., 1969.

[10] J. W. Ross, Jr., 'Solid-state and liquid membrane ion selective electrodes', chapter 2 of ref. [9].

[11] J. N. Butler, 'Thermodynamic studies', chapter 5 of ref. [9].

[12] R. P. Buck, paper presented at the International Workshop on Ion and Enzyme Electrodes in Biology and Medicine, Schloss Reisensburg (near Ulm), Germany, September 15-18, 1974.

[13] A. K. Covington, 'Heterogeneous membrane electrodes', chapter 3 of ref [9].

[14] G. J. Moody and J. D. R. Thomas, Selective Ion Sensitive Electrodes, Merrow Publishing Co., Watford, 1971.

[15] G. Kahr, Beitrag zum elektromotorischen Verhalten von ionenselektiven Festkörpermembranelektroden, Dissertation ETH, Juris, Zürich, 1972.

[16] J. Koryta, Anal. Chim. Acta 61, 329 (1972).

[17] J. Koryta, Ion-Selective Electrodes, Cambridge University Press, Cambridge, 1975.

[18] R. P. Buck, Anal. Chem. 44, 270R (1972); 46, 28R (1974).

[19] R. P. Buck, Anal. Chem. 48, 23R (1976).

[20] R. P. Buck, Anal. Chem. 50, 17R (1978).

[21] R. P. Buck, Crit. Rev. Anal. Chem. 5, 323 (1975).

[22] R. P. Buck, in Ion-Selective Electrodes in Analytical Chemistry (H. Freiser, ed.), Plenum Press, New York, 1978.

[23] K. Cammann, Das Arbeiten mit ionenselektiven Elektroden, Springer-Verlag, Berlin, 1973.

[24] P. L. Bailey, Analysis with Ion-Selective Electrodes, Heyden International Topics in Science (L. C. Thomas, ed.), Heyden, London, 1976.

[25] J. Vesely, O. J. Jensen, and B. Nicolaisen, Anal. Chim. Acta 62, 1 (1972).

[26] J. Vesely, Coll. Czech. Chem. Comm. 36, 3364 (1971).

[27] J. Siemroth, I. Hennig, and R. Claus, in Ion-Selective Electrodes (E. Pungor, ed.), Akadémiai Kiadó, Budapest, 1977, p. 185.

[28] H. Hirata, K. Higashiyama, and K. Date, Anal. Chim. Acta 51, 209 (1970).

[29] H. Hirata and K. Higashiyama, Anal. Chim. Acta 54, 415 (1971).

[30] H. Hirata and K. Higashiyama, Z. Anal. Chem. 257, 104 (1971).

[31] E. H. Hansen, C. G. Lamm, and J. Růžička, Anal. Chim. Acta 59, 403 (1972).

[32] M. Mascini and A. Liberti, Anal. Chim. Acta 64, 63 (1973).

[33] J. Vesely, Coll. Czech. Chem. Comm. 39, 710 (1974).

[34] K. Tóth, Ph. D. Thesis, Veszprém, 1964.

[35] W. Jaenicke, Z. Elektrochem. 55, 648 (1951); W. Jaenicke and M. Haase, Z. Elektrochem. 63, 521 (1959).

[36] R. P. Buck, Anal. Chem. 40, 1432 (1968).

[37] E. Pungor and K. Tóth, Analyst 95, 625 (1970); Hung. Sci. Instruments 18, 1 (1970).

[38] W. E. Morf, D. Ammann, E. Pretsch, and W. Simon, Pure Appl. Chem. 36, 421 (1973).

[39] G. P. Bound, B. Fleet, H. von Storp, and D. H. Evans, Anal. Chem. 45, 788 (1973).

[40] F. G. K. Baucke, Electrochim. Acta 17, 851 (1972).

[41] J. Havas, IUPAC International Symposium on Selective Ion-Sensitive Electrodes, paper 30, Cardiff, 1973.

[42] W. E. Morf, G. Kahr, and W. Simon, Anal. Chem. 46, 1538 (1974).

[43] H.-R. Wuhrmann, W. E. Morf, and W. Simon, Helv. Chim. Acta 56, 1011 (1973).

[44] R. P. Buck, in Ion and Enzyme Electrodes in Biology and Medicine (M. Kessler et al., eds.), Urban & Schwarzenberg, Munich, 1976.

[45] J. Bagg and G. A. Rechnitz, Anal. Chem. 45, 271 (1973).

[46] Handbook of Chemistry and Physics, 52th ed., Chem. Rubber Publishing Co., Cleveland, Ohio, 1970.

[47] L. R. Sillén and A. E. Martell, Stability Constants of Metal-Ion Complexes, Spec. Publ. No. 17, The Chem. Soc., London, 1964.

[48] R. A. Durst, 'Analytical techniques and application of ion selective electrodes', chapter 11 of ref. [9].

[49] W. Simon, presented at the Euchem Conference on Trace Analysis in Complex Mixtures, Baden bei Wien (Austria), May 9-11, 1978.

[50] R. A. Durst, private communication.

[51] J. D. R. Thomas, presented at the International Conference on Ion-Selective Electrodes, Budapest, 5-9 September, 1977.

[52] J. Gulens and B. Ikeda, Anal. Chem. 50, 782 (1978).

[53] Analytical Methods Guide, 7th ed., Orion Research Inc., Cambridge, Mass., 1975.

[54] V. Rothmund and G. Kornfeld, Z. Anorg. Allgem. Chem. 103, 129 (1918).

[55] G. Eisenman, ed., Glass Electrodes for Hydrogen and Other Cations, Dekker, New York, 1967.

[56] H. Garfinkel, 'Cation-exchange properties of dry silicate membranes', in Membranes, Vol. 1 (G. Eisenman, ed.), Dekker, New York, 1972.

[57] H. A. Klasens and J. Goossen, Anal. Chim. Acta 88, 41 (1977).

[58] A. Hulanicki and A. Lewenstam, Talanta 23, 661 (1976); 24, 171 (1977).

[59] Ion-Selective, Solid-State Electrodes for Iodide, Bromide, Chloride and Cyanide Type IS 550, Philips, Eindhoven.

[60] W. Nernst, Z. Phys. Chem. 47, 52 (1904).

[61] B. Fleet and H. von Storp, Anal. Chem. 43, 1575 (1971).

Chapter 11

Liquid-Membrane Electrodes Based on Liquid Ion-Exchangers

Porous membranes with fixed, electrically charged ion-exchange sites were developed especially in the nineteen twenties and thirties as models for biological membranes (for a review, see Michaelis [1, 2] and Sollner [3, 4]). Such membranes exhibit a strong tendency towards permselectivity, which means that they are easily permeable for counterions (ions attracted by the ion-exchange sites) but poorly permeable for the oppositely charged coions (ions repelled by the ion-exchange sites). The theory of ion transport across porous membranes was pioneered by Teorell [5-7], Meyer and Sievers [8, 9], and was later generalized by Schlögl [10-13] and by Helfferich [13, 14] (see also Chapters 4 and 7). More recently, a modern theory of the irreversible thermodynamics of membrane transport has been developed [15-19] that provides a fundamental background for the physics of porous membranes (for a review, see [20-23]).

Liquid ion-exchange membranes are homogeneous phases, formed from water-immiscible organic liquids so as to be devoid of aqueous pores, and based on mobile ionic or ionogenic components, such as hydrophobic acids, bases, or salts. Such systems were first studied systematically in 1933 by Beutner [24]. In his important work, he already drew the remarkable conclusion that liquid membranes might respond as electrodes to changes in external solution conditions. The potential response was postulated to be due to the phase boundary potentials which reflect the partition equilibria between the aqueous solutions and the membrane phase. An extension of the work of Beutner was performed by Kahlweit et al. [25-28] who applied the fixed-site concepts of the classical Teorell-

211

Meyer-Sievers theory to describe the potential response of liquid ion-exchange membranes, thereby also taking into account the membrane-internal transport processes. However, this Teorell-Meyer-Sievers approach of liquid membranes, for evident reasons, constitutes a correct description only for fortuitous situations where the basically mobile charged sites can be regarded as "fixed", i. e. as trapped and evenly distributed within the membrane phase. An adequate theoretical formalism that is applicable to the membrane systems used in analytical devices became available only in 1967 owing to the remarkable contribution by Sandblom, Eisenman, and Walker [29]. These authors first developed rigorous relationships for the electrical properties and transport processes in liquid ion-exchange membranes and also took into account the possibility for association or complex formation between ionic sites and counterions. Part of the following discussion will be devoted to the original theory by Sandblom et al. [29, 30]. We shall demonstrate, however, that the key results of this theory are all easily deducible from the rather universal formalism of membrane transport elaborated in the preceding part (see also [31, 32]).

11.1. MEMBRANE MATERIALS AND OBSERVED ION SELECTIVITIES

Most of the exciting early work in the field of ion-exchange membranes was carried out on solid (porous) membranes. As of the later sixties, when the first ion-selective liquid-type membranes were introduced [33-41], this class has become a focus of attention. Whereas the porous membrane is at best capable of producing permselectivity for a given class of ions (cations or anions, depending on the charge of the ion-exchange sites involved), it is the compact or liquid counterpart that may exhibit increased selectivity or even specificity for a given sort of ions. In fact, it follows from elementary theoretical arguments that the selectivity of the membrane medium (solvent) in the extraction of counterions becomes operative only in the absence of aqueous pores, and that the ion-binding specificity of the incorporated sites or ligands can be exploited only if an adequate mobility of these charge carriers is guaranteed (see also below).

It goes without saying that the development of the first ion-selective electrodes based on liquid ion-exchangers was accomplished more or less intuitively. All the same, a number of these membrane systems (see Table 11.1) have withstood attempts at displacement by superior formulations and are still attractive components for potentiometric sensors. Three groups of liquid ion-exchanger electrodes have attained special analytical significance:

a) Divalent-ion-sensitive electrodes based on dialkyl-phosphates or diarylphosphates dissolved in appropriate solvents. Such membrane electrodes with considerable specificity for calcium ions (membrane solvent: dioctylphenylphosphonate), respectively with comparable selectivities for calcium and magnesium (solvent: 1-decanol) were introduced in 1967 by Ross [34, 35]. Significant improvements of the calcium electrode were realized by Moody et al. [52] and particularly by Růžička et al. [56] (see Table 11.2). A review on the membrane compositions and reported selectivities of different calcium electrodes has been given elsewhere [57]. A recently described electrode [58], which also shows selectivity for magnesium and calcium ions, is based on different principles.

b) Anion-sensitive electrodes based on organic ammonium ions (Sollner and Shean [33], Coetzee and Freiser [38]), respectively on organic complexes of nickel(II) and iron(II) (Ross [35]). The selectivities of such systems among anions depend to a large extent on the membrane solvent used, on the membrane preparation [59], and also on the external solution conditions (see Sections 11.2 - 11.4). The selectivity sequence, however, is commonly the same and follows the Hofmeister lyophilic series [33, 60], i. e.

$$R^- > ClO_4^- > I^- > NO_3^- > Br^- > Cl^- > F^-$$

where R^- symbolizes highly lipophilic organic anions. As a rule, the discrimination of liquid anion-exchange membranes between different counterions is less pronounced (when compared, e. g., with the silver compound solid-state membranes), which makes them appropriate as sensors for various anions, such as nitrate, chloride, or perchlorate [35, 38, 44, 61]. Because of their lack of ion specificity, these electrodes can be applied only for analyses of solutions that meet a number of restrictive criteria.

c) Cation-sensitive electrodes based on tetra(p-chlorophenyl)-borate or similar anions dissolved in nitroaromatic solvents (Baum and coworkers [40, 48], Scholer and Simon [41]). The selectivity behavior is analogous to that of the anion-exchangers summarized above. Thus, the selectivity sequence is

$$R^+ > Cs^+ > Rb^+ > K^+ > Na^+ > Li^+ \ .$$

Membrane electrodes of this type were introduced as sensors for potassium ions [40] and have found acceptance in biological applications [61-63] although they suffer from an immense selectivity for large organic cations. Indeed, such electrodes can be used for the direct potentiometric determination of acetylcholine [48], tubocurarin [41], and other lipophilic quaternary ammonium ions [41] in the presence of ionic backgrounds corresponding to blood serum.

A deeper understanding of the ion-selectivity behavior of membrane electrodes with liquid ion-exchangers is obtained from theoretical results which are discussed in the following.

Table 11.1. Properties of some classical liquid ion-exchange membrane electrodes

Membrane composition				
Counterion (Primary ion)	Ion-exchange site	Solvent	Selectivity sequence	Reference
Sensors for anions				
Not given	Lauryl(trialkylmethyl) ammonium ion (Amberlite LA-2)	Benzene or xylene or nitrobenzene	$SCN^- \gg Cl^-,\ I^- > Br^-$	33
NO_3^- } ClO_4^-	$Ni(o\text{-}phen)_3^{2+}$ } $Fe(o\text{-}phen)_3^{2+}$ }	Not given (nitrobenzene?)	$ClO_4^- > I^- > NO_3^- > Br^- > Cl^- \sim F^-$	35, 42, 43
Cl^-	Dimethyldistearyl-ammonium ion	Not given (decanol?)	$ClO_4^- > I^- > NO_3^- > Br^- > Cl^- > F^-$	35
Various anions	Methyltricapryl-ammonium ion (Aliquat 336S)	1-Decanol	$SCN^- > ClO_4^- > I^- > NO_3^- > Br^- > Cl^-$	38, 44
Various anions	Tetraheptyl-ammonium ion	Benzene	$PdCl_4^{2-},\ ZnCl_4^{2-} > NO_3^- > Br^- > Cl^-$	45
ClO_4^-	Dodecyloctyl-methylbenzyl-ammonium ion	Nitrobenzene		42

(Continued)

I^-, NO_3^-, Br^-	Tetraalkyl-ammonium ion	Chlorobenzene		46
Sulfonates (anionic detergents)	Hexadecyl-pyridinium ion	Nitrobenzene		47

Sensors for monovalent cations

H^+	Oleate	Nitrobenzene	$Cs^+ > Rb^+ > K^+ > Na^+ > Li^+$	39
K^+	Tetra(p-chloro-phenyl)borate	3-Nitro-o-xylene/p-hexylnitro-benzene	$K^+ > Na^+$	40
Acetyl-choline	Tetra(p-chloro-phenyl)borate	3-Nitro-o-xylene	Acetylcholine > choline	48
Various cations	Tetraphenylborate or dipicrylaminate/thenoyltrifluoro-acetone	2-Nitro-p-cymene	$R^+ > Cs^+ > Rb^+ > K^+ > Na^+ > Li^+$ R^+: organic "onium" ions, e.g. alkaloids, surfactants,etc.	41
Quaternary ammonium ions or picrate	Tetraphenylborate	Nitrobenzene		49
K^+	Dipicrylaminate	Xylene		50
Initially no ions present in the membrane	Dioctyl-phthalate		$(C_6H_{13})_4N^+ > (C_5H_{11})_4N^+ > (C_4H_9)_4N^+ > (C_3H_7)_4N^+ > K^+ > H^+$	51

(Continued)

Sensors for divalent cations

Ca^{2+}	Dioctylphosphate or didecylphosphate	Dioctylphenyl-phosphonate	$H^+ >> Zn^{2+} > Ca^{2+} \sim Fe^{2+} \sim Pb^{2+} > Cu^{2+} > Ni^{2+} > Sr^{2+} \sim Mg^{2+} \sim Ba^{2+} > Na^+$	34, 35, 52
Ca^{2+} & Mg^{2+}	Dialkylphosphate	1-Decanol	$Zn^{2+} \sim Fe^{2+} \sim Cu^{2+} > Ni^{2+} \sim Ca^{2+} = Mg^{2+} \sim Ba^{2+} > Sr^{2+} > Na^+$	35
Ca^{2+}	Thenoyltrifluoro-acetone	Tributyl-phosphate	$Ca^{2+} > Mg^{2+} > Na^+$	36, 37
Heavy metal ions	Dithizonate or 2-(alkylthio)-acetate	Dipentyl-phthalate, xylene, or other solvents	Various selectivity sequences	35, 53-55

Table 11.2. Selectivities of Ca^{2+} electrodes based on liquid ion-exchangers [56][a]

Selective membrane components[b]	log K$_{CaM}^{Pot}$ for ion M^{z+} =						
	Na$^+$	K$^+$	Mg^{2+}	Sr^{2+}	Ba^{2+}	H$^+$	Zn^{2+}
DDP$^-$/DOPP	-3.50[c]	-4.00	-1.85	-1.77	-2.00	8.00[d]	0.51
DOPP$^-$/DOPP	-5.20[c] -4.70[e]	-5.70	-3.60	-1.77	-3.60	4.20[d]	-1.22

[a] Reported selectivities as obtained from the separate solution method, except where footnotes c-e apply. For a detailed discussion, see Reference 57.

[b] DDP$^-$: didecylphosphate; DOPP$^-$: di(p-octylphenyl)phosphate; DOPP: dioctylphenylphosphonate.

[c] Fixed interference method, using Ca^{2+}-buffered solutions.

[d] Fixed primary ion method (varying pH).

[e] Value estimated from the basic calibration curve in Reference 56.

11.2. IMPLICATIONS OF THE GENERAL MEMBRANE THEORY

In this section we make use of some results of the general membrane theory, worked out in Chapters 3-7 and summarized in Chapter 9, that are basic to the rationalization of liquid-membrane electrodes.

11.2.1. Potential Response of Ion-Exchange Membranes to Monovalent Counterions

As the first example we consider a cation-exchange liquid membrane in contact with aqueous solutions of two cations I^+ and J^+. The predominant species existing within such a membrane phase are the free sites S^- and the free counterions I^+ and J^+ (case of complete dissociation of ions), respectively electrically neutral complexes of the type IS and JS (case of nearly complete association). Of these species, only the charged forms exert a direct influence on the electrical potential, whereas the neutral complexes have a more or less regulationary effect on the population of free ions in the membrane. Hence we may immediately substitute a relationship of the form (9.7) to describe the membrane-internal diffusion potential for the present case:

$$E_D = (1-\tau)\frac{RT}{F}\ln\frac{a_i(0) + a_j(0)}{a_i(d) + a_j(d)} - \tau\frac{RT}{F}\ln\frac{a_s(0)}{a_s(d)} \qquad (11.1)$$

where:

$$\tau = \frac{u_s}{u_i + u_s} \qquad (11.2)$$

As was shown in Chapter 4, this expression is based on the

primary assumptions of a zero-current steady-state, of electro-neutrality, and of equal mobilities of comparable ionic forms within the membrane. The boundary potential contribution can be formulated, according to Eq. (9.4), as follows:

$$E_B = \frac{RT}{F} \ln \frac{w_i k_i a_i' + w_j k_j a_j'}{w_i a_i(0) + w_j a_j(0)} - \frac{RT}{F} \ln \frac{w_i k_i a_i'' + w_j k_j a_j''}{w_i a_i(d) + w_j a_j(d)} \qquad (11.3)$$

where w_i and w_j are appropriate weighting factors (see below). Assuming conservation of ion-exchange sites within the membrane, we may use the following approximation to eliminate the unknown quantities $a_s(0)$ and $a_s(d)$ in Eq. (11.1):

$$a_s(x) + a_{is}(x) + a_{js}(x) = X = const(x) \qquad (11.4)$$

where X denotes the total activity of sites. Combination of Eqs. (11.1) and (11.3) then leads to a straightforward description of the potential response and the ion-selectivity behavior of the membrane.

In the case of nearly complete dissociation between sites and counterions, such as obtains for non-complexing ionic compounds in comparatively polar membrane media [64], we have

$$a_s(0) = a_s(d) = X \qquad (11.5)$$

and as a consequence of the electroneutrality assumption:

$$a_i(0) + a_j(0) = a_i(d) + a_j(d) \qquad (11.6)$$

220

The diffusion potential of membranes that meet the mentioned requirements therefore approximates zero, as becomes evident from Eq. (11.1). Finally, one obtains the following expressions for the membrane potential E_M (see Eq. (11.3) with $w_i = w_j = 1$), respectively for the emf E of the membrane electrode assembly:

$$E_M = \frac{RT}{F} \ln \frac{k_i a_i' + k_j a_j'}{k_i a_i'' + k_j a_j''} \qquad (11.7)$$

and

$$E = E_i^o + \frac{RT}{F} \ln \left[a_i' + K_{ij}^{Pot} a_j' \right] \qquad (11.8)$$

where

$$K_{ij}^{Pot} = \frac{k_j}{k_i} \qquad (11.9)$$

Equation (11.9) demonstrates that the ion selectivity of dissociated ion-exchanger membrane systems is dictated almost exclusively by the extraction properties of the membrane solvent used. A convincing example for such a relationship is given in Figure 11.1. The documented correlation between the observed selectivity coefficients and the known ratios of single-ion distribution constants [41, 64] is astounding and clearly corroborates the basic theoretical results. In addition, it nicely illustrates the aforementioned marked preference of corresponding membranes for large, lipophilic counterions. The differences in the selectivity factors in Figure 11.1 indeed amount to several orders of magnitude.

221

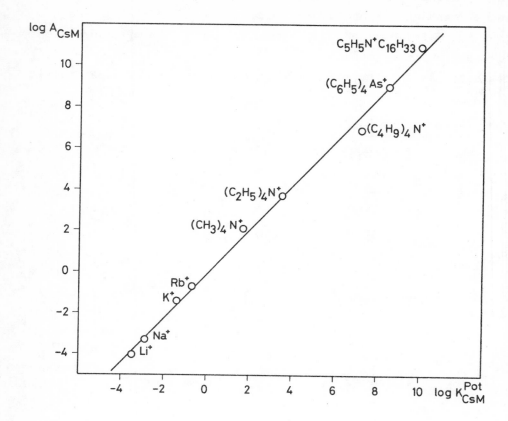

<u>Figure 11.1.</u> Correlation between the potentiometric selectivity factors, K_{CsM}^{Pot}, and the ion-exchange equilibrium constants, $A_{CsM} = k_M/k_{Cs}$, for a liquid membrane electrode based on the solvent 2-nitro-p-cymene [41]. Tetraphenylborate anions were incorporated as dissociated exchanger sites for cations M^+.

An easy explanation for such behavior may be obtained through the definition of the ionic distribution coefficient k_j:

$$k_j = \exp\left[\frac{\Delta G^O_{H,j} - \Delta G^O_{S,j}}{RT}\right] \qquad (11.10)$$

$\Delta G^O_{H,j}$ is the free energy of hydration of species J, and $\Delta G^O_{S,j}$ is the free energy of solvation for the same species in the membrane medium. Since $\Delta G^O_{H,j}$ is normally the decisive energy term, we conclude that poor hydration of ions favors their extraction into the liquid membrane and vice versa. In some cases, it is possible to even get a qualitative correlation between the logarithms of the selectivity coefficients and the corresponding hydration energies [44, 64, 65]. Figures 11.2 and 11.3 show such correlations for a cation-selective electrode based on tetraphenylborate sites, respectively for anion-sensors based on quaternary ammonium ions. Although in the latter case association between ion-exchange sites and counterions may become important, it obviously has little influence on the ion-selectivity behavior of the system [38, 44, 45, 65-67]. Thus, the selectivity sequence of different anion-sensitive liquid membrane electrodes is typically the same and follows the lyophilic series (see Figure 11.3 and Table 11.1). The range of available ion specificity, on the other hand, is mainly a function of the membrane solvent used. A comparatively high discrimination between different counter-ions is obtained when using nitroaromatic solvents (Figures 11.2 and 11.3 and Table 11.3), whereas modest selectivities are reported for membranes with decanol which is found appropriate for sensors for chloride.

When using nonpolar solvents as membrane liquids and/or strong complexing agents as incorporated sites, it is possible to encounter situations where association (ion pair formation

223

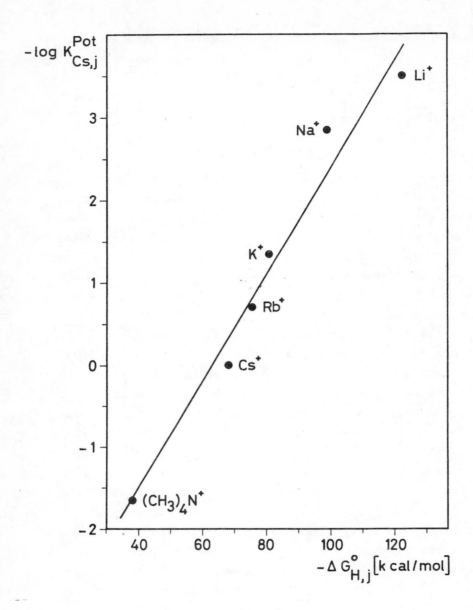

Figure 11.2. Dependence of the observed selectivities of a cation-exchange membrane (Figure 11.1) on the free energies of hydration of cations.

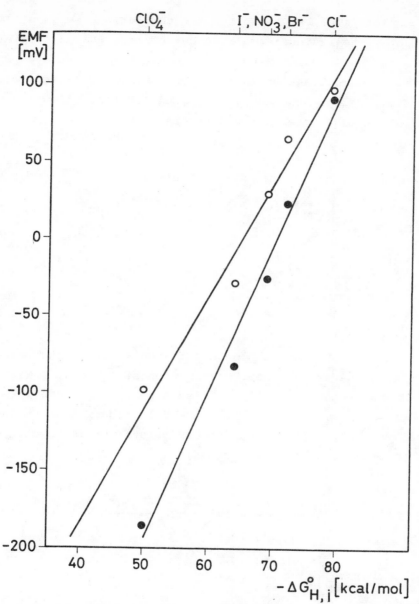

<u>Figure 11.3.</u> Correlation between the EMF response of anion-sensitive liquid membrane electrodes to 10^{-2}M solutions of different anions and the corresponding free energies of hydration [65]. The membrane contained quaternary ammonium ions in o-nitrophenyl octyl ether (●), respectively in methyl decyl ketone (o).

or complexation) between sites and counterions becomes almost complete. The residual small number of free charge carriers in the membrane, nevertheless, remains decisive for the formation of the transmembrane potential. Hence Eqs. (11.1) and (11.3) are still valid. Equation (11.4) leads to the following approximation for associated ion-exchangers:

$$a_{is}(0) + a_{js}(0) = a_{is}(d) + a_{js}(d) = X \qquad (11.11)$$

or

$$\frac{a_s(d)}{a_s(0)} = \frac{K_{is}a_i(0) + K_{js}a_j(0)}{K_{is}a_i(d) + K_{js}a_j(d)} \qquad (11.12)$$

where K_{is} and K_{js} are the stability constants of the respective complexes in the membrane. Combination of Eq. (11.1) with expressions of the type (11.3) (inserting $w_i = K_{is}$ and $w_j = K_{js}$, respectively $w_i = w_j = 1$) is now easily accomplished. The final result for membranes with associated cation-exchangers and monovalent counterions reads:

$$E_M = (1-\tau) \frac{RT}{F} \ln \frac{k_i a_i' + k_j a_j'}{k_i a_i'' + k_j a_j''} + \tau \frac{RT}{F} \ln \frac{K_{is}k_i a_i' + K_{js}k_j a_j'}{\mathring{K}_{is}k_i a_i'' + K_{js}k_j a_j''} \qquad (11.13)$$

or, after some reduction:

$$E = E_i^o + (1-\tau) \frac{RT}{F} \ln \left[a_i' + K_{ij}^{(1)} a_j' \right] + \tau \frac{RT}{F} \ln \left[a_i' + K_{ij}^{(2)} a_j' \right] \qquad (11.14)$$

226

Table 11.3. Selectivities of ClO_4^- electrodes based on liquid ion-exchangers

Anion X^-	$\log K_{ClO_4X}^{Pot}$		
	Electrode in Fig. 11.3 [65][a]	ORION electrode [35][b]	Theoretical values[c]
ClO_4^-	0.0	0.0	0.0
I^-	-1.9	-1.9	-1.9
NO_3^-	-3.0	-3.0	-
Br^-	-3.9	-3.3	-
Cl^-	-5.1	-3.7	-3.8

[a] N,N-dimethyl-N-hexadecyl-N-(1-decyloxycarbonyl)ethyl ammonium salt in o-nitrophenyl octyl ether.

[b] 1:3 complexes of iron (II) with substituted o-phenanthroline in nitrobenzene (see Table 11.1).

[c] Values $\log(k_X/k_{ClO_4})$ obtained from single-ion extraction constants k [64] for the system water-nitrobenzene at $25^\circ C$.

where the selectivity factors are evidently given by

$$K_{ij}^{(1)} = \frac{k_j}{k_i} \tag{11.15}$$

$$K_{ij}^{(2)} = \frac{K_{js}k_j}{K_{is}k_i} \tag{11.16}$$

In contrast to the Nicolsky-type potential function (11.8), as obtained for the situation with dissociated exchanger sites, Eq. (11.14) constitutes a three-parameter description of the electrode response and the involved ion selectivity. The additional parameters are τ, which is the electrical transference number of the free sites in the membrane (see Eq. (11.2)), and a second selectivity factor, reflecting the ion-binding properties of the sites or ligands. The latter term can be identified with the equilibrium constant of the following ion-exchange reaction:

$$IS(m) + J^+(aq) \rightleftharpoons JS(m) + I^+(aq) \qquad (11.17)$$

By subdividing this reaction into a number of suitable steps, one may derive an alternative formulation of the site-induced selectivity (see also Chapter 12):

$$K_{ij}^{(2)} = \frac{K_{js}^w}{K_{is}^w} \frac{k_{js}}{k_{is}} \approx \frac{K_{js}^w}{K_{is}^w} \qquad (11.18)$$

where the complex stability constants K_{is}^w and K_{js}^w refer to aqueous solution. The selectivity term $K_{ij}^{(2)}$ thus turns out to be approximately independent of the membrane solvent since the distribution coefficients k_{is} and k_{js} of the involved neutral, isosteric complexes are presumed to be quite similar. Inspection of Eq. (11.14) makes clear, however, that the ion specificity inherent to certain electrically charged complexing agents can never be fully exploited in ion-selective membrane electrodes. Evidently, the potentiometric selectivity behavior of such devices usually depends in a rather complicated way on both the membrane solvent and the incorporated sites. An emf-response function of the familiar type (11.8)

228

can be deduced only for situations where $\tau = 0$ (i. e. for fixed sites) or where $K_{is} = K_{js}$ (nonspecific sites) [*]. However, under such conditions the sites are seen to be no longer capable of modifying the membrane's selectivity between counterions. The desirable condition for an exclusive selectivity control by the ligand sites would be $\tau = 1$ (infinitely mobile sites) which cannot be fulfilled in reality. This could explain the fiasco experienced when attempts were made to utilize negatively charged ionophores, such as nigericin or monensin [68], as components in highly selective sensors for alkali ions. In contrast to this, a wide range of ion specificities is accessed by the incorporation of electrically neutral ionophores into liquid membranes, as will be demonstrated in Chapter 12.

11.2.2. Potential Response of Ion-Exchange Membranes to Divalent Counterions

The large values of $-\Delta G_H^o$ ($>10^3$ kJ/mol) characteristic of divalent cations make it improbable that these species are extracted as free ions into liquid ion-exchange membranes (see Eq. (11.10)). Therefore, it is realistic to assume that divalent-ion-selective membranes based on negatively charged ligands predominantly form complexes of the type IS_2 and JS_2, and that the only charge-carrying species to be taken into account are IS^+, JS^+, and S^-. The membrane-internal diffusion potential will then be determined in a way analogous to Eq. (11.1) with all the terms referring to free cations being replaced by the corresponding ones for the cationic 1:1 complexes. Hence we can write:

$$E_D = (1-\tau) \frac{RT}{F} \ln \frac{a_{is}(0) + a_{js}(0)}{a_{is}(d) + a_{js}(d)} - \tau \frac{RT}{F} \ln \frac{a_s(0)}{a_s(d)} \qquad (11.19)$$

---

[*] The latter case seems to be realized in certain anion-sensitive liquid-membrane electrodes [38, 44, 45, 65-67].

where:

$$\tau = \frac{u_s}{u_{is} + u_s} \tag{11.20}$$

Recalling the stability constants of the positively charged 1:1 complexes:

$$K_{is} = \frac{a_{is}(0)}{a_i(0) \; a_s(0)} = \frac{a_{is}(d)}{a_i(d) \; a_s(d)} \tag{11.21a}$$

$$K_{js} = \frac{a_{js}(0)}{a_j(0) \; a_s(0)} = \frac{a_{js}(d)}{a_j(d) \; a_s(d)} \tag{11.21b}$$

and adapting Eq. (11.12) to the present case (β_{iss} and β_{jss} are the cumulative stability constants of the predominant electrically neutral 1:2 complexes):

$$\left[\frac{a_s(d)}{a_s(0)} \right]^2 = \frac{\beta_{iss} \; a_i(0) + \beta_{jss} \; a_j(0)}{\beta_{iss} \; a_i(d) + \beta_{jss} \; a_j(d)} \tag{11.22}$$

we get the following:

$$E_D = (1-\tau) \; \frac{RT}{F} \; \ln \frac{K_{is}a_i(0) + K_{js}a_j(0)}{K_{is}a_i(d) + K_{js}a_j(d)}$$

$$+ \; (\tau - \frac{1}{2}) \; \frac{RT}{F} \; \ln \frac{\beta_{iss}a_i(0) + \beta_{jss}a_j(0)}{\beta_{iss}a_i(d) + \beta_{jss}a_j(d)} \tag{11.23}$$

The boundary potential term (Eq. (9.4)) is given for two divalent cations by

$$E_B = \frac{RT}{2F} \ln \frac{w_i k_i a_i' + w_j k_j a_j'}{w_i a_i(0) + w_j a_j(0)} - \frac{RT}{2F} \ln \frac{w_i k_i a_i'' + w_j k_j a_j''}{w_i a_i(d) + w_j a_j(d)} \qquad (11.24)$$

By using the trivial relationship

$$E_B = (1-\tau) \, 2E_B + (\tau - \frac{1}{2}) \, 2E_B \qquad (11.25)$$

and by inserting expressions of the type (11.24) with $w_i = K_{is}$ and $w_j = K_{js}$, respectively with $w_i = \beta_{iss}$ and $w_j = \beta_{jss}$, we succeed in combining terms for E_D and E_B. Finally we obtain the following fundamental result that describes the emf-response of divalent-ion sensors based on liquid ion-exchange membranes:

$$E = E_i^o + (1-\tau) \, \frac{RT}{F} \ln \left[a_i' + K_{ij}^{(1)} a_j' \right] + (\tau - \frac{1}{2}) \, \frac{RT}{F} \ln \left[a_i' + K_{ij}^{(2)} a_j' \right] \qquad (11.26)$$

The potentiometrically exhibited selectivity of such electrode assemblies for ions J^{2+} relative to the primary ions I^{2+} is generally controlled, apart from τ, by the following two selectivity parameters:

$$K_{ij}^{(1)} = \frac{K_{js} k_j}{K_{is} k_i} \qquad (11.27)$$

and

$$K_{ij}^{(2)} = \frac{\beta_{jss}k_j}{\beta_{iss}k_i} = \frac{K_{jss}}{K_{iss}} K_{ij}^{(1)} \tag{11.28}$$

The most interesting new aspect of Eq. (11.26), as compared to the corresponding expression (11.14) valid for monovalent counterions, is its tendency to approximate a conventional Nicolsky-type response function for realistic situations. Exactly speaking, it is reasonable to assume that the mobilities in the membrane phase are nearly the same for the free ligand and its 1:1 complex because both forms are single-charged and of comparable dimensions. This corresponds to the assumption that $\tau \cong 0.5$, for which case Eq. (11.26) immediately reduces to

$$E = E_i^o + \frac{RT}{2F} \ln \left[a_i' + K_{ij}^{Pot} a_j' \right] \tag{11.29}$$

with

$$K_{ij}^{Pot} \cong K_{ij}^{(1)}$$

$$= \frac{K_{js}k_j}{K_{is}k_i} = \frac{K_{js}^W k_{js}}{K_{is}^W k_{is}} \qquad [\tau \cong 0.5] \tag{11.30}$$

Equations (11.29) and (11.30) offer for the first time a justification of the concentration-independence suggested for the selectivity coefficients K_{ij}^{Pot} of divalent-ion-exchange membrane electrodes. The applicability of such devices as "water hardness electrodes" [35] is evidently due to the

232

fortuitous fact that $K_{CaMg}^{(1)} \cong 1$ and, correspondingly, $K_{CaMg}^{Pot} \cong 1$ (independent of a_{Ca}' and a_{Mg}') which enables the total activity of calcium and magnesium ions to be determined potentiometrically. Another important conclusion may be drawn from Eq. (11.30). The selectivity of the ion sensors in question among divalent cations is recognized to be controlled mainly by the formation reactions of the 1:1 cation/ligand complexes (terms K_{is}^{W} and K_{js}^{W} in Eq. (11.30)) and by the extraction reactions of these species (k_{is} and k_{js}). The distribution coefficients k of the positively charged complexes involved depend heavily on the membrane solvent and also on the nature of the cation. Therefore, again it is expected that the membrane solvent is a major factor determining the potentiometric selectivity of the membrane electrode assembly. This indeed agrees with the experimental facts mentioned in Section 11.1, example a).

11.2.3. Mixed Potential Response to Divalent and Monovalent Counterions (Origin of "Potential Dips")

The formalism applied in the foregoing sections also permits a straightforward description of cation-exchange liquid membrane electrodes in the presence of divalent ions I^{2+} and monovalent ions J^{+}. For such systems, Eqs. (11.1) and (11.19) have to be replaced by the following one:

$$E_D = (1-\tau) \frac{RT}{F} \ln \frac{a_{is}(0)+a_j(0)}{a_{is}(d)+a_j(d)} - \tau \frac{RT}{F} \ln \frac{a_s(0)}{a_s(d)} \qquad (11.31)$$

with:

$$\tau = \frac{u_s}{u_{is}+u_s} \qquad (11.20)$$

233

The steady-state condition in respect to total ligand leads, in analogy to Eq. (11.11), to the additional relation:

$$2\, a_{iss}(x) + a_{js}(x) = 2K_{iss}K_{is}a_i(x)\, (a_s(x))^2 + K_{js}a_j(x)\, a_s(x)$$

$$= X \tag{11.32}$$

or (with $\beta_{iss} = K_{iss}\, K_{is}$):

$$a_s(x) = \frac{\sqrt{8X\,\beta_{iss}a_i(x) + (K_{js}a_j(x))^2} - K_{js}a_j(x)}{4\beta_{iss}a_i(x)} \tag{11.33}$$

which holds, e. g., for the locations x=0 and x=d inside the membrane. The last expression allows us to eliminate in Eq. (11.31) the activities of the free ligand and its complexed form. We then obtain:

$$E_D = (1-\tau)\,\frac{RT}{F}\,\ln\frac{\sqrt{8X\,\beta_{iss}a_i(0)+(K_{js}a_j(0))^2} - K_{js}a_j(0)+4K_{iss}a_j(0)}{\sqrt{8X\,\beta_{iss}a_i(d)+(K_{js}a_j(d))^2} - K_{js}a_j(d)+4K_{iss}a_j(d)}$$

$$+\,\tau\,\frac{RT}{F}\,\ln\frac{\sqrt{8X\,\beta_{iss}a_i(0)+(K_{js}a_j(0))^2} + K_{js}a_j(0)}{\sqrt{8X\,\beta_{iss}a_i(d)+(K_{js}a_j(d))^2} + K_{js}a_j(d)} \tag{11.34}$$

Recalling the conditions (3.9a,b) for the interfacial equilibria, we may replace in Eq. (11.34) all the ionic activities referring to the membrane phase by the corresponding products of outside activities and ionic distribution coefficients, thereby arriving at a relationship for the total membrane potential E_M. Finally, the emf of the membrane electrode assembly comes out to be

$$E = E_i^o + (1-\tau) \frac{RT}{F} \ln \left[\sqrt{a_i' + \frac{1}{4} K_{ij}^{(2)} a_j'^2} - \sqrt{\frac{1}{4} K_{ij}^{(2)} a_j'^2} + \sqrt{K_{ij}^{(1)} a_j'^2} \right]$$

$$+ \tau \frac{RT}{F} \ln \left[\sqrt{a_i' + \frac{1}{4} K_{ij}^{(2)} a_j'^2} + \sqrt{\frac{1}{4} K_{ij}^{(2)} a_j'^2} \right] \qquad (11.35)$$

where the monovalent/divalent ion selectivity terms are defined as

$$K_{ij}^{(1)} = \left(\frac{2K_{iss}}{K_{js}} \right)^2 K_{ij}^{(2)} \qquad (11.36)$$

$$K_{ij}^{(2)} = \frac{(K_{js} k_j)^2}{2X \beta_{iss} k_i} \qquad (11.37)$$

These equilibrium parameters are representative for the
following ion-exchange reactions:

$$K_{ij}^{(1)}: \quad 2IS^+ (m) + 2J^+ (aq) \rightleftharpoons 2J^+ (m) + IS_2 (m) + I^{2+} (aq) \quad (11.38)$$

$$K_{ij}^{(2)}: \quad IS_2 (m) + 2J^+ (aq) \rightleftharpoons 2JS (m) + I^{2+} (aq) \qquad (11.39)$$

of which the second one is recognized to be controlled to a
large extent by the ion-binding properties of the ligands used
as membrane components. The theoretical emf-response function
(11.35) is clearly at variance with the empirical formalism
(11.40) that is usually preferred to describe the present case:

235

$$E = E_i^o + \frac{RT}{2F} \ln \left[a_i' + K_{ij}^{Pot} a_j'^2 \right] \tag{11.40}$$

Since it is conceivable that the two basic selectivity factors $K_{ij}^{(1)}$ and $K_{ij}^{(2)}$ assume widely different values, the gross selectivity coefficient K_{ij}^{Pot} determined from (11.40) will generally show large variations, even for one and the same system of membrane and counterions. The most interesting consequence of Eq. (11.35), as compared to (11.40), is that the plot of the potential response E versus log a_j' at a constant value of a_i' may exhibit a minimum. Such "potential dips" are well known in practice and are characteristic of the pH interference in calcium-selective electrodes based on liquid ion-exchangers [30, 35, 56]. Although it might be supposed that such phenomena were due to certain nonidealities in the experimental procedure, Figure 11.4 unmistakably demonstrates that the exhibited emf-minima are systematic effects of pH interference. The calculations according to Eq. (11.35) reveal that the conditions for the appearance of "potential dips" are $\tau < 0.5$ and $K_{CaH}^{(2)} \gg K_{CaH}^{(1)}$. The latter circumstance is probably due to the substantial basicity exhibited by the negatively charged ligand dialkylphosphate in the membrane (see Eq. (11.36)). It may finally be noted that potential minima of the type in question cannot be rationalized on the basis of the familiar Nicolsky equation (11.40), except when negative local values of the selectivity coefficient K_{CaH}^{Pot} are accepted.

11.3. THEORY OF SANDBLOM, EISENMAN, AND WALKER, AND ITS EXTENSIONS

The original theory of Sandblom, Eisenman, and Walker [29] describes with somewhat greater accuracy membrane systems with ionic sites S^-, two monovalent counterions I^+ and J^+, as well

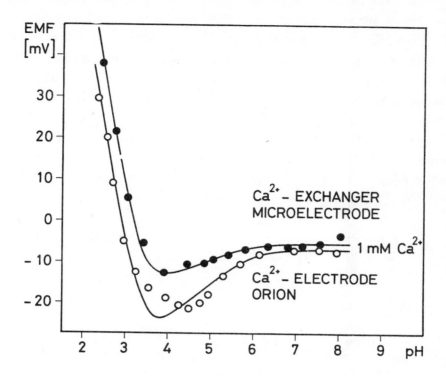

Figure 11.4. Interference of hydrogen ions in calcium-selective micro- and macroelectrodes based on dialkylphosphate/dioctyl-phenylphosphonate. The experimental EMF data obtained for 1 mM Ca^{2+} and a varying pH were taken from Reference 30. The solid curves were calculated from Eq. (11.35) with $K_{CaH}^{(1)}=10^{-1.5}M^{-1}$ $K_{CaH}^{(2)}=10^{9}\ M^{-1}$, $\tau=0.465$ and $E_{Ca}^{O}=-5.5$ mV (upper curve), respectively $\tau=0.429$ and $E_{Ca}^{O}=-7$ mV (lower curve).

as their associated forms IS and JS. The principal difference
to the preceding treatment is that the individual diffusion
mobilities of all these particles are explicitly taken into
account. On the other hand, the generalized results of this
theory cannot be simply deduced from the formulas compiled in
Chapter 9 but have to be derived ab initio.

The mass fluxes of the species in the membrane, which is
treated here as an ideal phase, are given by the Nernst-
Planck equation as follows (see also Chapter 4):

$$J_i = - RT u_i \frac{dc_i}{dx} - u_i c_i F \frac{d\phi}{dx} \qquad (11.41)$$

$$J_j = - RT u_j \frac{dc_j}{dx} - u_j c_j F \frac{d\phi}{dx} \qquad (11.42)$$

$$J_s = - RT u_s \frac{dc_s}{dx} + u_s c_s F \frac{d\phi}{dx} \qquad (11.43)$$

$$J_{is} = - RT u_{is} \frac{dc_{is}}{dx} \qquad (11.44)$$

$$J_{js} = - RT u_{js} \frac{dc_{js}}{dx} \qquad (11.45)$$

where u and c are the mobilities and local concentrations of
the indicated species in the membrane, respectively, and ϕ
is the local value of the electrical potential. With the
assumptions of electroneutrality and of a state of local
chemical equilibria, we can use the additional relations:

$$c_s = c_i + c_j \tag{11.46}$$

$$c_{is} = K_{is}c_i c_s \; ; \; \frac{dc_{is}}{dx} = K_{is} \frac{dc_i}{dx} c_s + K_{is}c_i \frac{dc_s}{dx} \tag{11.47}$$

$$c_{js} = K_{js}c_j c_s \; ; \; \frac{dc_{js}}{dx} = K_{js} \frac{dc_j}{dx} c_s + K_{js}c_j \frac{dc_s}{dx} \tag{11.48}$$

The electrical current density j and the total ligand flux J_s^{tot} are defined as follows:

$$j/F = J_i + J_j - J_s \tag{11.49}$$

$$J_s^{tot} = J_s + J_{is} + J_{js} \tag{11.50}$$

By imposing the conditions of a zero-current steady-state and of conservation of ligand in the membrane, we can write:

$$J_i + J_j - J_s = 0 \tag{11.49a}$$

$$J_s + J_{is} + J_{js} = 0 \tag{11.50a}$$

Insertion of Eqs. (11.41) - (11.48) into (11.49a) and (11.50a) and subsequent elimination of the term dc_s/dx leads to a fundamental relationship between the electrical potential gradient and the concentration profiles of free cations [29]:

$$- \frac{F}{RT} \frac{d\phi}{dx} \left[(u_s + u_{is}K_{is}c_i + u_{js}K_{js}c_j)(u_ic_i + u_jc_j) \right.$$

$$\left. + u_sc_s(u_{is}K_{is}c_i + u_{js}K_{js}c_j) \right]$$

$$= (u_s + u_{is}K_{is}c_i + u_{js}K_{js}c_j) \left[\frac{d}{dx} \ u_ic_i + u_jc_j \right]$$

$$+ u_sc_s \frac{d}{dx} \left[u_{is}K_{is}c_i + u_{js}K_{js}c_j \right] \tag{11.51}$$

Although a numerical evaluation of this general result is feasible by means of computer simulation [69], explicit solutions for the integrated membrane-internal potential difference E_D can be obtained only for the limiting cases discussed earlier.

The case of dissociated ion-exchangers corresponds here to the situation:

$$u_sc_s \gg u_{is}K_{is}c_ic_s + u_{js}K_{js}c_jc_s \approx 0 \tag{11.52}$$

Accordingly, Eq. (11.51) may readily be integrated to yield

$$E_D = \frac{RT}{F} \ln \frac{u_ic_i(0) + u_jc_j(0)}{u_ic_i(d) + u_jc_j(d)} \tag{11.53}$$

and after combination with (11.3) (see also Eq. (9.16)):

$$E_M = \frac{RT}{F} \ln \frac{u_i k_i a_i' + u_j k_j a_j'}{u_i k_i a_i'' + u_j k_j a_j''} \qquad (11.54)$$

This result is equivalent to Eq. (11.7), which obviously holds for comparable ionic mobilities, and its implications have already been discussed exhaustively in Section 11.2.1.

The case of nearly complete association between sites and counterions is characterized by

$$u_s c_s \ll u_{is} K_{is} c_i c_s + u_{js} K_{js} c_j c_s \qquad (11.55)$$

Here, Eq. (11.51) reduces to

$$-\frac{F}{RT} \frac{d\phi}{dx} \left[\left((u_i + u_s) c_i + (u_j + u_s) c_j \right) \left(u_{is} K_{is} c_i + u_{js} K_{js} c_j \right) \right]$$

$$= (u_{is} K_{is} c_i + u_{js} K_{js} c_j) \frac{d}{dx} \left[u_i c_i + u_j c_j \right]$$

$$+ u_s (c_i + c_j) \frac{d}{dx} \left[u_{is} K_{is} c_i + u_{js} K_{js} c_j \right] \qquad (11.51a)$$

According to Sandblom, Eisenman, and Walker, this expression may be transformed into the following one:

$$-\frac{F}{RT} \frac{d\phi}{dx} = (1-\tau) \frac{d}{dx} \ln \left[(u_i + u_s) c_i + (u_j + u_s) c_j \right]$$

$$+ \tau \frac{d}{dx} \ln \left[u_{is} K_{is} c_i + u_{js} K_{js} c_j \right] \qquad (11.56)$$

where the parameter τ turns out to be independent of the position x in the membrane:

$$\tau = \frac{u_s(u_{is}K_{is}-u_{js}K_{js})}{(u_j+u_s)u_{is}K_{is}-(u_i+u_s)u_{js}K_{js}} \tag{11.57}$$

Integration of (11.56) is now easily accomplished and yields:

$$E_D = (1-\tau)\frac{RT}{F}\ln\frac{(u_i+u_s)c_i(0)+(u_j+u_s)c_j(0)}{(u_i+u_s)c_i(d)+(u_j+u_s)c_j(d)}$$
$$+ \tau\frac{RT}{F}\ln\frac{u_{is}K_{is}c_i(0)+u_{js}K_{js}c_j(0)}{u_{is}K_{is}c_i(d)+u_{js}K_{js}c_j(d)} \tag{11.58}$$

After combination with expressions of the type (11.3), the final result reads:

$$E_M = (1-\tau)\frac{RT}{F}\ln\frac{(u_i+u_s)k_ia_i'+(u_j+u_s)k_ja_j'}{(u_i+u_s)k_ia_i''+(u_j+u_s)k_ja_j''}$$
$$+ \tau\frac{RT}{F}\ln\frac{u_{is}K_{is}k_ia_i'+u_{js}K_{js}k_ja_j'}{u_{is}K_{is}k_ia_i''+u_{js}K_{js}k_ja_j''} \tag{11.59}$$

This description of the membrane potential differs from the former simplified Eqs. (11.13) and (11.2) insofar as here the individual mobilities of all species in the membrane are taken into account. Nevertheless, it leads to a relationship for the emf-response of the membrane electrode that is formally identical to Eq. (11.14) of Section 11.2.1 ! Thus, from the practical point of view, the Sandblom-Eisenman-Walker formalism offers no advantages except that the selectivity-controlling parameters will be defined in a more general sense, namely:

$$K_{ij}^{(1)} = \frac{u_j + u_s \; k_j}{u_i + u_s \; k_i} \qquad\qquad (11.60)$$

$$K_{ij}^{(2)} = \frac{u_{js} \; K_{js} k_j}{u_{is} \; K_{is} k_i} \qquad\qquad (11.61)$$

$$\tau = \frac{u_s}{u_j + u_s} \; \frac{K_{ij}^{(1)}}{K_{ij}^{(1)} - K_{ij}^{(2)}} - \frac{u_s}{u_i + u_s} \; \frac{K_{ij}^{(2)}}{K_{ij}^{(1)} - K_{ij}^{(2)}} \qquad\qquad (11.62)$$

The last expression implies that - at least hypothetically - the parameter τ may accept values of even >1 or <0, which is in contrast to the simplified relation (11.2). However, this definitely does not give rise to abnormities in the emf-response curves. The calculation of "potential dips" claimed by Sandblom and Orme [30] appears to be an artifact (see, however, Fig. 11.4).

It was stated that the basic relations of the theory of Sandblom et al. "can easily be extended ho higher valences and to mixtures of univalent and divalent electrolytes, and that the conclusions reached for univalent ions will remain valid for mixtures, but the form of the expressions will be slightly different" [30]. However, no such extensions were ever performed explicitly, to the best of our knowledge. A rigorous description, along the lines of the Sandblom-Eisen-man-Walker theory, becomes available for divalent-ion-exchangers only when using the conceptual framework set forth in Sections 11.2.2 and 11.2.3. The basic assumptions were that the membrane phase contains predominantly uncharged complexes or associates between exchanger sites and counterions, and that the minority of charge-carrying species is restricted to single-charged

243

forms. This stratagem allows to deduce explicit expressions for the diffusion potential which are analogous to Eq. (11.58) [70]. To avoid lengthy derivations, we shall present here the key results only.

The extension of the Sandblom-Eisenman-Walker theory to divalent counterions I^{2+} and J^{2+} leads to the following relationship for the diffusion potential:

$$E_D = (1-\tau) \frac{RT}{F} \ln \frac{(u_{is}+u_s)c_{is}(0)+(u_{js}+u_s)c_{js}(0)}{(u_{is}+u_s)c_{is}(d)+(u_{js}+u_s)c_{js}(d)}$$

$$+ \tau \frac{RT}{F} \ln \frac{u_{iss}K_{iss}c_{is}(0)+u_{jss}K_{jss}c_{js}(0)}{u_{iss}K_{iss}c_{is}(d)+u_{jss}K_{jss}c_{js}(d)} \quad (11.63)$$

where:

$$\tau = \frac{u_s(u_{iss}K_{iss}-u_{jss}K_{jss})}{(u_{js}+u_s)u_{iss}K_{iss}-(u_{is}+u_s)u_{jss}K_{jss}} \quad (11.64)$$

K_{iss} and K_{jss} are the equilibrium constants of formation of the electrically neutral 1:2 complexes from the corresponding cationic 1:1 complexes. The steady-state assumption in respect to total ligand implies that $J_{iss}+J_{jss} \cong 0$, or in terms of mobilities and concentrations:

$$u_{iss}c_{iss}(x)+u_{jss}c_{jss}(x) = (u_{iss}K_{iss}c_{is}(x)+u_{jss}K_{jss}c_{js}(x))c_s(x)$$

$$= (u_{iss}\beta_{iss}c_i(x)+u_{jss}\beta_{jss}c_j(x))(c_s(x))^2 = \text{const}(x) \quad (11.65)$$

Equation (11.65) shares common features with (11.22) used in Section 11.2.2 and permits rewriting Eq. (11.63) in a form equivalent to (11.19) or (11.23). Hence the same relationship, Eq. (11.26), will be derived for the emf of the cell, as was the basis of discussions in Section 11.2.2. The only difference consists in the definition of the selectivity-controlling parameters, which are obtained here in the following form:

$$K_{ij}^{(1)} = \frac{u_{js}+u_s}{u_{is}+u_s} \frac{K_{js}k_j}{K_{is}k_i} \tag{11.66}$$

$$K_{ij}^{(2)} = \frac{u_{jss}}{u_{iss}} \frac{\beta_{jss}k_j}{\beta_{iss}k_i} \tag{11.67}$$

$$\tau = \frac{u_s}{u_{js}+u_s} \frac{K_{ij}^{(1)}}{K_{ij}^{(1)}-K_{ij}^{(2)}} - \frac{u_s}{u_{is}+u_s} \frac{K_{ij}^{(2)}}{K_{ij}^{(1)}-K_{ij}^{(2)}} \tag{11.68}$$

These expressions are to be compared with Eqs. (11.27), (11.28), and (11.20) in 11.2.2. It may be recognized that the two theoretical approaches offered become identical in all their consequences if $u_{is} = u_{js}$ and $u_{iss} = u_{jss}$.

The potential difference generated in ion-exchange membranes by mixtures of divalent and monovalent counterions, I^{2+} and J^+, may be described in analogy to Eqs. (11.58) and (11.63) as follows [70]:

$$E_D = (1-\tau) \frac{RT}{F} \ln \frac{(u_{is}+u_s)c_{is}(0)+(u_j+u_s)c_j(0)}{(u_{is}+u_s)c_{is}(d)+(u_j+u_s)c_j(d)}$$
$$+ \tau \frac{RT}{F} \ln \frac{2u_{iss}K_{iss}c_{is}(0)+u_{js}K_{js}c_j(0)}{2u_{iss}K_{iss}c_{is}(d)+u_{js}K_{js}c_j(d)} \tag{11.69}$$

with

$$\tau = \frac{u_s(2u_{iss}K_{iss}-u_{js}K_{js})}{(u_j+u_s)2u_{iss}K_{iss}-(u_{is}+u_s)u_{js}K_{js}} \qquad (11.70)$$

The former steady-state conditions (11.32) and (11.65) are here replaced by

$$2u_{iss}c_{iss}(x) + u_{js}c_{js}(x) = u_{iss}X = const(x) \qquad (11.71)$$

where X is a measure for the total amount of exchanger sites (involved as ligands in electrically neutral cation-complexes) in the membrane. Again, we can apply the same procedure as in Section 11.2 to derive expressions for the total membrane potential and the resulting emf-response function. As expected, the validity of Eq. (11.35) is finally confirmed. The refinements of the membrane model presented in this section add up to two slight modifications of the theoretical ion-selectivity parameters [70]: an additional factor $[u_{iss}(u_j+u_s)/u_{js}(u_{is}+u_s)]^2$ appears in Eq. (11.36), and $(u_{js}/u_{iss})^2$ in Eq. (11.37). However, these effects are immaterial for practical purposes.

11.4. INTERPRETATION OF THE APPARENT SELECTIVITY BEHAVIOR OF LIQUID MEMBRANES

It has been shown in the preceding sections that the theoretical potential-response functions of liquid ion-exchange membrane electrodes may differ significantly from the formalism (11.72) accepted in practice:

246

$$E = E_i^o + \frac{RT}{z_iF} \ln \left[a_i' + K_{ij}^{Pot} a_j'^{z_i/z_j} \right] \qquad (11.72)$$

These inconsistencies in the mathematical descriptions may
be responsible for considerable variations of the practical
selectivity coefficients K_{ij}^{Pot}. However, even in cases where
theory strictly predicts a behavior according to Eq. (11.72),
it is often observed that the apparent selectivity factors
have a tendency to converge on unity with decreasing activities of the sample solution [44, 71]. A rationalization of
this phenomenon has been given informally by Hulanicki and
Lewandowski [44] and more recently in a semiquantitative
approach by Jyo and Ishibashi [72]. It was recognized that
concentration gradients build up in the electrolyte solutions
("concentration polarization" [72]) which are a consequence
of the ion-exchange and diffusion processes reaching into the
membrane. Hence the ionic composition of the aqueous film
adhering to the surface of the membrane electrode is generally
not identical to that of the bulk sample solution. This means
that the activities a_i' and a_j' sensed by the ion-selective
electrode may differ to some extent from the intrinsic sample
activities a_i and a_j (see also Chapter 10). This gives rise
to characteristic discrepancies between the theoretical ion
selectivity of the membrane material and the apparent ion
selectivity of the measuring system, the latter parameter
being determined in practice from the following relationship:

$$E = E_i^o + \frac{RT}{z_iF} \ln \left[a_i + K_{ij}^{Pot}(app.) a_j^{z_i/z_j} \right] \qquad (11.73)$$

For extreme situations, the electrode measures the primary
ions eluted from the membrane rather than those coming from
the sample solution, which fact is evinced by an apparent

selectivity coefficient K_{ij}^{Pot} (app.) of nearly unity (see below).

In their treatment of liquid ion-exchange membranes which takes into account such diffusion effects, Jyo and Ishibashi [72] made the following simplifying assumptions:

a) Considerations are restricted to monovalent counterions $(z_i = z_j = 1$ or $-1)$.

b) The mobilities are assumed the same for the free counterions $(u_i = u_j)$, as well as for the neutral complexes $(u_{is} = u_{js})$.

c) Association between exchanger sites and counterions should be either negligible $(K_{is}, K_{js} \rightarrow 0;$ e. g. in membranes prepared from tetraphenylborate and nitrobenzene [41, 64]) or nonspecific $(K_{is} \approx K_{js};$ e. g. in certain anion-exchangers [44, 45, 72]).

If these requirements are met, the theoretical emf-response function of the membrane electrode assembly will definitely assume the familiar form (11.72) (see Sections 11.2.1 and 11.3), and the theoretical ion-selectivity coefficient comes out to be

$$K_{ij}^{Pot} = \frac{k_j}{k_i} \tag{11.9}$$

For an idealized membrane, this parameter corresponds to the equilibrium constant of the basic ion-exchange reaction and, therefore, characterizes the following law of mass action:

$$K_{ij}^{Pot} = \frac{a_i' \, c_j(0)}{a_j' \, c_i(0)} \qquad \text{, for a membrane with pre-} \tag{11.74a}$$
dominantly free ions,

respectively:

$$K_{ij}^{Pot} = \frac{a_i' \, c_{js}(0)}{a_j' \, c_{is}(0)} \quad , \text{ for a membrane with pre-} \qquad (11.74b)$$

, for a membrane with pre-dominantly associated ions and $K_{is} = K_{js}$.

Another consequence of the assumptions a) to c) is that the fluxes of counterions will simply obey Fick's law. This may be verified using Eqs. (11.41) - (11.50a). Thus, the flux of species I at steady-state may be approximated by

$$J_i = -D_i' \, \frac{a_i' - a_i}{\delta'} = -D_i \, \frac{c_i(\delta) - c_i(0)}{\delta} \quad \text{[dissociated system]} \quad (11.75a)$$

or

$$J_i = -D_i' \, \frac{a_i' - a_i}{\delta'} = -D_{is} \, \frac{c_{is}(\delta) - c_{is}(0)}{\delta} \quad \text{[associated system]} \quad (11.75b)$$

where $D_i = u_i RT$ and $D_{is} = u_{is} RT$ are the diffusion coefficients of ionic forms in the membrane, D_i' is the mean diffusion co-efficient of counterions in the sample solution, δ is the thickness of the Nernstian diffusion layer existing inside the membrane boundary, and δ' is the corresponding value for the aqueous diffusion layer. From Eq. (11.75), we obtain the following relationship:

$$a_i' = a_i + C(x_i(\delta) - x_i(0)) \qquad (11.76)$$

where x_i denotes the fraction of the total concentration of exchanger sites, X, that is occupied by counterions I, and the parameter C is defined as

$$C = \frac{D_i}{D_i'} \frac{\delta'}{\delta} X \quad \text{or} \quad \frac{D_{is}}{D_i'} \frac{\delta'}{\delta} X \qquad\qquad (11.77)$$

An analogous expression is found for the ions J:

$$a_j' = a_j + C(x_j(\delta) - x_j(0)) \qquad\qquad (11.78)$$

Since it is assumed that the ion-exchanger was initially prepared in or converted to the form containing the primary ion I,
it holds that

$$x_i(\delta) = 1-x_j(\delta) \cong 1 \qquad\qquad (11.79)$$

The ionic composition of the membrane surface, on the other
hand, is determined by the ion-exchange equilibrium (Eq.
(11.74)) as follows:

$$x_i(0) = 1-x_j(0) = \frac{a_i'}{a_i' + K_{ij}^{Pot} a_j'} \qquad\qquad (11.80)$$

Equations (11.76) - (11.80) allow to calculate the unknown
boundary values of the sample activities, a_i' and a_j', in terms
of the given bulk activities a_i and a_j, the theoretical selectivity coefficient K_{ij}^{Pot}, and the basic experimental parameter C. After lengthy but trivial algebra, one arrives at
the following result for the total ion activity as sensed
by the ion-selective electrode (see Eqs. (11.72) and (11.73)
with $z_i = z_j$):

$$a_i' + K_{ij}^{Pot} a_j' = a_i + K_{ij}^{Pot}(app.)a_j$$

$$= \frac{1}{2} (a_i + K_{ij}^{Pot} a_j - K_{ij}^{Pot} C)$$

$$+ \frac{1}{2} \sqrt{(a_i + K_{ij}^{Pot} a_j - K_{ij}^{Pot} C)^2 + 4 K_{ij}^{Pot} C(a_i + a_j)} \qquad (11.81)$$

To demonstrate the systematic differences between the theoretical selectivities K_{ij}^{Pot} and the apparent values K_{ij}^{Pot} (app.), we shall discuss in more detail the limiting cases obtained for $a_i \rightarrow 0$. These were the cases considered by Jyo and Ishibashi [72]. Equation (11.81) predicts ideal selectivity behavior towards pure solutions of an interfering ion J only if the sample activities are sufficiently high, i. e.:

$$E = E_i^o + \frac{RT}{z_j F} \ln [K_{ij}^{Pot} a_j] \ , \ \text{for } a_j >> C \text{ and } K_{ij}^{Pot} a_j >> C \qquad (11.82)$$

In contrast, an apparent selectivity coefficient of approximately unity will be found for very diluted sample solutions:

$$E = E_i^o + \frac{RT}{z_j F} \ln a_j \ , \ \text{for } a_j << C \text{ and } a_j << K_{ij}^{Pot} C \qquad (11.83)$$

This means, of course, that a region of transient selectivities must appear at intermediate sample activities. In this activity range, the experimental calibration curve E vs. log a_j for any ion other than the primary ion will exhibit a non-linear section. Maximal distortions from a Nernstian response of the electrode are predicted for $a_j = C$ at which position

the potential value and the slope of the calibration curve
are calculated as follows:

$$E = E_i^o + \frac{RT}{z_j F} \ln \left[\sqrt{K_{ij}^{Pot}} \, a_j \right]$$
$$\left. \right\}$$
(11.84)

for $a_j = C$

$$\frac{dE}{d\ln a_j} = \frac{RT}{z_j F} \frac{\sqrt{K_{ij}^{Pot}} + 1}{2}$$
(11.85)

The last expression describes a highly super-Nernstian region
("break") of the response function for ions preferred by the
membrane electrode (i. e. for $K_{ij}^{Pot} \gg 1$), and a region with
half of the theoretical slope for ions discriminated by the
sensor ($K_{ij}^{Pot} \ll 1$). These theoretical predictions are in excellent qualitative agreement with experimental findings.
Figure 11.5 documents the response pattern of anion-selective
liquid-membrane electrodes based on methyltricapryl ammonium
salts (Aliquat 336S). The values on the left represent experimental data, quoted from tabulated selectivity coefficients [44], whereas the right-hand curves were drawn according to Eq. (11.81). An even better agreement is obtained
when the calculated plots are compared with the original calibration curves given in [44].

It becomes evident that, in practice, liquid-membrane
electrodes often exhibit an ideally linear response function
only for the primary ions. As a rule, the membrane type selected for an electrode application should therefore incorporate as counterions essentially the ions to be determined.
This fact is well known to the manufacturers of ion-selective
electrodes who offer various devices based on similar membrane
compositions but with differing primary ions (see Table 11.1).
On the other hand, nearly the same favorable performance can

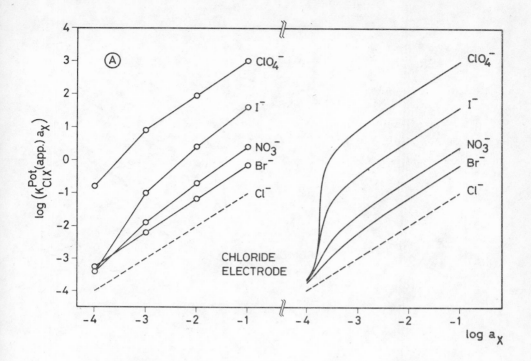

Figure 11.5. Response pattern of anion-sensitive liquid membrane electrodes based on quaternary ammonium salts (Aliquat 336S). A: chloride electrode, B: bromide electrode, C: perchlorate electrode. The shown functions are related to the apparent calibration plots for primary ions (dashed lines) and interfering ions (solid curves, see Eq. (11.73)). Left traces: illustration of experimental selectivities, as given in table 1 of Ref. 44. Right traces: corresponding theoretical curves, calculated from Eqs. (11.81) and (11.9) with $C = 1.8 \cdot 10^{-4}$ M and k_{ClO_4}: k_I: k_{NO_3}: k_{Br}: k_{Cl} = 10000: 400: 25: 7: 1.

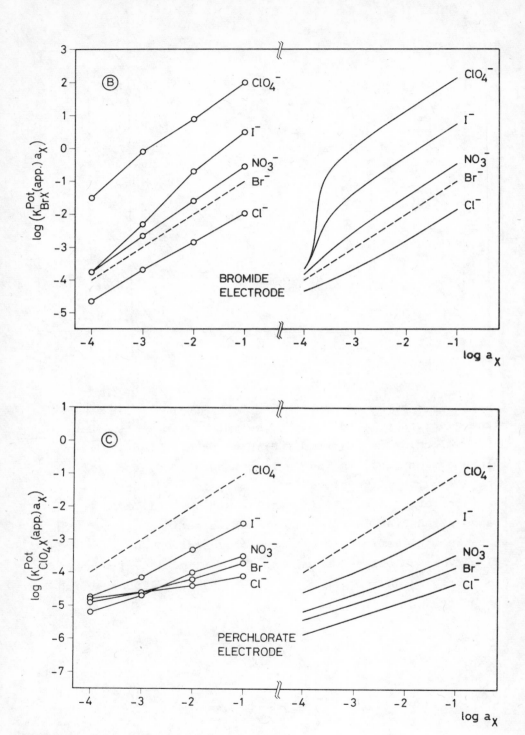

Figure 11.5. (continued)

254

be achieved by simply soaking a given membrane electrode in a
concentrated solution of the ion to be measured, whereby the
liquid ion-exchanger is converted into the appropriate form.
A so-called chloride electrode, for example, can be transformed into a perfect perchlorate sensor if it is exposed to
perchlorate solutions for a sufficiently long time [44].
Figure 11.5 C indicates that the new electrode modification
thus formed will retain its capability of responding to
chloride activities, but the slope of the calibration plot is
drastically reduced and the potentials are shifted relative
to the initial response curve (Figure 11.5 A). A similar
situation may be created unintentionally when a chloride
electrode is in continuous use in blood serum or other biological samples, containing significant traces of highly extractable lipophilic interferents. In fact, a clearly sub-
Nernstian response is typical for such biomedical applications
of liquid ion-exchange membrane electrodes (see, e. g.,
figure 9 in [73]).

The variations of the apparent selectivities documented in
Figure 11.5 may elucidate the reasons for the low precision
or poor reproducibility of selectivity specifications found
in the literature. In addition to the mode of preparation or
pretreatment of the membrane, it is the experimental parameter C that is decisive for the extent of ion specificity
exhibited by the electrode at a given activity level of the
sample solution. This parameter includes several experimental
aspects, namely (a) the viscosity of the liquid membrane phase
and the bulkiness of the interdiffusing species, which, according to Stokes-Einstein, determines the value of the diffusion
coefficient D, (b) the shape and dimensions of the electrode
surface and the stirring rate of the aqueous solution, which
is decisive for the film thickness δ', (c) the site concentration X of the membrane, and (d) the parameter δ which
roughly corresponds to the mean free path of diffusion in

the membrane. A more rigorous model shows that the last quantity is actually time-dependent and has to be replaced by $\delta \cong \sqrt{\pi Dt}$ for $\delta \ll d$ (see also Chapter 14). Figure 11.5 suggests that a fairly low value of C is required for realizing in practice the ion specificity dictated by the membrane material. In this respect, the use of nearly solid PVC membrane electrodes in flow-through cells (δ' of less than 10^{-2}cm, D on the order of $10^{-9} - 10^{-10}$ cm^{2}s^{-1} or even lower [74]), or of conventional liquid-membrane microelectrodes (tip diameter and adhering film (δ') of around 10^{-4}cm, D of about 10^{-6} cm^{2}s^{-1} in the absence of polymeric components) is especially attractive. A high value of C, on the other hand, leads to a near collapse of the apparent ion selectivity. This effect may be exploited to some extent, e. g. in sensors for chloride ions where the interference by more lipophilic anions could be reduced to a minimum (Figures 11.5 A and 11.6).

The results of this section clearly demonstrate that the practical ion selectivity, as exhibited by liquid membranes in well defined potentiometric experiments, must not necessarily be identical to the intrinsic ion specificity of these sensor materials. One may even come to the unorthodox conclusion that most of the modifications or improvements claimed in the literature for liquid ion-exchange membrane electrodes did not really involve systematical changes in ion selectivity, but were largely a consequence of diffusion-induced artifacts. Figure 11.6 gives some support to this opinion. Apparently, the selectivity data K_{ClX}^{Pot} reported for different chloride electrodes and different activities a_X can be rationalized perfectly by postulating a single set of basic selectivity coefficients and assuming individual values for the experimental parameter a_X/C.

It should finally be noted that the variations of apparent

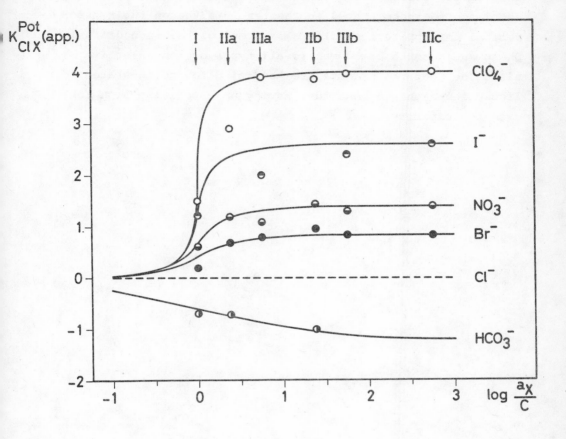

Figure 11.6. Reported selectivities for different chloride electrodes in comparison with theoretical expectations (Figure 11.5A). I: liquid membrane electrode (filter paper matrix) based on dimethyldistearyl ammonium ions [35]. II: PVC membrane electrode based on methyltridodecyl ammonium chloride; (a) 10^{-2}M, (b) 10^{-1}M solutions [59, 73]. III: PVC membrane electrode based on methyltricapryl ammonium chloride (Aliquat 336S); (a) 10^{-3}M, (b) 10^{-2}M, (c) 10^{-1}M solutions [44].

selectivity coefficients discussed in this section bear resemblance to those stated in Chapter 10 although their theoretical interpretation differs in some points. Such diffusion
phenomena seem to be common to all types of ion-selective
electrodes but the effects are magnified for solid-state and
liquid ion-exchange membranes having an especially high ion-
exchange capacity.

REFERENCES

[1]   L. Michaelis, <u>Kolloid-Z.</u> <u>62</u>, 2 (1933).

[2]   L. Michaelis and A. Fujita, <u>Biochem. Z.</u> <u>158</u>, 28 (1925).

[3]   K. Sollner, <u>Z. Elektrochem.</u> <u>36</u>, 36 (1930).

[4]   K. Sollner, <u>J. Macromol. Sci.-Chem.</u> A <u>3</u>, 1 (1969).

[5]   T. Teorell, <u>Trans. Faraday Soc.</u> <u>33</u>, 1053 (1937).

[6]   T. Teorell, <u>Z. Elektrochem.</u> <u>55</u>, 460 (1951).

[7]   T. Teorell, <u>Prog. Biophys. Biophys. Chem.</u> <u>3</u>, 305 (1953).

[8]   K. H. Meyer and J. F. Sievers, <u>Helv. Chim. Acta</u> <u>19</u>,
      649, 665 (1936).

[9]   K. H. Meyer and J. F. Sievers, <u>Trans. Faraday Soc.</u> <u>33</u>,
      1073 (1937).

[10] R. Schlögl, <u>Z. Phys. Chem. (Frankfurt am Main)</u> <u>1</u>, 305
      (1954).

[11] R. Schlögl, Zum Materietransport durch Porenmembranen,
      Habilitationsschrift, Georg-August-Universität,
      Göttingen, 1957.

[12] R. Schlögl, <u>Stofftransport durch Membranen</u>, Steinkopff,
      Darmstadt, 1964.

[13] F. Helfferich and R. Schlögl, <u>Discussions Faraday Soc.</u>
      <u>21</u>, 133 (1956).

[14] F. Helfferich, <u>Ion Exchange</u>, McGraw-Hill, New York, 1962.

[15] A. J. Staverman, <u>Rec. Trav. Chim.</u> <u>70</u>, 344 (1951);
      <u>Trans. Faraday Soc.</u> <u>48</u>, 176 (1952).

[16] J. G. Kirkwood, in <u>Ion Transport across Membranes</u>
      (H. T. Clarke, ed.), Academic, New York, 1954, p. 119.

[17] K. S. Spiegler, <u>Trans. Faraday Soc.</u> <u>54</u>, 1408 (1958).

[18] O. Kedem and A. Katchalsky, <u>Biochim. Biophys. Acta</u> <u>27</u>,
      229 (1958).

[19] O. Kedem and A. Katchalsky, <u>Trans. Faraday Soc.</u> <u>59</u>,
      1918, 1931, 1941 (1963).

[20] A. Katchalsky and P. F. Curran, Nonequilibrium Thermo-
      dynamics in Biophysics, Harvard University Press, Cam-
      bridge, Mass., 1965.

[21] G. Eisenman, ed., Membranes - A Series of Advances,
      Vol. 1, Marcel Dekker, New York, 1972.

[22] C. P. Bean, 'The physics of porous membranes - -
      neutral pores', chapter 1 of ref. [21].

[23] P. Meares, J. F. Thain, and D. G. Dawson, 'Transport
      across ion-exchange resin membranes: The frictional
      model of transport', chapter 2 of ref. [21].

[24] R. Beutner, Physical Chemistry of Living Tissues and
      Life Processes, Williams and Wilkins, Baltimore, 1933.

[25] K. F. Bonhoeffer, M. Kahlweit, and H. Strehlow, Z. Physik.
      Chem. (Frankfurt) 1, 21 (1954).

[26] M. Kahlweit, H. Strehlow, and C. S. Hocking, Z. Physik.
      Chem. (Frankfurt) 4, 212 (1955).

[27] M. Kahlweit and H. Strehlow, Z. Elektrochem. 58, 658
      (1954).

[28] M. Kahlweit, Arch. Ges. Physiol. 271, 139 (1960).

[29] J. Sandblom, G. Eisenman, and J. L. Walker, Jr.,
      J. Phys. Chem. 71, 3862, 3971 (1967).

[30] J. Sandblom and F. Orme, 'Liquid membranes as electrodes
      and biological models', chapter 3 of ref. [21].

[31] W. E. Morf, Anal. Chem. 49, 810 (1977).

[32] W. E. Morf and W. Simon, in Ion-Selective Electrodes in
      Analytical Chemistry (H. Freiser, ed.), Plenum, New
      York, 1978.

[33] K. Sollner and G. M. Shean, J. Am. Chem. Soc. 86, 1901
      (1964).

[34] J. W. Ross, Science 156, 1378 (1967).

[35] J. W. Ross, in Ion-Selective Electrodes (R. A. Durst, ed.),
      National Bureau of Standards Special Publication 314,
      Washington, 1969.

[36] A. Shatkay, Anal. Chem. 39, 1056 (1967).

[37] R. Bloch, A. Shatkay, and H. A. Saroff, Biophys. J. 7, 865 (1967).

[38] C. J. Coetzee and H. Freiser, Anal. Chem. 40, 2071 (1968); 41, 1128 (1969).

[39] G. Eisenman, Anal. Chem. 40, 310 (1968).

[40] G. Baum and W. M. Wise, Ger. Offen. 2024636 (1970); G. Baum and M. Lynn, Anal. Chim. Acta 65, 393 (1973).

[41] R. Scholer and W. Simon, Helv. Chim. Acta 55, 1801 (1972).

[42] N. Ishibashi and H. Kohara, Anal. Lett. 4, 785 (1971).

[43] J. E. W. Davies, G. J. Moody, and J. D. R. Thomas, Analyst 97, 87 (1972).

[44] A. Hulanicki and R. Lewandowski, Chemia Analityczna 19, 53 (1974).

[45] G. Scibona, L. Mantella, and P. R. Danesi, Anal. Chem. 42, 844 (1970); P. R. Danesi, F. Salvemini, G. Scibona, and B. Scuppa, J. Phys. Chem. 75, 554 (1971).

[46] A. L. Grekovich, E. A. Materova, and T. I. Pron'kina, Elektrokhimiya 7, 436 (1971).

[47] C. Gavach and C. Bertrand, Anal. Chim. Acta 55, 385 (1971).

[48] G. Baum, Anal. Letters 3, 105 (1970).

[49] C. Gavach and P. Seta, Anal. Chim. Acta 50, 407 (1970).

[50] J. Růžička and J. C. Tjell, Anal. Chim. Acta 49, 346 (1970).

[51] T. Higuchi, C. R. Illian, and J. L. Tossounian, Anal. Chem. 42, 1674 (1970).

[52] G. J. Moody, R. B. Oke, and J. D. R. Thomas, Analyst 95, 910 (1970).

[53] S. Lal and G. D. Christian, Anal. Chem. 43, 410 (1971).

[54] G. L. Stucky, Ger. Offen. 2057114 (1971).

[55] J. Růžička, C. G. Lamm, and J. C. Tjell, Ger. Offen. 2034686 (1971).

[56] J. Růžička, E. H. Hansen, and J. C. Tjell, <u>Anal. Chim. Acta</u> <u>67</u>, 155 (1973).

[57] W. Simon, D. Ammann, M. Oehme, and W. E. Morf, <u>Ann. N. Y. Acad. Sci.</u> <u>307</u>, 52 (1978).

[58] D. Erne, W. E. Morf, S. Arvanitis, Z. Cimerman, D. Ammann, and W. Simon, <u>Helv. Chim. Acta</u> <u>62</u>, in press (1979).

[59] M. Oehme, Dissertation ETH, Juris, Zürich, 1977; K. Hartman, S. Luterotti, H. F. Osswald, M. Oehme, P. C. Meier, D. Ammann, and W. Simon, <u>Mikrochim. Acta</u> <u>1978II</u>, 235.

[60] J. A. Kitchener, <u>Ion-Exchange Resins</u>, Wiley, New York, 1957; R. Kunin and R. J. Myers, <u>Ion Exchange Resins</u>, Wiley, New York, 1950.

[61] J. Koryta, <u>Ion-Selective Electrodes</u>, Cambridge University Press, Cambridge, 1975.

[62] W. M. Wise, M. J. Kurey, and G. Baum, <u>Clin. Chem.</u> <u>16</u>, 103 (1970).

[63] J. L. Walker, <u>Anal. Chem.</u> <u>43</u>, 89A (1971).

[64] J. Rais, <u>Coll. Czech. Chem. Comm.</u> <u>36</u>, 3080, 3253 (1971).

[65] A. V. von Rechenberg, Dissertation ETH, Juris, Zürich, 1974.

[66] H. J. James, G. P. Carmack, and H. Freiser, <u>Anal. Chem.</u> <u>44</u>, 853 (1972).

[67] S. Bäck, <u>Anal. Chem.</u> <u>44</u>, 1696 (1972).

[68] Orion Research Inc. <u>Newsletter</u> <u>2</u>, 14 (1970).

[69] F. S. Stover and R. P. Buck, <u>Biophys. J.</u> <u>16</u>, 753 (1976).

[70] W. E. Morf, unpublished results.

[71] K. Srinivasan and G. A. Rechnitz, <u>Anal. Chem.</u> <u>41</u>, 1203 (1969).

[72] A. Jyo and N. Ishibashi, private communication of unpublished results (<u>Anal. Chem.</u>, in preparation).

[73] P. C. Meier, D. Ammann, W. E. Morf, and W. Simon, in
Medical and Biological Applications of Electrochemical
Devices (J. Koryta, ed.), Wiley, in press.

[74] U. Oesch and W. Simon, Helv. Chim. Acta 62, 754 (1979).

Chapter 12

Liquid-Membrane Electrodes Based on Neutral Carriers

Electrically neutral, lipophilic ion-complexing agents of rather small relative molar mass are known to behave as ionophores or ion carriers [1]. They have the capability to selectively extract ions from aqueous solutions into a hydrophobic membrane phase and to transport these ions across such barriers by carrier translocation (see also Chapter 8). Neutral carriers are therefore predestined to be incorporated as the working principle in natural or artificial ion-transport systems, as well as in ion-selective liquid-membrane electrodes where the ion specificity inherent to the carriers can be exploited to a large extent.

Although Moore and Pressman's discovery of the effect of naturally occurring neutral carrier antibiotics in biological membrane systems dates to 1964 [2], their fundamental property, namely their role as highly selective complexing agents for alkali metal ions was recognized only two years later by Štefanac and Simon [3]. Some of the molecules studied, such as valinomycin (1) and the macrotetrolides (2-6, see Figure 12.1), show a striking differentiation between Na^+ and K^+. This fact led Štefanac and Simon to the pioneering idea of utilizing such neutral ionophores as ion-selective components in liquid-membrane electrodes. The first specimens of carrier-based electrodes were generally considered to be "exotic systems" [4], but in the meantime the analytical potential of these ion-selective sensors has been recognized. In the event, the valinomycin-based membrane electrode [5 - 13] became one of the most important ion sensors because of its unsurpassed specificity for potassium ions, the sodium ions being discriminated by a factor K_{KNa}^{Pot} of around 10^{-4}. Similar ion-selective properties, although less pronounced, are

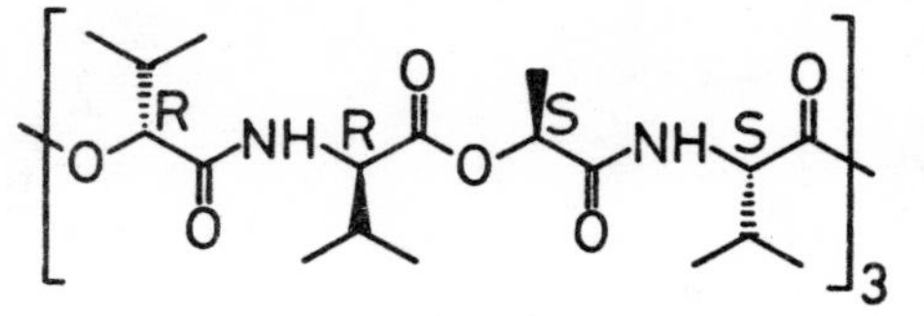

VALINOMYCIN 1

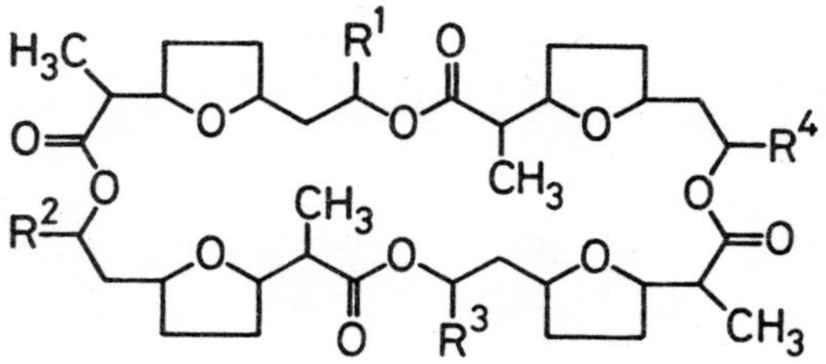

$R^1 = R^2 = R^3 = R^4 = CH_3$		NONACTIN	2
$R^1 = R^2 = R^3 = CH_3$	$R^4 = C_2H_5$	MONACTIN	3
$R^1 = R^3 = CH_3$	$R^2 = R^4 = C_2H_5$	DINACTIN	4
$R^1 = CH_3$	$R^2 = R^3 = R^4 = C_2H_5$	TRINACTIN	5
$R^1 = R^2 = R^3 = R^4 = C_2H_5$		TETRANACTIN	6

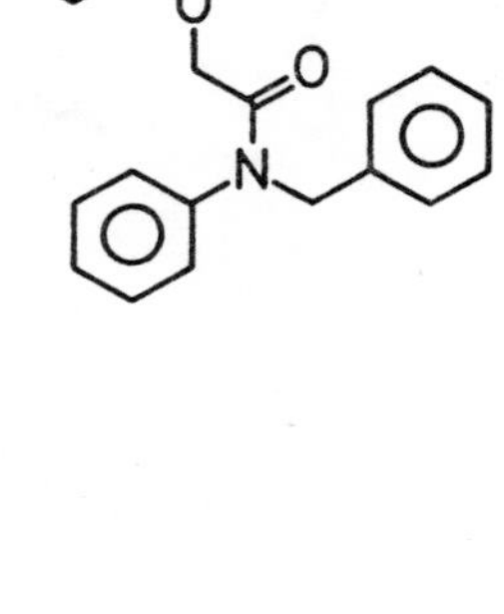

7

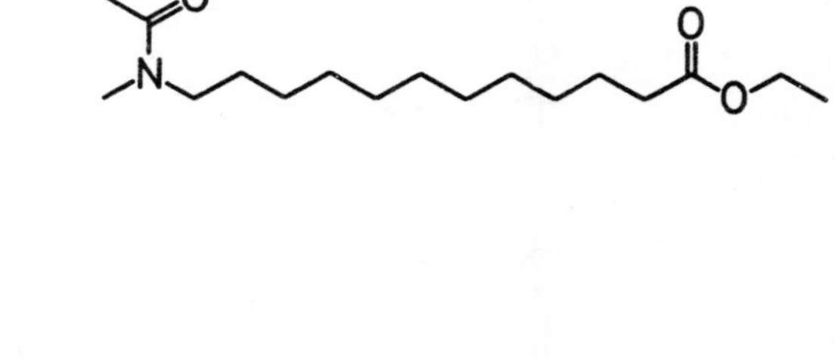

8

Figure 12.1. Structure of the carriers discussed.

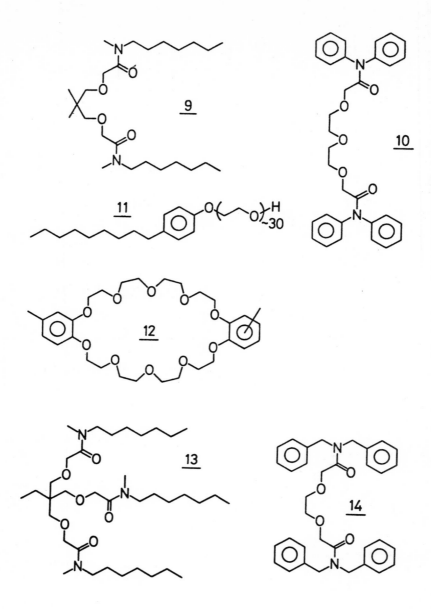

Figure 12.1. (continued)

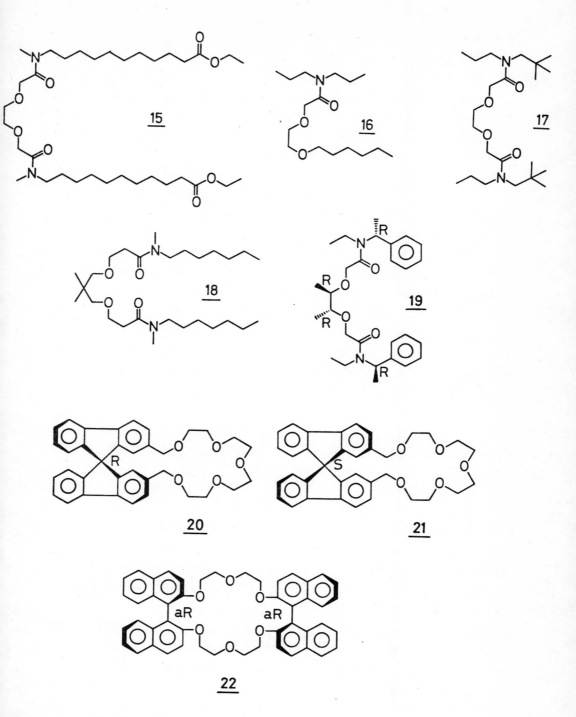

Figure 12.1. (continued)

267

mimicked by the synthetic macrocyclic polyethers (crown compounds) pioneered by Pedersen [14]. Some of these compounds (e. g. 12 in Figure 12.1) have also been considered for electrode applications [15]. The macroheterobicyclic ligands first synthesized by Lehn [16 - 18] show extremely interesting complexation selectivities for A-cations but, unfortunately, do not behave as carriers in membranes. This was ascribed to their low lipophilicity and especially to their slow exchange kinetics. A wide range of remarkable ion selectivities for alkali and alkaline earth metal cations may, however, be induced in membranes by nonmacrocyclic synthetic ion carriers (7 - 10, 13 - 19). Such molecules were first introduced in 1972 by Ammann et al. [19]; the synthesis was prompted by model calculations carried out by Morf and Simon [7, 20 - 22]. Out of several hundred carrier molecules prepared by Simon's group, the ligands 8, 7 (or 13), 9, 10, and 18 (Fig.12.1) are - so far - the most attractive ones and constitute the active components of electrodes specific for Ca^{2+}, Na^+, Li^+, Ba^{2+}, and UO_2^{2+}, respectively [8, 23 - 27].

12.1. CHARACTERISTICS OF NEUTRAL CARRIERS AND REPORTED SELECTIVITIES FOR MEMBRANE ELECTRODES

In the meantime a large number of macrocyclic and non-macrocyclic neutral complexing agents for cations are accessible, and many of them have been used as carriers in ion-selective electrode applications (for a review, see References 24 - 33). The structures of a selection of ligands are given in Figure 12.1. In addition to carriers for group 1A cations (e. g., 1 - 7, 9, 12 - 14) and 2A cations (8, 10, 11, 15 - 17), ionophores for transition metals [34], for uranyl [35] (e. g. 18), and even for a given enantiomer of chiral ammonium ions [36-38] (19-22, discrimination by chiral recognition) have been introduced. The latter molecules were designed especially in view of their use for selective ion separations.

A ligand that behaves as an ionophore for A-cations must meet the following requirements [21, 22, 33]:

a) The carrier molecule should be composed of polar and non-polar groups.
b) The carrier should be able to assume a stable conformation that provides a cavity, surrounded by the polar groups, suitable for the uptake of a cation, while the nonpolar groups form a lipophilic shell around the coordination sphere. These groups must ensure sufficiently large lipid-solubility for ligand and complex. This is one reason why the classical electrically charged complexing agents such as EDTA do not behave as carriers in membrane systems.
c) Among the polar groups of the ligand sphere, there should be preferably 5-8, but not more than 12 coordinating sites such as oxygen atoms.
d) High selectivities are achieved by locking the coordinating sites into a rigid arrangement around the cavity. Such rigidity can be enhanced by the presence of bridged struc-tures, e. g. hydrogen bonds. Within one group of the perio-dic system, the cation that best fits into the offered ca-vity is preferred. Ideally, all cations should be forced into accepting the same given number of coordinating groups.
e) Notwithstanding requirement d), the ligand should be flexible enough to allow a sufficiently fast ion-exchange. This is possible only with a stepwise substitution of the solvent molecules by the ligand groups. Thus a compromise between stability (d) and exchange rate (e) has to be found.
f) To guarantee adequate mobility, the overall dimensions of a carrier should be rather small but still compatible with high lipid-solubility.

269

Table 12.1. Neutral carrier liquid-membrane electrodes

Ion	Ligand [Figure 12.1]	Membrane material	References
Li^+	9	PVC; TEHP	8,24,32,39
Na^+	7	PVC; DBS	8,24
Na^+	13	PVC; o-NPOE or DBS	40
K^+	1	filter paper; DPE	5
K^+	1	filter paper; aromatic solvent	6
K^+	1	PVC; DNP	7,8
K^+	1	PVC; DOA	9
K^+	1	SR	10
K^+	1	cellulose acetate; plasticizer	11
K^+	1 ?	collodion ?	12
K^+	1	PBC copolymer	13
K^+	12	PVC; DPP	15
NH_4^+	2/3	filter paper; TEHP	41
NH_4^+	2/3 ?	collodion ?	42
NH_4^+	2/3	cellulose acetate; plasticizer	11
NH_4^+	2/3	PVC; TEHP	7,8
Ca^{2+}	8	PVC; o-NPOE	8,24,32,43
Sr^{2+}	11	filter paper; ENB	44
Ba^{2+}	11	filter paper; ENB	45
Ba^{2+}	11	PVC; DNPE	46
Ba^{2+}	10	PVC; o-NPOE	8,24,47

PVC: polyvinyl chloride; SR: silicone rubber; DOA: dioctyl adipate; ENB: 4-ethylnitrobenzene; DNPE: di-2-nitrophenyl ether; TEHP: tris(2-ethylhexyl)phosphate; DBS: dibutyl sebacate; DNP: dinonyl phthalate; o-NPOE: o-nitrophenyl octyl ether; DPE: diphenyl ether; DPP: dipentyl phthalate; PBC: poly(bisphenol-A carbonate)

The underlying principles involved in points a) to e) will be shortly discussed in Section 12.5.

For applications in ion-selective electrodes, the carrier molecules are incorporated into a water-immiscible membrane and usually interposed between the sample and an inner reference solution. In the classical approach, the membrane consists simply of the electrically neutral carrier dissolved in an appropriate organic solvent (e. g., a plasticizer), the solution being held by a preferably inert matrix (e. g. PVC, see Table 12.1). Such bulk membranes, as a rule, function as cation-selective electrodes and often exhibit a nearly Nernstian response to the primary ions. The potentiometric selectivity of these sensors among different ions is dictated mainly by the complexation specificity of the carrier molecules involved, but it may also be influenced to some extent by the membrane solvent and by other parameters (see Sections 12.2 - 12.5).

For a series of relevant membrane electrode systems the observed selectivity coefficients are compiled in Table 12.2. The specifications given in Table 12.2 clearly show that the K^+-electrode based on valinomycin $\underline{1}$ and the Ca^{2+}-electrode based on the synthetic carrier $\underline{8}$ are far superior to the corresponding sensors based on classical liquid ion-exchangers (see Section 11.1 and Figure 12.2). The outstanding selectivities realized for these carrier membrane electrodes opened attractive applications in analytical and clinical chemistry, such as the measurement of ion activities in blood serum or whole blood. An extensive review on this subject has been given elsewhere [25 - 27, 49].

A deeper understanding of the potentialities and the limitations of neutral carrier membrane electrodes can be obtained only by a thorough theoretical treatment. The following

Table 12.2. Reported selectivity factors, log K_{IJ}^{Pot}, for PVC-membrane electrodes based on neutral carriers

Ligand of the membrane system	J = Li+	Na+	K+	Rb+	Cs+	NH4+	H+	Mg2+	Ca2+	Sr2+	Ba2+
I=Li+											
Ligand 9 [39]	0	-1.3	-2.2	-2.5	-2.7	-1.4	-0.1	-3.8	-3.3	-3.4	-3.8
I=Na+											
Ligand 7 [8]	-1.5	0	-0.3	-0.8	-1.1	-0.8	-0.3	-3.2	-2.9	-2.2	-2.4
Ligand 13 [40]	0.5	0	-2.4	-2.4	-2.5	-2.3	-0.7	-2.8	-0.5	-1.0	-1.4
I=K+											
Ligand 1 [8]	-	-5.5	0	0.5	-0.4	-1.8	-5.0	-5.3	-4.6	-3.3	-3.4
Ligand 12 [15]	-2.3a	-2.4b / -2.7c	0	-0.1a	-0.6a	-1.2a	-	-4.0a	-3.5a	-	-4.0a
I=NH4+											
Ligand 2,3 [8]	-2.7	-2.7	-0.8	-1.3	-2.4	0	-1.6	-5.0	-4.5	-4.0	-4.5
I=Ca2+											
Ligand 8 [43]	-2.8	-5.0c / -6.1d	-5.2c	-4.0	-4.0	-5.1	-0.1c / -4.4	-5.1c	0	-2.1	-3.2
I=Sr2+											
Ligand 11e [44]	-2.7	-2.7	-2.1	-	2.3	-2.7	-3.3	-3.2	-2.7	0	2.5
I=Ba2+											
Ligand 11f [45]	-3.7	-3.7	-2.1	-	-	-3.2	-3.7	-4.0	-4.0	-2.7	0
Ligand 10 [47]	-3.2	-2.4	-1.6	-2.0	-2.4	-2.3	-1.3	-5.1	-3.7	-1.5	0

Reported selectivities as obtained from the separate solution method using 10⁻¹M solutions, except for: a Separate solution method, 10⁻²M solutions. b Fixed interference method, no metal buffers. Background concentration of the interfering ions: 10⁻²M. c Fixed interference method, no metal buffers. Background concentration of the interfering ions: 10⁻¹M. d Fixed interference method, metal-buffered solutions. Background concentration of the interfering ions: 10⁻¹M. e Sr²⁺-complex of ligand 11. f Ba²⁺-complex of ligand 11.

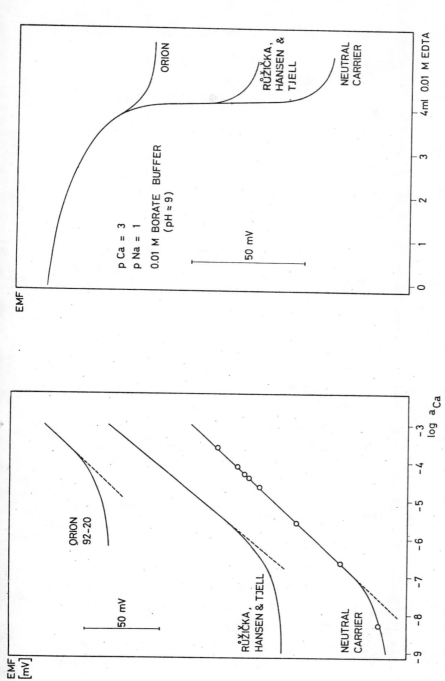

Figure 12.2. Comparison of the EMF-response of a neutral carrier PVC membrane electrode [43] (ligand **8** in Fig. 12.1.) with the published response of other Ca^{2+}-sensors [48] (liquid ion-exchangers given in Table 11.2.).

Left: response to metal-buffered aqueous solutions of Ca^{2+} containing 0.1 M Na$^+$.

Right: titration of 0.001 M CaCl$_2$ with EDTA at pH 9 in the presence of 0.1 M Na$^+$.

sections are meant to give an introduction to the response mechanism, the potentiometric selectivity behavior, as well as the molecular aspects of ion selectivity of such systems.

12.2. MECHANISM OF CATION SPECIFICITY (PERMSELECTIVITY) OF NEUTRAL CARRIER MEMBRANES

The predominant problem of the rationalization of the behavior of neutral carrier membrane electrodes is to explain their permselectivity for cations, as is observable in both potentiometric and ion-transport studies. Such experiments generally confirm that a carrier membrane tends to be permeable for cations only and, accordingly, its electrical properties are scarcely influenced by sample anions such as chloride. Experimental evidence for cation permselectivity of all the analytically relevant neutral-carrier-based liquid membrane electrodes is given in Table 12.3. Throughout, a cation transference number of close to 1.0 is found in electrodialysis experiments, which means that practically no anions are transported across such membranes when an electrical potential gradient is applied. This agrees perfectly with the nearly Nernstian slope of the emf-response function observed for the same membranes (Table 12.3), indicating cation specificity of these sensors.

In the history of neutral-carrier-based membrane electrodes, different theories and views were called upon to explain the origin of cation permselectivity (zero anion flux).

a) The classical thin membrane model by Ciani, Eisenman, and Szabo [51 - 53] did not stipulate electroneutrality, and cationic carrier complexes therefore were assumed to be the only charged species existing within the membrane. Hence, permselectivity for cations was explained by a complete

Table 12.3. Transport numbers and slopes of the electrode response of cation-permselective neutral carrier membranes [50]

| Cation studied | Membrane composition | | | Electrolytes | | Transference number for cations studied a | Slope of electrode response in percent of theoretical slope |
	Ligand; wt.%	Solvent;d wt.%	Matrix; wt.%	Anode compartment	Cathode compartment		
Ca^{2+}	15; 3	o-NPOE; 65	PVC; 32	10^{-3}M $CaCl_2$	10^{-3}M KCl	0.99 ± 0.08	94
Ca^{2+}	15; 3	DBS; 65	PVC; 32	10^{-3}M $CaCl_2$	10^{-3}M KCl	1.00 ± 0.105	--
Ca^{2+}	15; 3	o-NPOE; 65	PVC; 32	$5 \cdot 10^{-4}$M $CaCl_2$ / $5 \cdot 10^{-4}$M $MgCl_2$	10^{-3}M KCl	0.99 ± 0.08	94
Ca^{2+}	15; 3	o-NPOE; 65	PVC; 32	$5 \cdot 10^{-4}$M $CaCl_2$ / $5 \cdot 10^{-4}$M NaCl	10^{-3}M KCl	0.99 ± 0.02	94
Ca^{2+}	15; 3	o-NPOE; 65	PVC; 32	10^{-4}M $CaCl_2$	10^{-4}M KSCN	0.995 ± 0.025	94
Na^{+}	14; 3	DBS; 65	PVC; 32	10^{-3}M NaCl	10^{-3}M KCl	0.92 ± 0.08	--
Na^{+}	14; 3	DMK; 65	PVC; 32	10^{-3}M NaCl	10^{-3}M KCl	0.92 ± 0.061	--
Na^{+}	14; 3	o-NPOE; 65	PVC; 32	10^{-3}M NaCl	10^{-3}M KCl	0.90 ± 0.075	96
Li^{+}	9; 5.8	TEHP; 62.8	PVC; 31.4	10^{-2}M LiCl	10^{-2}M KCl	0.97 ± 0.11	97
Li^{+}	9; 5.8	TEHP; 62.8	PVC; 31.4	10^{-3}M LiCl	10^{-3}M KCl	1.02 ± 0.21	97
Li^{+}	9; 5.8	TEHP; 62.8	PVC; 31.4	10^{-4}M LiCl	10^{-4}M KCl	0.98 ± 0.10	97
K^{+}	1; 3	DPP; 67	PVC; 30	10^{-2}M KCl	10^{-2}M HCl	1.08 ± 0.07	--
K^{+}	1; 5	---	Silicone rubber;95	10^{-2}M KCl	10^{-2}M HCl	1.1 ± 0.15	--

Table <u>12.3.</u> (continued)

			10^{-2} M KCl	10^{-2} M HClO$_4$		
K$^+$	<u>1</u>; 5	---	Silicone rubber; 95		1.1 ± 0.15	--
K$^+$	<u>1</u>; 1 DOA; 66	PVC; 33	$9 \cdot 10^{-4}$ M KCl	$9 \cdot 10^{-4}$ M KCl	1.02 ± 0.04	100
PEAH$^+$ [b]	<u>20/21</u>;1 DOA; 65	PVC; 34	$4 \cdot 10^{-3}$ M PEAHCl	$4 \cdot 10^{-3}$ M PEAHCl	0.95 ± 0.04	(120) [c]

[a] Measured on the membrane specified in the region of ohmic behavior of the current-voltage curve (at low voltage); 95% confidence limits. [b] PEAH$^+$: ^{14}C-α-phenylethylammonium cation. [c] From electrode response in buffered 10^{-1} M and 10^{-2} M solutions of the ammonium salt; activity coefficients unknown. [d] o-NPOE: o-nitrophenyloctyl ether; DBS: dibutyl sebacate; DMK: decylmethyl ketone; TEHP: tris-(2-ethyl-hexyl)-phosphate; DPP: dipentyl phthalate; DOA: dioctyl adipate.

exclusion of free hydrophilic anions from the lipid membrane. This theory was extended to thick membranes by Boles and Buck [29, 54] who still assumed large deviations from electroneutrality to occur within the membrane interior.

b) A more recent suggestion made informally by Buck [29] was that slow anion interfacial kinetics permit near-Nernstian response to cations.

The remaining theories c) to e) of thick electroneutral carrier membranes may be summarized under the general assumption that the anions present within the membrane are rather immobile. The reasons for such a behavior may be as follows:

c) The membrane contains permanent anions that are chemically bound to the supporting material, as was suggested by Kedem, Perry, and Bloch [55] (see also [54]).

d) The membrane contains permanent anions that are immobilized because of their poor water-solubility. Lipophilic anions, such as tetraphenylborate, were intentionally introduced into certain carrier membranes to improve their response characteristics (Morf et al. [26, 56, 57]; Seto et al. [58]).

e) The membrane extracts anions from the sample solution but the integral mobility of these species across the membrane is low as compared to the cationic complexes. This interpretation of zero anion flux was first given by Wuhrmann et al. [59] (see also [22, 25, 26, 60 - 62]).

The actual mechanism for permselectivity in thick neutral-carrier-based membranes was revealed only in 1977. To this end electrodialytic transport experiments (as well as inter-

diffusion studies) were carried out on 0.02-cm solvent polymeric membranes, using ^{14}C-labeled valinomycin and millimolar
aqueous solutions of $^{42}K^+$ and $^{36}Cl^-$ [50]. The concentration
profiles thus determined for ionophore and ions in a stack of
five membrane sections are displayed in Figure 12.3. It is
shown that an ionophore concentration gradient builds up
during electrodialysis which is in agreement with theoretical
predictions [60] (see also Chapter 8). This is unambiguous
proof for carrier translocation accompanying the cation transport and it is a requirement for the back-diffusion of free
carriers to end up in a steady-state [60]. According to Figure 12.3, cations as well as anions enter the membrane phase.
Although this was expected, it is most remarkable that the
overall concentration of cations K^+ exceeds by far (factor 85)
the one of the sample anions Cl^-. This fact can be explained
only when assuming either dramatic deviations from electroneutrality or the presence of anionic species other than Cl^-
within the membrane phase. The rather high cation concentration
level (about 0.5 mmol l^{-1}, see Figure 12.3) as well as the
relatively low electrical resistance of only about 10^6 Ω cm^2
found for the same membranes were demonstrated [50] to be at
variance with the space-charge theory a) mentioned above, however. Therefore, anionic species distinct from Cl^- must definitely reside in the membrane phase. The uneven distribution
of these species throughout the membrane, which was initially
composed of chemically identical sections, clearly rules out
the predominance of bound anionic sites, as suggested by
hypothesis c). The apparent anion deficiency found in Figure
12.3 is around 0.5 mmol l^{-1}; such high levels of impurities
(i. e., lipophilic anions corresponding to case d)) are
excluded by select methods of membrane preparation.

It was therefore concluded, and corroborated by further
evidence [50], that the additional anions in the membrane
phase must originate from the aqueous system, which situation

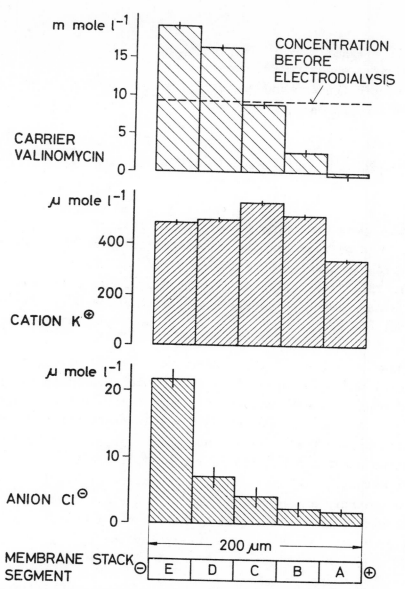

Figure 12.3. Ionophore and ion concentration profiles within a valinomycin-based liquid membrane (stack) in contact with aqueous KCl solutions [50]. Results represent the local concentrations of valinomycin, potassium (mainly as complexes), and chloride immediately after interrupting a steady-state electrodialysis; the 95% confidence limits are given by vertical bars. The dashed line indicates the concentration level of valinomycin established under zero-current conditions.

conforms in principle to theory e). It was suggested that anionic sites are generated according to the following carrier-induced ion-exchange reaction, involving water present in the membrane phase:

$$K^+(aq) + Val(m) + H_2O(m) \rightleftharpoons KVal^+(m) + OH^-(m) + H^+(aq) \qquad (12.1)$$

Because of a possible formation of water clusters involving the OH^- formed by reaction (12.1), the mobility of these anionic sites is likely to be low*), as is demanded by hypothesis e). This "constellation" in the end allows a neutral-carrier-based solvent polymeric membrane to perform as cation-specific electrode.

Mechanisms of type (12.1), however without the participation of carriers, may also explain the heretofore unresolved behavior of other representatives of liquid membrane electrodes. Correspondingly, we can understand why solvent-polymeric membranes without any real ion-exchange sites (e. g., the system dioctylphthalate/PVC mentioned in Table 11.1) are capable of a perfect Nernstian response to highly extractable cations such as quaternary ammonium ions [63]. In a similar way, the classical cation-exchange liquid membranes may respond to lipophilic sample anions such as tetraphenylborate or perchlorate [64].

---

*) A direct electric transference of such sites occurs only under extreme conditions. In fact, a transference number for K^+ of nearly 1.0 was observed in electrodialysis experiments up to voltages of about 30 V. This value dropped to 0.77 only at 70 V, the remaining 23% of the current being contributed by ions other than Cl^- [50].

The presented facts [50] lead to the conclusion that, among the available theories of carrier membranes, the model developed in Reference 60 (hypothesis e)) gives the most satisfying description of the observed electromotive and transport behavior. This model indeed assumes anions of low mobility to exist within neutral carrier membranes. To facilitate the theoretical description, these species were treated as fixed anionic sites of approximately constant concentration, similarly to the Teorell-Meyer-Sievers concept worked out earlier [54, 55]. Then one can apply, for example, the following expression (Eq. (9.10)) to describe the zero-current diffusion potential produced within valinomycin-based membranes, containing potassium complexes $KVal^+$ and anions Cl^- as mobile charged species:

$$E_D = \frac{u_{KVal} - u_{Cl}}{u_{KVal} + u_{Cl}} \frac{RT}{F} \ln \frac{u_{KVal} a_{KVal}(0) + u_{Cl} a_{Cl}(0)}{u_{KVal} a_{KVal}(d) + u_{Cl} a_{Cl}(d)} \qquad (12.2)$$

Since it holds that

$$a_{KVal}(0) \approx a_{KVal}(d) \approx X \qquad (12.3a)$$

$$a_{Cl}(0) \approx a_{Cl}(d) \approx 0 \qquad (12.3b)$$

where X is the mean activity of immobile anionic sites, Eq. (12.2) leads to the approximation

$$E_D = 0 \qquad (12.4)$$

The boundary potential of the liquid membranes in question is determined as follows (see Eq. (9.3)):

$$E_B = \frac{RT}{F} \ln \frac{k_K a'_K}{a_K(0)} - \frac{RT}{F} \ln \frac{k_K a''_K}{a_K(d)} \qquad (12.5)$$

where a_K are the activities of free potassium ions, and k_K is the distribution coefficient of these species between aqueous solutions (' resp. ") and membrane (x = 0 resp. d). The predominant partitioning of cations between the outside solutions and the respective complexes in the membrane (stability constant β_{KVal}) can be characterized by overall distribution coefficients [25, 26, 60] which are defined as

$$K'_K = k_K \frac{a_{KVal}(0)}{a_K(0)} = \beta_{KVal} k_K a_{Val}(0) \qquad (12.6a)$$

$$K''_K = k_K \frac{a_{KVal}(d)}{a_K(d)} = \beta_{KVal} k_K a_{Val}(d) \qquad (12.6b)$$

This allows to transform Eq. (12.5) into the following one:

$$E_B = \frac{RT}{F} \ln \frac{K'_K a'_K}{a_{KVal}(0)} - \frac{RT}{F} \ln \frac{K''_K a''_K}{a_{KVal}(d)} \qquad (12.7)$$

Recalling Eqs. (12.3a) and (12.4) we arrive at the following result for the total membrane potential, $E_M = E_B + E_D$:

$$E_M = \frac{RT}{F} \ln \frac{K'_K a'_K}{K''_K a''_K} \qquad (12.8)$$

Equation (12.8) demonstrates that neutral carrier membranes must definitely meet a second requirement, in addition to cation permselectivity, to bring out an ideal emf-response to cations (see also Eq. (12.6)). A reduction of Eq. (12.8) to a Nernstian expression is possible only as long as the activity of free carrier ligands is constant throughout the membrane, that is

$$a_{Val}(0) = a_{Val}(d) = a_{Val} \qquad (12.9)$$

respectively:

$$K_K' = K_K'' = K_K \qquad (12.10)$$

These conditions are usually fulfilled if no lipophilic anions are present in the outside solutions; such species would strongly facilitate the extraction of electrolyte into the organic membrane, thereby leading to both a loss of cation permselectivity and a decrease of the free ligand activity (see Section 12.4).

An alternative theory suited for the description of neutral carrier membrane electrodes is based on the Planck formalism evolved in Reference 65. This membrane model, treated in the Chapters 4 - 6, does not stipulate ideally fixed anionic sites. According to Eqs. (9.7) and (9.8), Planck's result for the diffusion potential of the valinomycin-based membranes discussed here may be written as[*]:

---

[*] For simplicity the same activity coefficients were used for all ionic species in the membrane ($KVal^+$, OH^-, Cl^-). The electroneutrality condition then yields the approximation $a_{KVal}(x) = a_{OH}(x) + a_{Cl}(x)$, which was inserted into Eqs. (12.11) and (12.12).

$$E_D = \frac{u_{KVal} - \bar{u}_X}{u_{KVal} + \bar{u}_X}\, \frac{RT}{F}\, \ln \frac{a_{KVal}(0)}{a_{KVal}(d)} \qquad (12.11)$$

where

$$\bar{u}_X = u_{KVal} \cdot \frac{[u_{OH}a_{OH}(d) + u_{Cl}a_{Cl}(d)]\, e^{-\frac{F}{RT}E_D} - [u_{OH}a_{OH}(0) + u_{Cl}a_{Cl}(0)]}{u_{KVal}a_{KVal}(d)\, e^{-\frac{F}{RT}E_D} - u_{KVal}a_{KVal}(0)}$$

$$(12.12)$$

The findings in this section have shown, however, that the following relations apply to the present case:

$$a_{Cl}(x) \ll a_{KVal}(x), \text{ i.e. } u_{Cl}a_{Cl}(x) \ll u_{KVal}a_{KVal}(x) \qquad (12.13)$$

$$u_{OH} \ll u_{KVal}, \text{ i.e. } u_{OH}a_{OH}(x) \ll u_{KVal}a_{KVal}(x) \qquad (12.14)$$

Hence, the mean mobility $\bar{u}_X$, representative for all anions in the membrane, becomes very low as compared to the mobility u_{KVal} of the cationic species. This conclusion offers for the first time a clear-cut explanation for the apparently poor mobility of sample anions in carrier membranes, as was postulated more intuitively in earlier theoretical approaches [22, 25, 26, 59 - 62]. It follows:

$$E_D = \frac{RT}{F}\, \ln \frac{a_{KVal}(0)}{a_{KVal}(d)} \qquad (12.15)$$

The last expression can easily be combined with Eq. (12.7). The final result for the membrane potential E_M turns out to be identical to Eq. (12.8). This implies that the Planck formalism, assuming low-mobility anionic sites, and the Teorell-Meyer-Sievers concept of fixed anionic sites constitute equivalent descriptions of neutral carrier membrane electrodes. Both models are referred to in the following sections.

12.3. CATION SELECTIVITY OF CARRIER MEMBRANE ELECTRODES

In this section we focus on the potentiometric selectivity behavior of ideally cation-sensitive liquid membranes based on neutral carriers. Correspondingly, the following restrictions are a priori imposed: a) cation permselectivity, and b) constant activity of free carriers (see Sections 12.2 and 12.4). The formal discussion, initiated above for the example of the valinomycin-based membranes, is extended here insofar as carrier complexes of different cations and of different stoichiometries (including the 1:0 complexes corresponding to the minor fraction of free cations) are taken into account. All these cationic forms have to be considered as separate permeating species.

The relative population of a given 1:n complex between the cation M^{z+} and ionophores S in the membrane is controlled by the individual distribution parameter $K_{m,n}$, defined in analogy to Eq. (12.6):

$$K_{m,n} = k_m \frac{a_{ms,n}(x)}{a_m(x)} = \beta_{ms,n} \, k_m \, a_s^n \qquad (12.16)$$

where: $\beta_{ms,n}$: stability constant of the complex MS_n^{z+} in the membrane (for free cations, $\beta_{ms,0} \equiv 1$)

$a_{ms,n.}$: activity of the complex MS_n^{z+} in the membrane

a_s : activity of the free carrier S in the membrane

This parameter evidently constitutes a measure for the fundamental equilibrium (12.17):

$$M^{z+} \text{ (aqueous)} + n \text{ S (membrane)} \xrightleftharpoons{K_{m,n}} MS_n^{z+} \text{ (membrane)} \qquad (12.17)$$

An alternative, and very useful, formulation of the distribution coefficient $K_{m,n}$ is obtained by breaking down this reaction into the following steps:

1. Transfer of free ligands from the membrane into the boundary layer of the outside solution:

$$S \text{ (membrane)} \xrightleftharpoons{1/k_s} S \text{ (aqueous)} \qquad (12.18)$$

2. Formation of cationic complexes in the aqueous phase:

$$M^{z+} \text{ (aqueous)} + n \text{ S(aqueous)} \xrightleftharpoons{\beta_{ms,n}^w} MS_n^{z+} \text{ (aqueous)} \qquad (12.19)$$

3. Transfer of complexes into the membrane:

$$MS_n^{z+} \text{ (aqueous)} \xrightleftharpoons{k_{ms,n}} MS_n^{z+} \text{ (membrane)} \qquad (12.20)$$

From this consideration, it follows:

$$K_{m,n} = \beta_{ms,n}^w \, k_{ms,n} (a_s/k_s)^n \qquad (12.21)$$

The advantage of Eq. (12.21), as compared to (12.16), is that membrane-independent quantities, $\beta_{ms,n}^{w}$, are used to describe the complexation specificity of the carrier ligands. Equation (12.21) is also applicable to free cations, which case is evidently characterized by the values $n = 0$, $\beta_{ms,0}^{w} \equiv 1$, and $k_{ms,0} \equiv k_{m}$.

12.3.1. Selectivity Between Cations of the Same Charge

In cells containing primary ions I^{z+} as well as interfering ions J^{z+}, both of the same charge $z=1, 2, \ldots.$, all the different 1:n carrier complexes formed by these species are involved in the generation of the membrane potential. Recalling Eqs. (9.16) and (12.16), we therefore obtain the relationship [25, 26, 65 - 67]:

$$E_M = \frac{RT}{zF} \ln \frac{\sum\limits_{n} u_{is,n} K_{i,n} a_i' + \sum\limits_{n} u_{js,n} K_{j,n} a_j'}{\sum\limits_{n} u_{is,n} K_{i,n} a_i'' + \sum\limits_{n} u_{js,n} K_{j,n} a_j''} \qquad (12.22)$$

In cases where the individual mobilities of cationic forms are not significantly different, one can use a simplified description:

$$E_M = \frac{RT}{zF} \ln \frac{\sum\limits_{n} K_{i,n} a_i' + \sum\limits_{n} K_{j,n} a_j'}{\sum\limits_{n} K_{i,n} a_i'' + \sum\limits_{n} K_{j,n} a_j''} \qquad (12.23)$$

Both expressions are applicable to carrier-modified bilayer membranes (exclusion of anions) or to bulk membranes (stationary or poorly mobile anions) since no assumption concerning the mechanism of cation permselectivity is needed for the derivation of the basic Eq. (9.11). For the emf-response of such idealized membrane electrode assemblies, an equation of the well-known Nicolsky type can readily be derived:

$$E = E_i^o + \frac{RT}{zF} \ln [a_i' + K_{ij}^{Pot} a_j'] \tag{12.24}$$

The selectivity is given by [25, 26, 33, 66, 67]

$$K_{ij}^{Pot} \cong \frac{\sum\limits_{n} K_{j,n}}{\sum\limits_{n} K_{i,n}} = \frac{\sum\limits_{n} \beta_{js,n}^{W} k_{js,n} (a_s/k_s)^n}{\sum\limits_{n} \beta_{is,n}^{W} k_{is,n} (a_s/k_s)^n} \tag{12.25}$$

and characterizes the following ion-exchange reaction:

I^{z+} (complexed, membrane) $+ J^{z+}$ (free, aqueous) $\rightleftharpoons$

J^{z+} (complexed, membrane) $+ I^{z+}$ (free, aqueous) $\tag{12.26}$

It becomes evident that the cation selectivity of carrier-based ion sensors may depend on various factors:
1. The selectivity behavior of the carrier ligands used, which can be fully characterized by the values $\beta_{ms,n}^{W}$.
2. The activity a_s of free ligands in the membrane (given by the membrane composition), this dependency being expressed by the relation

$$\frac{\partial \log K_{ij}^{Pot}}{\partial \log a_s} \cong \bar{n}_j - \bar{n}_i \tag{12.27}$$

where $\bar{n}$ is the mean degree of complexation of the subscripted ion in the membrane.
3. The extraction properties of the membrane solvent which are decisive for the magnitude of the ratios $k_{ms,n}/k_s^n$. The influence of the membrane solvent on K_{ij}^{Pot} may be estimated using an electrostatic model (see Section 12.5). According to such considerations, the free energy of transfer of

cationic complexes from water into the membrane (as compared to the uncomplexed ligands) is given by

$$- RT \ln \frac{k_{ms,n}}{k_s^n} = N \frac{(z_m e)^2}{2r_C} \left(\frac{1}{\varepsilon} - \frac{1}{78.5} \right) + \text{const} \qquad (12.28)$$

where ε is the dielectric constant of the membrane solvent,
$z_m e$ is the charge (in electrostatic units) and $2r_C$ is the
overall diameter of the cationic forms, and N is Avogadro's
number.

A very simple relationship can be derived from Eq. (12.25)
for neutral carriers that predominantly form 1:1 complexes
with cations, as is the case for most of the natural ionophores known to-date. Here, the selectivity becomes independent of the ligand concentration. Since the dimensions and
therefore the distribution coefficient of carrier-complexes
of given charge and stoichiometry are roughly independent of
the nature of the central ion (see Eq. (12.28)), the selectivity behavior can approximately be described as [22, 51 -
53, 59, 60, 68 - 70]

$$K_{ij}^{Pot} \approx \frac{\beta_{js}^w}{\beta_{is}^w} \qquad (12.29)$$

where the stability constants refer to the respective 1:1
complexes. Accordingly, the selectivity of corresponding
neutral carrier membrane electrodes among ions of the same
charge is scarcely influenced by the ion-selective behavior
of the membrane solvent used but is mainly given by the
complexation properties of the incorporated ligands. This
exclusive ligand control of ion selectivity, unrealizable

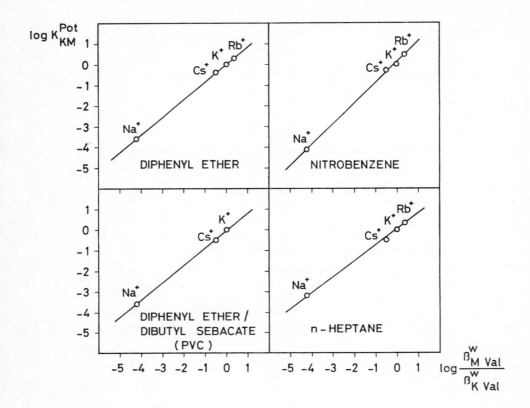

<u>Figure 12.4.</u> Correlation between theoretical and experimental selectivity factors for liquid-membrane electrodes based on the carrier valinomycin (<u>1</u> in Fig. 12.1) in different membrane solvents (from Reference 26).

for the classical ion-exchangers based on charged ligands, is
nicely demonstrated in Figure 12.4 for a series of liquid
membrane electrodes based on the carrier antibiotic valinomycin. The correlations found attest a good agreement between
the theoretical selectivity factors given in the simplified
form (12.29) and the experimental values (see also [22, 59,
68 - 70]). No such simple correlations are obtained, however,
if cation-carrier complexes of different stoichiometries are
involved [67] and/or the central ion remains partly solvated
[59, 70 - 72] (incomplete covering by the ligand shell). Several molecular aspects of the ion specificity exhibited by
natural ionophores and synthetic model compounds will be
summarized in Section 12.5.

12.3.2. Monovalent/Divalent Cation Selectivity

It is of some practical importance to also give a description of the selectivity behavior of carrier membrane electrodes
towards cations of different charge, for instance towards a
mixed solution of divalent ions I^{2+} and monovalent ions J^+. A
theoretical approach to this problem has already been worked
out in Part A and is based on a Teorell-Meyer-Sievers concept
of the carrier membrane (fixed anionic sites of activity X).
If the cation-exchange reaction at the phase boundary is the
dominant factor determining the membrane potential, the emf-
response of the corresponding electrode cell is given by
Eq. (9.19):

$$E = E_i^o + \frac{RT}{F} \ln \left[\sqrt{a_i' + \tfrac{1}{4} K_{ij}^M a_j'^2} + \sqrt{\tfrac{1}{4} K_{ij}^M a_j'^2} \right] \qquad (12.30)$$

Since the extraction properties of carrier membranes have
previously been characterized by overall distribution coefficients $K_{m,n}$, however, we have to replace the parameters k_m

291

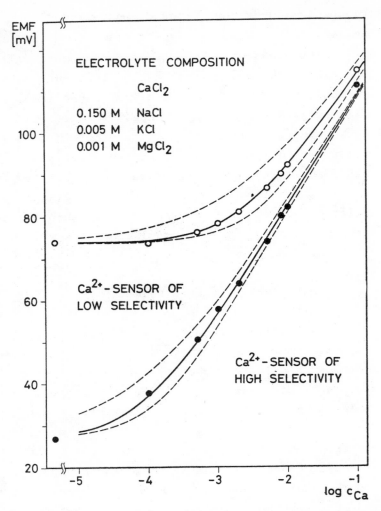

Figure 12.5. Comparison between calculated and measured EMF-response curves for different Ca^{2+}-sensitive carrier-based PVC membrane electrodes ($25^{\circ}C$)[70]. The electrolyte composition is comparable to blood serum, the Ca^{2+}-level being variable. Solid lines: calculated according to Eq. (12.30). Lower dashed lines: calculated according to Eq. (12.32) where K_{ij}^{Pot} is assumed to be activity-independent. Upper dashed lines: calculated according to Eq. (9.23) with constant K_{ji}^{Pot}-values. Upper traces: ligand 15; lower traces: ligand 17 (Figure 12.1).

occurring in Eq.(9.20). Thus, we obtain the following relationship for the monovalent/divalent ion selectivity of
idealized carrier membranes:

$$K_{ij}^M \cong \frac{(\sum\limits_n K_{j,n})^2}{2 \times \sum\limits_n K_{i,n}} = \frac{(\sum\limits_n \beta_{js,n}^W \ (k_{js,n}/k_s^n) \ a_s^n)^2}{2 \times \sum\limits_n \beta_{is,n}^W \ (k_{is,n}/k_s^n) \ a_s^n} \qquad (12.31)$$

An emf-behavior according to Eq. (12.30) was in fact observed for real ion-sensor systems [70]. Figure 12.5, illustrating the determination of Ca^{2+} in mixed electrolyte solutions
comparable to blood serum, shows impressively that the one-
parameter equation permits a perfect interpolation of emf-
values over the whole Ca^{2+}-activity range. Obviously, the
common Nicolsky equation (9.21)

$$E = E_i^o + \frac{RT}{2F} \ln \left[a_i' + K_{ij}^{Pot} \ a_j'^2 \right] \qquad (12.32)$$

may also be used to give a relatively close fit of experimental values, which is in agreement with usual findings. Thus,
the theoretical selectivity factor K_{ij}^M corresponds to the
practical K_{ij}^{Pot}-value, as obtained from the separate solution
method:

$$K_{ij}^M \approx K_{ij}^{Pot} \qquad (12.33)$$

A discussion of these selectivity parameters is rather involved since Eq. (12.31) obviously cannot be reduced to a
simple form comparable to (12.29). Nevertheless, Eq. (12.31)
reveals some important rules which have to be considered in

designing ion-selective sensors (see also Section 12.5). First of all, the selectivity coefficient K_{ij}^M strongly reflects the complexation behavior of the carriers used. In addition, it depends again on the free ligand activity, that is

$$\frac{\partial \log K_{ij}^M}{\partial \log a_s} \simeq 2\,\bar{n}_j - \bar{n}_i \qquad\qquad (12.34)$$

Another selectivity-determining factor is the mean activity of anionic sites in the membrane. This parameter can be varied systematically by the additional incorporation of permanent anions into the membrane (Section 12.4.2), which procedure generally leads to an increased preference for divalent over monovalent cations. The remaining terms in Eq. (12.31) are the ratios of distribution coefficients, describing the transfer of charged complexes relative to the transfer of free ligands. From Eq. (12.28), these ratios are readily seen to decrease with increasing charge of the cation and decreasing dielectric constant of the membrane phase. When assuming the same stoichiometry 1:n for all complexes in the membrane, one may get the following relationship (Eqs. (12.31) and (12.28)):

$$\frac{\partial \log K_{ij}^M}{\partial (1/\varepsilon)} \simeq \frac{2\,N\,e^2}{2.303\,RT \cdot 2r_C} = \frac{487}{2r_C (\text{in } \mathring{A})} \qquad (12.35)$$

Accordingly, the preference of a sensor for monovalent over divalent cations is efficiently improved when reducing the polarity of the membrane solvent, and vice versa. Such a relationship was indeed observed for neutral carrier membranes in potentiometric measurements (Figure 12.6 and [73, 74]) as well as in ion transport experiments [74].

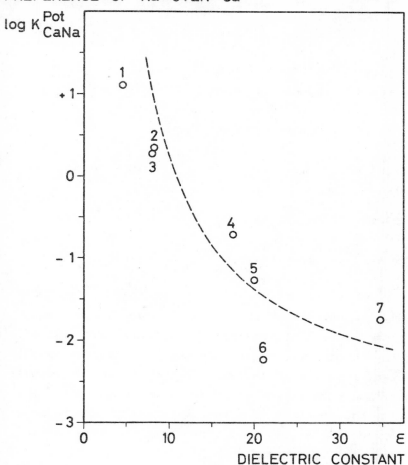

PREFERENCE OF Na$^+$ OVER Ca^{2+}

$\log K^{Pot}_{CaNa}$

DIELECTRIC CONSTANT

Figure 12.6. Dependence of the monovalent/divalent ion selectivity on the polarity (as described by the dielectric constant)
of the membrane solvent used (ligand 15). The solvents are: 1:
dibutylsebacate; 2: tris-(2-ethylhexyl)-phosphate; 3: 1-deca-
nol; 4: acetophenone; 5: 2-nitro-p-cymene; 6: p-nitroethylben-
zene; 7: nitrobenzene. The selectivity factors were obtained
from the EMF-values measured on 0.1 M chloride solutions [22].
The curve was calculated from Eq. (12.35) with $2r_C = 15$ Å.

12.4. ANION EFFECTS IN CARRIER MEMBRANE ELECTRODES

12.4.1. Anion Interference in Conventional Carrier Membranes

Whereas carrier membrane electrodes usually show a nearly ideal cation sensitivity when exposed to chloride solutions, the presence of lipophilic anions in the sample may lead to significant nonidealities in the emf-behavior [25 - 30, 54 - 58, 60 - 62, 64, 75]. Species that are easily soluble in the organic membrane phase, e. g. thiocyanate or perchlorate, facilitate the extraction of electrolyte according to the reaction

$$I^{z+}(aq) + z\ Y^-(aq) + n\ S(m) \underset{\phantom{K_{ex}}}{\overset{K_{ex}}{\rightleftharpoons}} IS_n^{z+}(m) + z\ Y^-(m) \qquad (12.36)$$

This reaction, competing with the "normal" process analogous to (12.1), has a twofold effect. First, the activity of mobile anions in the membrane is increased substantially, which involves a clear loss of cation permselectivity and consecutive distortions from a pure cation response of the electrode. Simultaneously, the activity of free ligands at the membrane-solution interface decreases; near consumption of free carriers is signaled by conversion of the electrode response into a pure anionic function (see below).

The interference in neutral carrier membrane electrodes by lipophilic sample anions[*)] was studied by several authors

----

[*)] The sensitivity to lipophilic, easily extractable ions constitutes a possible limitation of all liquid membrane electrodes. Such interferences were also reported for the classical ion-exchanger electrodes [64, 76]. In the case of neutral carrier membranes, the observed effects increase with increasing lipophilicity of the interfering anions, that is according to the Hofmeister series $Cl^- < Br^- < NO_3^- < SCN^- \sim ClO_4^- < R^-$ [54, 54a, 56, 61].

[26, 29, 54, 56 - 58, 60 - 62]. The experimental results usually show a region of cation-dominant response at low levels of interfering anions, a maximum in potential, and a decrease in response at high sample activities (Figures 12.7 and 12.8). Different theoretical treatments were offered (see hypotheses a) to e) in Section 12.2) that succeeded in explaining most of the observable effects. First, Boles and Buck [29, 54] postulated that space-charge control of potential, i. e., exclusion of anions from the membrane, occurs at the low bathing activities and that the maximum and decrease of potential appear in an activity region where the membrane becomes electroneutral. It was claimed that consumption of free carriers does not occur because of the presumed small formation constants for ion-carrier complexing and ion pairing [29]. In the theory proposed by Morf et al. [26, 56, 57, 60], in agreement with later experimental findings [62] (see also Figures 12.7 and 12.8), the distinct effects of anion interference are mainly a consequence of high salt extraction and the concomitant consumption of free carriers. The assumption of low anion mobility in electroneutral carrier membranes was required for rationalizing the region of cation-dominant potential response. Although there is no a priori reason why anions should have low mobilities [29], this approach has now been underpinned by results from the more sophisticated Planck-type description outlined in Section 12.2. An alternative description of anion effects, along the lines of the Teorell-Meyer-Sievers theory, was initiated by Kedem, Perry, and Bloch [55] and others [54, 56 - 58]. Taking into account the in-depth information on carrier membranes available in the meantime, it seems necessary, however, that all the mentioned theories be revised or completed.

For simplicity, we focus first on carrier membrane systems where the two species IS_n^{z+} and Y^- are the predominant permeating species. When choosing a fixed-site model for

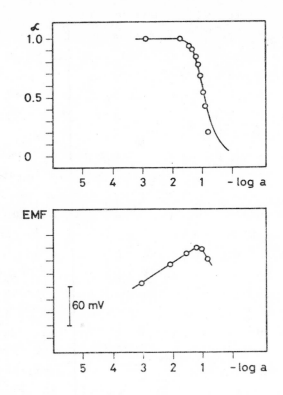

Figure 12.7. Correlation between salt extraction and potential response of a PVC membrane based on the Ca^{2+}-carrier $\underline{16}$ (Figure 12.1) in o-nitrophenyl octyl ether for $Ca(SCN)_2$ solutions as sample. The percentage of uncomplexed carriers ($\alpha \cdot 100\%$) was obtained from ^{13}C-n.m.r. spectroscopic studies. The solid lines were calculated according to theory [62].

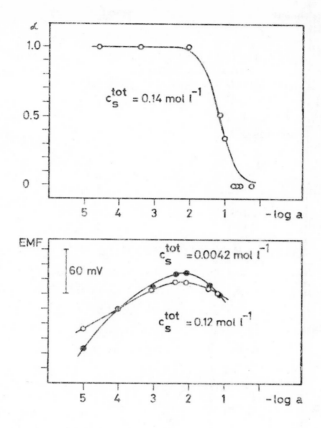

Figure 12.8. Correlation between salt extraction and potential response of a PVC membrane based on nonactin (2 in Fig. 12.1) in o-nitrophenyl octyl ether for KSCN solutions as sample [62].

the carrier membrane, we may apply in principle the former Eqs. (12.2) and (12.7) to describe the diffusion potential

$$E_D = \frac{u_{is} - u_y}{z u_{is} + u_y} \frac{RT}{F} \ln \frac{z^2 u_{is} a_{is}(0) + u_y a_y(0)}{z^2 u_{is} a_{is}(d) + u_y a_y(d)} \qquad (12.37)$$

respectively, the boundary potential

$$E_B = \frac{RT}{zF} \ln \frac{K'_{i,n} a'_i a_{is}(d)}{K''_{i,n} a''_i a_{is}(0)} \qquad (12.38)$$

The parameters entering in Eqs. (12.37) and (12.38) are the individual mobilities u and the local activities a of cationic forms and sample anions. The ratio of overall distribution coefficients in (12.38), each characterizing the partitioning of cations at one membrane/solution interface, may be reduced as follows, witness Eq. (12.6a,b):

$$\frac{K'_{i,n}}{K''_{i,n}} = \frac{\beta_{is} k_i [a_s(0)]^n}{\beta_{is} k_i [a_s(d)]^n} = \left(\frac{a_s(0)}{a_s(d)}\right)^n \qquad (12.39)$$

This ratio is evidently affected by the extraction of cations since for carrier-based liquid membrane electrodes the total activity a_s^{tot} of all forms of ligand S trapped within the membrane is approximately constant (steady-state condition, replacing the corresponding assumption made in the older theories [51, 54]:

$$n \, a_{is}(x) + a_s(x) = a_s^{tot} \qquad (12.40)$$

Another relationship follows from the electroneutrality condition and the assumption of fixed anionic sites:

$$z \, a_{is}(x) = a_y(x) + X \tag{12.41}$$

Since the inner solution of membrane electrodes normally contains no lipophilic anions, we get:

$$a_y(d) \cong 0 \; ; \; a_{is}(d) \cong \frac{1}{z} X \; ; \; a_s(d) \cong a_s^{tot} - \frac{n}{z} X \tag{12.42}$$

Determination of the residual activities $a_{is}(0)$, $a_y(0)$, and $a_s(0)$ becomes possible after defining the equilibrium constant for the extraction reaction (12.36):

$$K_{ex} = \beta_{is} k_i k_y^z = \frac{a_{is}(0) [a_y(0)]^z}{a_i' a_y'^z [a_s(0)]^n} \tag{12.43}$$

Equations (12.37) - (12.43) offer an implicit but well-defined description of the potential response of carrier membrane electrodes to sample solutions containing interfering anions. The following limiting cases may be discerned for conventional membranes having $a_s^{tot} \gg X$ (see Figure 12.3).

1. At sufficiently low sample activities, it holds that

$$K_{ex}(a_s^{tot})^n \, a_i' \, a_y'^z \ll X^{z+1} \tag{12.44}$$

Here the extraction of lipophilic anions Y^- into the mem-

301

brane remains negligible and, therefore, these species exert no detectable influence on the membrane potential. A pure cation response is obtained, as described in Section 12.2:

$$E = E_i^o + \frac{RT}{zF} \ln a_i'$$

(12.45)

2. At intermediate sample activities, i. e.,

$$x^{z+1} \ll K_{ex}(a_s^{tot})^n a_i' a_y'^z \ll (a_s^{tot})^{z+1}$$

(12.46)

the extraction of ions is governed by reaction (12.36) and leads to comparable activity levels of cationic complexes and sample anions in the membrane. However, consumption of carriers does not yet come into play. In this case, the emf-response assumes the form (9.26):

$$E = E_{i,y}^o + \frac{u_{is}}{zu_{is}+u_y} \frac{RT}{F} \ln a_i' - \frac{u_y}{zu_{is}+u_y} \frac{RT}{F} \ln a_y'$$

(12.47)

Equation (12.47) indicates a loss of cation specificity, which is observable in Figure 12.8 by the relatively broad maximum of the response curves.

3. At comparatively high sample activities,

$$(a_s^{tot})^{z+1} \ll K_{ex}(a_s^{tot})^n a_i' a_y'^z$$

(12.48)

302

the activity of ionic forms residing inside the membrane
boundary exceeds by far that of free ligands. Since the activity of cationic complexes (and that of counterions) is
limited by the total ligand available, Eq. (12.43) predicts
proportionality between $[a_s(0)]^n$ and $1/a_i' a_y'^z$. Inserting
this dependency into Eq. (12.38) we can derive a Nernstian
response to anions:

$$E = E_y^o - \frac{RT}{F} \ln a_y' \qquad\qquad (12.49)$$

Here the cationic complexes of fixed activity evidently
behave as anion-exchangers. Exactly the same situation
holds for certain anion-selective membrane electrodes
using metal complexes as ion-exchange sites (e. g., complexes of transition metal cations with neutral ligands
of the phenanthroline type, see Table 11.1). However, these
systems cannot attain a cation-response region at low
sample activities because of the extremely high stability
of the involved metal complexes. It may be recognized that
neutral ligands must be sufficiently weak complex formers
in order to realize ionophoric behavior.

The experimental results presented in Figures 12.7 -
12.10 (see also [54, 54a, 56 - 58, 60 - 62]) illustrate
that the emf-response to cations in solutions containing
lipophilic anions may be far from the nearly Nernstian behavior obtained for chloride solutions. The observable
effects of anion interference are in agreement with the
theoretical predictions. For sensors responsive to monovalent cations, these nonidealities may be nearly eliminated by employing membrane components of low dielectric
constant [57, 77] (Figures 12.9 and 12.10), thereby re-

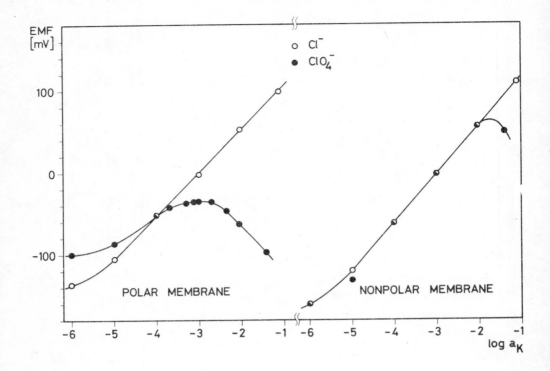

Figure 12.9. EMF response of valinomycin-based PVC membrane electrodes to K$^+$ in samples containing different anions (25°C). Left: a highly polar solvent (2-nitro-p-cymene) was used as membrane component. Right: a rather nonpolar solvent (dioctyl sebacate) was incorporated.

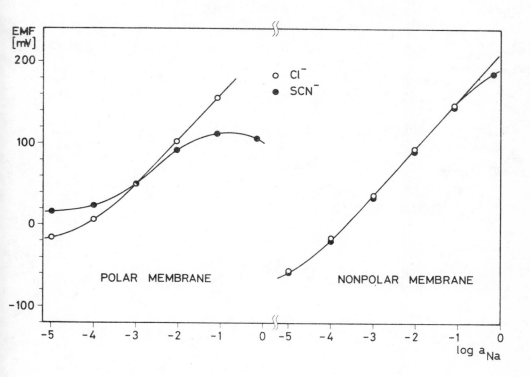

Figure 12.10. EMF response of PVC membrane electrodes based on the neutral carrier 14 (Fig. 12.1) to Na^+ in samples containing different anions ($25°C$) [57]. Left: o-nitrophenyl octyl ether was used as membrane solvent. Right: dibenzyl ether was used as membrane solvent.

ducing the value of the extraction constant K_{ex}. Polar membrane solvents, such as nitrobenzene and the higher homologs [6, 54, 75], are a rather poor choice in this respect. In contrast, these polar solvents cannot be replaced in carrier membrane electrodes selective for divalent cations, as has been explained in Section 12.3.2. An alternative method for improving the cation specificity is discussed later.

A more elegant description of anion effects in neutral carrier membrane electrodes, including those discussed in the next section, is based on the extended Planck formalism. Adapting the general result, Eqs. (9.27) – (9.29), to the present case, we can immediately write the following solution for the membrane potential:

$$E_M = \frac{u_{is}}{zu_{is}+\bar{u}_y} \frac{RT}{F} \ln \frac{a'_{is}}{a''_{is}} - \frac{\bar{u}_y}{zu_{is}+\bar{u}_y} \frac{RT}{F} \ln \frac{k_y a'_y + k_r a'_r}{k_y a''_y + k_r a''_r}$$

$$\text{(12.50)}$$

$$= \frac{u_{is}}{zu_{is}+\bar{u}_y} \frac{RT}{F} \ln \frac{K'_{i,n} a'_i}{K''_{i,n} a''_i} - \frac{\bar{u}_y}{zu_{is}+\bar{u}_y} \frac{RT}{F} \ln \frac{k_y a'_y + k_r a'_r}{k_y a''_y + k_r a''_r}$$

The mean mobility $\bar{u}_y$ of anions in the membrane is influenced by both the sample anions Y^- and the primary anionic sites, denoted here as R^-:

$$\bar{u}_y = \frac{(u_y k_y a''_y + u_r k_r a''_r) e^{-\frac{F}{RT} E_M} - (u_y k_y a'_y + u_r k_r a'_r)}{(k_y a''_y + k_r a''_r) e^{-\frac{F}{RT} E_M} - (k_y a'_y + k_r a'_r)}$$

$$\text{(12.51)}$$

All the activities appearing in these expressions refer to the aqueous boundaries that are in equilibrium with the adjoining membrane surfaces. For the conventional carrier membrane electrodes it holds that $u_r \approx 0$ and $k_y a_y'' << k_r a_r''$ (i.e., $a_y(d) << a_r(d)$, see Section 12.2). Case 1 of the preceding treatment corresponds here to the situation where

$$k_y a_y' << k_r a_r' \quad \text{resp.} \quad \bar{u}_y \approx 0 \qquad\qquad (12.52)$$

Cases 2 and 3 result again from an increased anion permeability, i. e.,

$$k_y a_y' >> k_r a_r' \ , \ k_r a_r'' \quad \text{resp.} \quad \bar{u}_y \approx u_y \qquad\qquad (12.53)$$

Evidently, the two theoretical models developed for carrier membranes lead to basically the same results. The choice between these two approaches depends on the more academic question whether the anionic sites existing in the membrane phase are to be considered as fixed or as immobile.

For many carrier membrane systems, especially those prepared with rather nonpolar membrane solvents, the effects of ionic association (ion pairing) are not to be overlooked [26, 54, 62]. Thus the formation of associates of the type $IS_n Y_z$ is predominant in media of low dielectric constant. Since these species are electrically neutral, however, they have no direct influence on the diffusion potential. Accordingly, the emf of associated membrane systems is still described by Eqs. (12.50) and (12.51) as far as the ions IS_n^{z+}, Y^-, and R^- are the only charge-carrying species. The only difference exists in the formal result obtained for the position of the emf-maximum separating cases 1 and 3 [26, 62]. In striking contrast, the formation of charged associates may give rise to entirely new response characteristics. In certain sensors

307

for divalent cations, for example, it is conceivable that the single-charged species IS_nY^+ is the predominant permeating ion besides the nearly immobile sites R^-. The membrane electrode will then approximate a Nernstian response to these permeating species, as may be inferred from Eq. (12.50) using $\bar{u}_y = 0$, $z = 1$, and the subscript isy instead of is:

$$E = E^o_{isy} + \frac{RT}{F} \ln a'_{isy} \tag{12.54}$$

Taking into account the expressions for ion-carrier complexing and ion pairing, we obtain:

$$E = \text{const} + \frac{RT}{F} \ln [K'_{i,n} \, a'_i] + \frac{RT}{F} \ln a'_y \tag{12.55}$$

Equation (12.55) reveals that anion interference here causes an over-Nernstian region of the emf-response curve, the actual slope being up to four times larger than the expected value of 29.6 mV (25^oC) which is otherwise characteristic of divalent-ion sensors. A slope of about 100 mV was in fact observed for different carrier membrane electrodes when exposed to $Ba(SCN)_2$ or $Ca(SCN)_2$ solutions [78]. It has been verified that both the divalent cation and the monovalent anion give a contribution of around 50 mV, which value roughly corresponds to the Nernstian slope for a monovalent cation. The reason is that both species are part of the permeating complex which indeed represents a monovalent cation. A similar situation was found for uranyl-selective electrodes [35]. These sensors give a highly selective, linear response to UO_2^{2+} ions with a slope of around 60 mV when the aqueous solution is buffered at pH 3. Here, the membrane evidently forms carrier complexes with ion pairs of the type UO_2OH^+, the activity of OH^- ions in the sample being kept constant.

308

12.4.2. Properties of Carrier-Based Liquid Membranes with Incorporated Ion-Exchangers

The preceding treatment has shown that a relatively high number of anionic sites immobilized within the bulk membrane phase is required for rendering a carrier membrane perm-selective and cation-specific. In 1973, Kedem, Perry, and Bloch [55] first reported on this subject and advocated the use of polymeric membrane matrices with fixed negative charges to improve the response characteristics of valinomycin-based membranes. The introduction of such chemically bound sites into the commonly used supporting materials PVC and silicone rubber, however, seems questionable. This led Morf et al. [56, 57] to the intentional creation of mobile anionic sites in carrier membranes, accomplished by the addition of ex-tremely lipophilic anions, such as tetraphenylborate, to the membrane phase. These species are expected to be virtually trapped within the organic phase because of their poor water-solubility; rather than compete with the neutral carriers in complexing the cations they should dislodge the sample anions from the membrane, thereby reducing or even eliminating the effects of anion permeation. In the meantime, this modified membrane type has replaced the conventional carrier membrane in practically all sensors for divalent cations and has led to significant improvements (see below). The incorporation of ionic components into carrier membranes has also proved beneficial in that the membrane resistance can be lowered considerably, which point became crucial for the development of workable microelectrodes based on neutral carriers [79 - 83].

For a theoretical description of carrier-based ion-ex-changer membranes (see also [26, 56, 57]) one can directly apply the former Eqs. (12.50) and (12.51) which account for any type of mobile anionic sites R^-. In contrast to the nearly immobile OH^- groups prevailing in the conventional carrier membranes, the tetraphenylborate or related exchanger ions introduced here can move easily but are confined to the membrane phase because of their high lipophilicity. This means that, ideally, the steady-state flux of mobile ions R^- becomes $J_r = 0$, resulting in a Boltzmann distribution of such components across the membrane. Hence:

$$\frac{k_r \, a'_r}{k_r \, a''_r} = e^{-\frac{F}{RT} E_M} \tag{12.56}$$

where a'_r and a''_r are the activity levels maintained in the aqueous boundaries by interfacial equilibrium. By matching the number of anionic sites in the membrane to that of cation-complexing carriers, it is possible to create situations where

$$a_r(x) = X \approx \frac{z}{n} \, a_s^{tot} \tag{12.57}$$

respectively:

$$\frac{k_y \, a'_y}{k_r \, a'_r} = \frac{a_y(0)}{a_r(0)} \approx 0 \; ; \quad \frac{k_y \, a''_y}{k_r \, a''_r} = \frac{a_y(d)}{a_r(d)} \approx 0 \tag{12.58}$$

The membrane potential is then found from Eqs. (12.50) and (12.56) to obey the simple relation:

$$E_M = \frac{RT}{zF} \ln \frac{K'_{i,n} \, a'_i}{K''_{i,n} \, a''_i} = \frac{RT}{zF} \ln \frac{a'_i}{a''_i} + n \frac{RT}{zF} \ln \frac{a_s(0)}{a_s(d)} \qquad (12.59)$$

A rigorous solution for the residual activities of free carriers can be deduced from Eqs. (12.40) - (12.43); details are given elsewhere [26].

Equation (12.59) clearly demonstrates that the additional incorporation of permanent anions into carrier membranes allows to realize a perfect Nernstian emf-response to cations:

$$E = E^o_i + \frac{RT}{zF} \ln a'_i \qquad (12.60)$$

as long as

$$K'_{i,n} \cong K''_{i,n} \qquad \text{resp.} \qquad a_s(0) \cong a_s(d) \qquad (12.61)$$

Here, any interference by sample anions can be successfully eliminated; the unfavorable region of high anion permeability found earlier (case 2 in Section 12.4.1) does evidently not appear because interfering anions are virtually excluded from the liquid cation-exchanger phase. Transition of the emf-response into an anionic function is obtained only when complete consumption of free carriers occurs, in analogy to the former case 3:

$$E = E^o_y - \frac{RT}{F} \ln a'_y \qquad (12.62)$$

for

$$K'_{i,n} \ll K''_{i,n} \quad \text{resp.} \quad a_s(0) \to 0 \qquad (12.63a)$$

However, a detailed study [26] shows that such a degeneration of the response behavior of carrier-based electrodes proceeds only at comparatively high levels of interfering anions, that is for

$$K_{ex}(a_s^{tot})^{-1} \, a'_i \, a'^z_y \gg (a_s^{tot} - \frac{n}{z} X)^{z-n} \qquad (12.63b)$$

Among other terms, the charge z and stoichiometry n of the predominant cationic complexes are surprisingly found to be the decisive parameters. It may be recognized that the permanent incorporation of anionic components such as tetra-phenylborate ($\frac{n}{z} X \to a_s^{tot}$, see Eq. (12.57)) is attractive for carrier membrane systems having n=z, and especially for n>z where distortions from the ideal cation response can be eliminated completely. These theoretical expectations were entirely confirmed by experiment. For liquid ion-exchanger membrane electrodes with carriers for K^+ (1:1 complexes), Ca^{2+} (1:2 complexes), and Na^+ (1:2 complexes), the observed emf-response was found to be in perfect agreement with the predicted function (see Figures 12.11 and 12.12 and References 56 and 57). These results clearly encourage the use of the modified carrier-membrane type, especially in sensors for divalent cations which require the presence of polar, highly solvating membrane media.

Through the addition of anionic components to carrier membranes, the selectivity behavior among cations is also

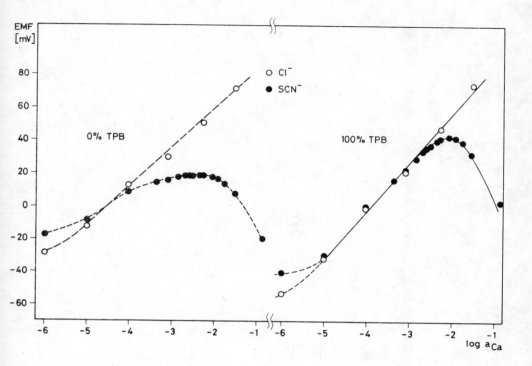

<u>Figure 12.11.</u> EMF response of neutral carrier electrodes to Ca^{2+} in samples containing different anions (25°C) [56]. Left: conventional membrane type. Right: membrane with tetraphenylborate (equivalent to 100 mol-% of the ligand concentration); solid curves according to theory [26, 56]. Carrier: <u>15</u> (Fig. 12.1); membrane solvent: o-nitrophenyl octyl ether.

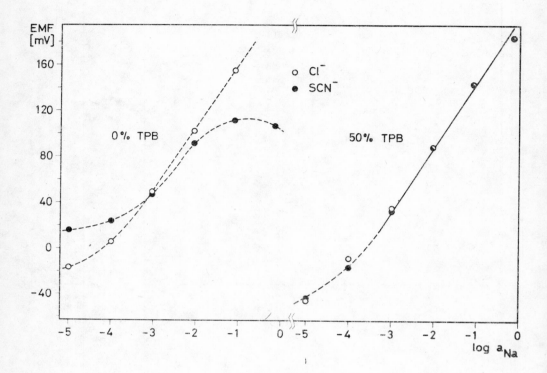

Figure 12.12. EMF response of neutral carrier electrodes to
Na$^+$ in samples containing different anions (25°C) [57]. Left:
conventional membrane. Right: membrane with tetraphenylborate
(equivalent to 50 mol-% of the ligand concentration); solid
line according to Eq. (12.60). Carrier 14 (Fig. 12.1); membrane
solvent: o-nitrophenyl octyl ether.

changed to some extent. By this treatment the activity X of
ionic charges in the membrane phase is directly increased and,
simultaneously, the activity a_s of free ligands is reduced.
It is readily seen from Eqs. (12.31) and (12.34) (assuming
$\bar{n}_i = \bar{n}_j$) that the preference of divalent over monovalent
cations is thereby increased substantially, whereas the selectivity between metal ions of the same charge remains unaffected. A pronounced reduction of the selectivity coefficient is
expected for hydrogen ions because the OH^- groups residing in
the conventional membranes are replaced here by tetraphenylborate or similar ions, having much lower basicity. These theoretical results compare favorably with the observed effects.
For example, Table 12.4 demonstrates that the introduction of
negatively charged membrane components leads to a perfection
of Ca^{2+}-selective electrodes based on neutral carriers. In
sensors for monovalent cations, similar improvements of
nearly all response characteristics can be realized by the
use of relatively nonpolar membrane materials.

12.5. MOLECULAR ASPECTS OF CATION-SELECTIVE CARRIERS

12.5.1. Molecular Basis of Ion Selectivity

In the end of the sixties, the known number of electrically
neutral ionophores was still restricted to a few naturally
occurring antibiotics and to the class of macrocyclic polyethers (crown compounds). Further attempts at imitating by
chance the structural principle of natural ionophores were
not successful [85]. Progress on the design of synthetic ion-
carriers was primarily due to the development of molecular-
level theories and models [16 - 18, 20 - 23, 33, 86, 87]. The
theory worked out by Morf and Simon [20 - 22] was successful
at reconstructing the observed selectivities of a variety of
antibiotics and model compounds, as well as ionic hydration

Table 12.4. Effect of the membrane component tetraphenyl-borate on the selectivity of Ca^{2+}-carrier membrane electrodes

Membrane composition[a]	Selectivity factors, $\log K_{CaM}^{Pot}$ [b]			
	M = Na^+	K^+	Mg^{2+}	H^+
ETH 1001 1% o-NPOE 66% PVC 33%	-4.6	-4.6	-5.2	1.2
ETH 1001 1% NaTPB 0.5% o-NPOE 64% PVC 34.5%	-5.0	-5.2	-5.1	0.0

[a] ETH 1001: carrier 8 in Figure 12.1; o-NPOE: o-nitrophenyl octyl ether; NaTPB: sodium tetraphenylborate.

[b] Determined by the fixed interference method (Na^+, K^+: 1.0 M; Mg^{2+}: 0.33 M; H^+: pH 2) [84].

energies, and was capable of predicting molecular requirements for electrically neutral ionophores. The selectivity behavior of carrier membranes was interpreted on the basis of specific effects on the free energies of transfer of cations:

$$- RT \ln (\beta^{W}_{is} k_{is}) = \Delta G^{O}(\text{transfer}) = \Delta G^{O}_{L} - \Delta G^{O}_{H} \qquad (12.64)$$

where: ΔG^{O}_{L}: free energy for transfer of a cation from the gas phase into the cavity formed by the ligand (in a given membrane medium)

ΔG^{O}_{H}: free energy of hydration, i. e., for transfer of a cation from the gas phase to water

Each of these free energies reflects the following contributions [20 - 22]:

a) Disengagement of the coordinating sites in the ligand or in water.

b) Electrostatic, dispersion, and repulsive interactions between the cation and the coordinating shell of ligand or water.

c) Polarization of the surrounding solvent medium by the charged complex.

d) Changes in the volume of solvent induced by the cation.

e) Correction for relating calculated quantities to standard states.

f) Adoption of a conformation of the ligand suitable for cation coordination.

g) Additional deformation of the ligand to optimally fit a given cation by the coordinating sites.

h) A statistical term depending on the number of coordinating sites in the ligand.

Among the various interactions existing between hard cations and electrically neutral ligands [20 - 22], the electrostatic bonds built up between the ionic charge and ligand dipoles are primarily responsible for the formation of stable complexes. For cations coordinated by n binding sites, these interactions are calculated as

$$- E = N \cdot n \frac{z \, e \, p}{r^2} - N \cdot b \frac{p^2}{r^3} \qquad (12.65)$$

where: - E: absolute energy value of ion-dipole interactions
(first term) and dipole-dipole interactions (second term), per mol of complex

ze: ionic charge, in electrostatic units

p: permanent dipole moment, in electrostatic units

r: distance between the centers of ion and dipole

b: numerical factor depending on the coordination geometry [20]

N: Avogadro's number

Calculated interaction energies are given in Figure 12.13 for idealized complexes of different cations. The data indicate that an increase in the dipole moment of the ligand sites generally increases the stability of the complexes. The effect is especially large for small and multiply charged cations. Therefore polar binding sites, such as amide groups, have to be introduced in carriers to compete with water molecules in the complexation of the small ions Ca^{2+}, Li^+, and Na^+ (see Section 12.5.2). High selectivities between different cations are to be expected for polydentate ligands that offer a predetermined coordination sphere. A coordination number of about eight is especially attractive for complexing Ca^{2+}, as shown in Figure 12.13.

318

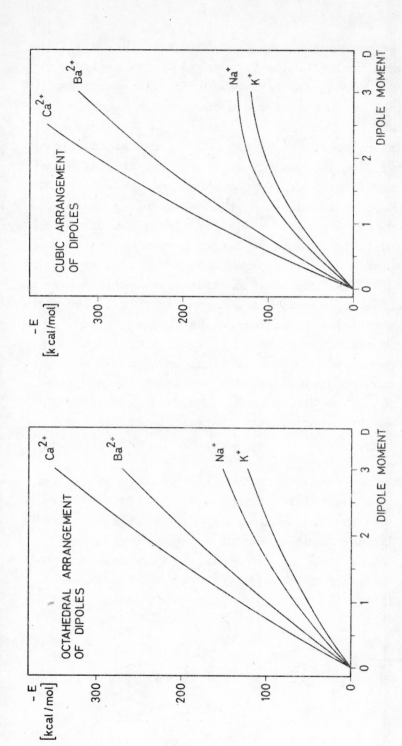

Figure 12.13. Energy of the electrostatic interactions between cation and coordinating ligand sites (van der Waals radius 1.40 Å; varying dipole moment). The values -E were calculated from Eq. (12.65) for complexes of Na$^+$ (ionic radius 0.98 Å), K$^+$ (1.33 Å), Ca^{2+} (1.06 Å), and Ba^{2+} (1.43 Å). Left: octahedral coordination. Right: cubic coordination.

319

The salient characteristic of most carriers, the ability to distinguish among the cations of one group of the periodic system, is largely due to intra-ligand steric interactions. First, the mutual repulsion of the n coordinating atoms a priori precludes their contact with cations smaller than a certain critical radius. This minimal cavity radius is determined as 0.6 and 1.0 Å for coordination spheres of six (octahedral) and eight (cubic) oxygen atoms, respectively [20 - 22]. Accordingly, an eight coordination can hardly be realized for Li^+ or Mg^{2+}. Second, the reduced flexibility of the ligand skeleton often entails a somewhat larger cavity which, as a rule, also increases with increasing number of coordinating sites [16, 21, 22]. The cation which best fits into the offered cavity is finally preferred since the complexation of any other ion involves an increase in conformational energy. A contrasting view on the selectivity of carrier antibiotics was suggested by Krasne and Eisenman [87] who came to the conclusion that carbonyl groups make for a selectivity sequence $K^+ > Rb^+ > Cs^+ > Na^+ > Li^+$. However, such binding sites are found in a wealth of synthetic carriers, the selectivities being quite different.

Another important contribution to the stability of carrier complexes in membranes results from the electrostatic interaction between the charged complexes and the surrounding membrane solvent. If the membrane medium is approximated by a structureless dielectric, one can apply the Born equation to describe the polarization induced by the complex ions (see also Eq. (12.28)):

$$-\Delta G_B = N \cdot \frac{(z\ e)^2}{2\ r_C} \left(1 - \frac{1}{\varepsilon}\right) \qquad (12.66)$$

$$r_C = r_{Ion} + s$$

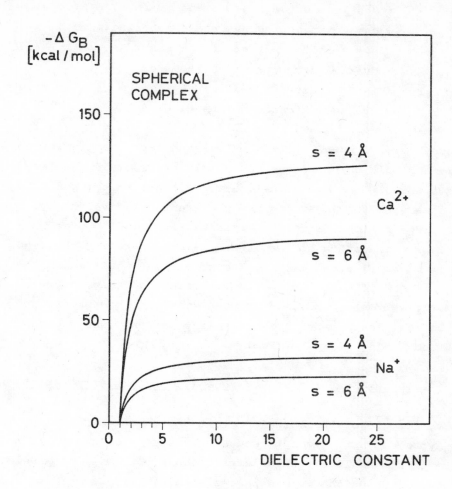

Figure 12.14. Free energy of the electrostatic interactions between cationic complex and membrane solvent. The values $-\Delta G_B$ were estimated for two metal ions of nearly the same size but of different charge, for two values of the ligand shell thickness s, and for a varying dielectric constant of the membrane medium (see Eq. (12.66)).

where: $-\Delta G_B$: absolute value of free energy for transfer of a
complex from the gas phase into the membrane
ϵ: dielectric constant of the membrane phase
$2\ r_C$: overall diameter of the complex
r_{Ion}: ionic radius
s: average thickness of the ligand layer

Figure 12.14 shows ΔG_B-values calculated for a varying ϵ; so-
dium- and calcium-complexes having diameters of 10 and 14 Å
were studied. The theoretical curves illustrate the point:
the stratagem for enhancing the preference of divalent over
monovalent cations of the same radius (e. g., Ca^{2+} over Na^+)
involves a reduction of the ligand layer separating the central
ion from the external medium. The effect is magnified in solvents of high dielectric constant. Correspondingly, nitroaromatic solvents were used as polar membrane components in
all carrier-based sensors for divalent cations (see Figure
12.6 and Tables 12.1 and 12.3).

12.5.2. Design Features of Membrane-Active Complexing Agents

The effects of the more important design parameters may
nicely be demonstrated by discussing the development of ionophore 8 (Figure 12.1); this carrier shows extremely high selectivity for Ca^{2+} ions in membrane systems [8, 43, 84, 88].
Such 3,6-dioxaoctanedioic diamides form 1:2 Ca^{2+}-carrier
complexes [56, 67, 89, 90]. As expected from theoretical considerations [20, 21, 33] calcium is coordinated by eight
oxygen atoms (two ether groups and two amide groups per ligand) [89, 90]; the ester carbonyl groups of the side chains
do not participate in the coordination of the cation [90, 91].
The arrangement of the four coordinating atoms in each ligand
allows the formation of five-membered chelate rings in the
complex. The high polarity of the amide carbonyl coordinating
sites ensures sufficient strength of interaction of the ligands

322

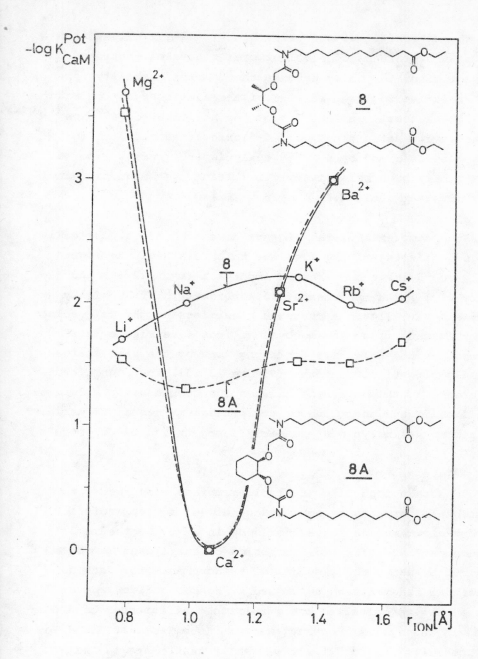

Figure 12.15. Influence of the average thickness of the ligand layer around the metal cation on the ion selectivity of liquid membrane electrodes.

with the cation. Simultaneously, highly polar binding sites
lead to a preference of divalent over monovalent cations of
the same size; they also enhance the complex stability for
small relative to large cations of the same charge (see Figure
12.13). The nearly perfect fit of the nonsolvated Ca^{2+} ion
in the cavity defined by two 3,6-dioxaoctanedioic diamides,
which was confirmed by a X-ray analysis [90], gives rise to
an extremely high selectivity for calcium over magnesium and
other divalent cations.

As the complexes of the ligands discussed are electrically
charged, effects arising from the thickness s of the ligand
shell around the cation become important (Eq. (12.66) and
Figure 12.14). An increase of this parameter in the highly
Ca^{2+}-selective ligand $\underline{8}$ causes a clear loss in the Ca^{2+} selectivity in respect to the monovalent ions (see Figure 12.15).
As expected the selectivities among ions of the same charge
are only slightly influenced (Figure 12.15). The significant
difference in the bulkiness between the complexes of calcium-
specific ligands and those of the potassium-selective carrier
antibiotic valinomycin are clearly demonstrated by CPK models
[33].

To ensure a long life of liquid membrane electrodes, the
distribution of complex and ligand between aqueous solution
and organic membrane phase should be in favor of the membrane.
A rough estimate of the distribution of the ligand may be
obtained by using the lipophilicity increments for various
structural fragments as described by Hansch and coworkers [92].
This allows to calculate the partition coefficient P of a compound in the system 1-octanol-water[*)] (see Table 12.5 and References 16, 33, and 92). It was shown that there is indeed

---

[*)] The parameter P is identical to k_s for membranes prepared
from 1-octanol.

Table 12.5. Lipophilicities (log P) calculated for various carriers

Ligand (Figure 12.1)		Lipophilicity, log P [a]
1	valinomycin	-0.4[b]
2	nonactin	7.8
7	ETH 157	8.3
8	ETH 1001	8.5
9	ETH 149	4.6
10	ETH 231	6.7
11	nonylphenoxy-polyethyleneglycol	6.9
12	dimethyldibenzo-30-crown-10	4.6
13	ETH 227	6.4

[a]Calculated from the increments given in table 6 of Reference 33.

[b]The actual lipophilicity of valinomycin is probably much higher because the polar ligand groups take part in intramolecular hydrogen bonds which force the macrocyclic ligand into a highly lipophilic conformation [95].

a correlation between the lipophilicity of a carrier, characterized by log P, and the observed lifetime of the corresponding PVC liquid membrane electrodes [93]. For the investigated
types of electrically neutral ligands, high lifetime of the
membrane electrodes was achieved by using carrier molecules
having log P values above about 4 (Table 12.5). Some compounds
with extremely high lipophilicity (log P $\geqslant$ 15), obtained by
simply increasing the length of the hydrocarbon side chains in
3,6-dioxaoctanedioic diamide-type ligands, did no longer induce
cation selectivity in membrane electrodes [94]. As these ligands and highly selective carriers show similar complexing
and ion-extracting properties, the lack of ion selectivity in
membranes is probably caused by kinetic effects [94]. The observed loss of ionophoric behavior and the resulting irrever-
sibility of the membrane electrode action may be ascribed to
substantial kinetic limitations in the interfacial reaction
(12.18).

The design of a series of further ligands selective for
Li$^+$, Na$^+$, and Ba^{2+} was achieved by a stepwise optimization of
the ligand structure. Correlations of the type given in Figure
12.16 can be applied to this end. Unfortunately, every change
of constitution of the ligand affects several molecular parameters simultaneously, such as dipole moment and polarizability
of the ligand groups, conformation of the molecule, and ligand
thickness. In Figure 12.16 selectivity factors for a number of
compounds are plotted in the order of decreasing polarity (decreasing with the expected polar substituent constant) of the
substituents attached to the carbon atoms carrying the two
ether oxygen atoms [88]. According to model calculations, an
increase of the dipole moment of the ligand groups[*] is expected

[*] This holds only for a given complex stoichiometry. It might
be expected that the basicities of the ether oxygen atoms of the
compounds shown in Figure 12.16 would increase from left to
right [88].

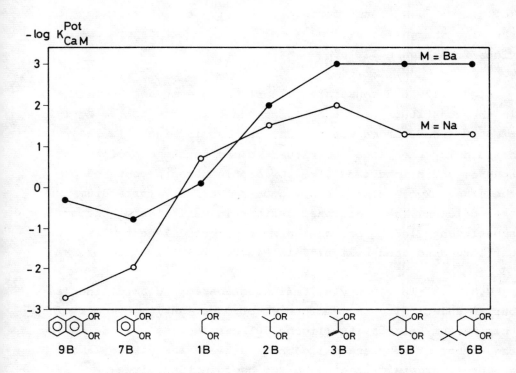

<u>Figure 12.16.</u> Influence of the constitution of ligands on the selectivity of the corresponding liquid membrane electrodes [88]. The ligand denoted here as 3B is identical to <u>8</u> in Figure 12.1.

to result in a selectivity enhancement for divalent relative
to monovalent cations when the two are of the same radius
(e. g., Ca^{2+}, Na^+) and for small relative to large cations of
the same charge (e. g., Ca^{2+}, Ba^{2+}); these trends are obvious from Figure 12.16.

Figure 12.16 demonstrates clearly that even slight changes
in the constitution of the ligand can shift the ion selectivities by orders of magnitude. One carrier in Figure 12.16, which
has a bulky phenylene substituent close to the coordination
sphere, and at the same time low dipole moments centered on
the ether oxygen atoms, shows good selectivity for sodium
over calcium. The additional systematic variation of the amide
substituents led to one of the most successful neutral sodium
carriers described so far, 7 in Figure 12.1 [8, 24].

N,N,N',N'-Tetraphenyl-3,6-dioxaoctanedioic diamide induces
barium selectivity in membrane electrodes. An improvement of
this barium selectivity (ligand 10) was achieved by increasing
the number of ethylene oxide units between the diphenylamide
groups (increase of the number of coordinating sites per ligand, see Figure 12.17).

Lithium-selective ligands were obtained by increasing the
chain length between the ether groups of 15 by one carbon
atom. The most promising representative of these 3,7-dioxa-
nonanedioic diamides is 9. This ligand forms only a 1:1 complex
with lithium with ethyl alcohol as solvent [39]. The reason
for the somewhat surprising lithium ion selectivity may reside in the more favorable arrangement of the four coordinating
oxygen atoms around the small lithium ion.

Carrier 13, which is based on the skeleton of the lithium-
selective ligands but contains two more potential coordinating
sites, shows an especially strong discrimination of large al-

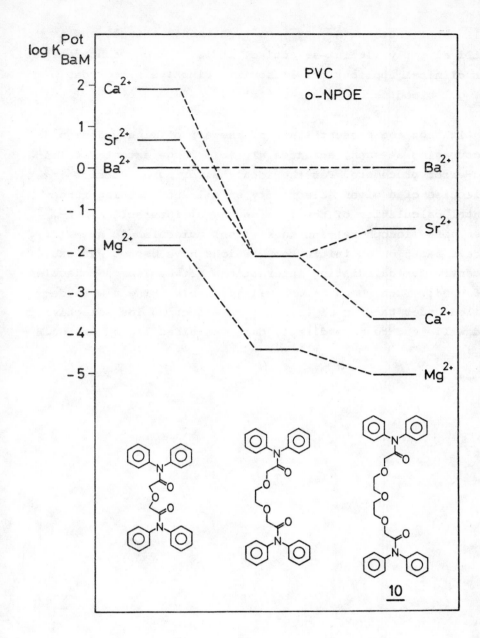

<u>Figure 12.17.</u> Influence of the number of ethyleneoxide units between the diphenylamide groups on the ion selectivity of the corresponding liquid membrane electrodes.

kali metal cations (K^+, Cs^+, Rb^+). Although this ligand is still slightly lithium-selective, it is a strong candidate for clinical applications in sodium-selective electrodes (e. g., blood serum measurements) [27, 40].

Although the present state of the art of ligand design is encouraging as such, advances are to a large extent a trial-and-error procedure. For the ideal tailoring of carrier molecules of a given selectivity an all-encompassing fundamental calculation of the free energy of interaction between host (ionophore) and guest (ion) molecules is necessary. Models based on ab initio computations have become available recently for calculating interactions between large molecules [96 - 98]. Such computations will probably bring a more detailed understanding of the factors affecting ion selectivity and will lead to more direct, computer-aided ligand design.

REFERENCES

[1] Yu. A. Ovchinnikov, V. T. Ivanov, and A. M. Shkrob, BBA Library 12, Membrane-Active Complexones, Elsevier, Amsterdam, 1974.

[2] C. Moore and B. C. Pressman, Biochem. Biophys. Res. Commun. 15, 562 (1964).

[3] Z. Štefanac and W. Simon, Chimia 20, 436 (1966); Microchem. J. 12, 125 (1967).

[4] R. A. Durst, in Ion-Selective Electrodes (R. A. Durst, ed.), Nat. Bur. of Standards Spec. Publ. 314, Washington, 1969.

[5] L. A. R. Pioda, V. Stankova, and W. Simon, Anal. Lett. 2, 665 (1969).

[6] M. S. Frant and J. W. Ross, Jr., Science 167, 987 (1970).

[7] W. E. Morf, Ch. U. Züst, and W. Simon, in Molecular Mechanisms of Antibiotic Action on Protein Biosynthesis and Membranes (E. Munoz, F. Garcia-Ferrandiz, and D. Vasquez, eds.), Proceedings of a Symposium held at the University of Granada (Spain), June 1 - 4, 1971, Elsevier Scientific Publishing Co., Amsterdam, 1972, p. 523.

[8] D. Ammann, R. Bissig, Z. Cimerman, U. Fiedler, M. Güggi, W. E. Morf, M. Oehme, H. Osswald, E. Pretsch, and W. Simon, in Ion and Enzyme Electrodes in Biology and Medicine (M. Kessler, L. C. Clark, Jr., D. W. Lübbers, I. A. Silver, and W. Simon, eds.), Urban & Schwarzenberg, Munich, Berlin, Vienna, 1976, p. 22.

[9] U. Fiedler and J. Růžička, Anal. Chim. Acta 67, 179 (1973).

[10] J. Pick, K. Tóth, E. Pungor, M. Vašák, and W. Simon, Anal. Chim. Acta 64, 477 (1973).

[11] O. Kedem, E. Loebel, and M. Furmansky, Ger. Offen. 2027128, Dec. 1970; O. Kedem, M. Furmanski, E. Loebel, S. Gordon, and R. Bloch, Isr. J. Chem. 7, 87p (1969).

[12] I. H. Krull, C. A. Mask, and R. E. Cosgrove, Anal. Lett. 3, 43 (1970).

[13] O. H. LeBlanc, Jr., and W. T. Grubb, Anal. Chem. 48, 1658 (1976).

[14] C. J. Pedersen, J. Am. Chem. Soc. 89, 2495 (1967); 89 7017 (1967); 92, 386 (1970); 92, 391 (1970).

[15] J. Petránek and O. Ryba, Anal. Chim. Acta 72, 375 (1974); J. Electroanal. Chem. 44, 425 (1973).

[16] J.-M. Lehn, Struct. Bonding 16, 1 (1973).

[17] J.-M. Lehn and J. P. Sauvage, Chem. Commun. 1971, 440.

[18] B. Dietrich, J.-M. Lehn, and J. P. Sauvage, Tetrahedron Letters 34, 2885 (1969).

[19] D. Ammann, E. Pretsch, and W. Simon, Tetrahedron Letters 24, 2473 (1972).

[20] W. E. Morf and W. Simon, Helv. Chim. Acta 54, 794 (1971).

[21] W. E. Morf and W. Simon, Helv. Chim. Acta 54, 2683 (1971).

[22] W. Simon, W. E. Morf, and P. Ch. Meier, Struct. Bonding 16, 113 (1973).

[23] E. Pretsch, D. Ammann, and W. Simon, Research/Development 25 (3), 20 (1974).

[24] W. Simon, E. Pretsch, D. Ammann, W. E. Morf, M. Güggi, R. Bissig, and M. Kessler, Pure Appl. Chem. 44, 613 (1975).

[25] W. E. Morf and W. Simon, Hung. Sci. Instr. 41, 1 (1977).

[26] W. E. Morf and W. Simon, in Ion-Selective Electrodes in Analytical Chemistry (H. Freiser, ed.), Plenum, New York, 1978.

[27] P. C. Meier, D. Ammann, W. E. Morf, and W. Simon, in Medical and Biological Applications of Electrochemical Devices (J. Koryta, ed.), Wiley, in press.

[28] R. P. Buck, Anal. Chem. 44, 270R (1972); 46, 28R (1974).

[29] R. P. Buck, Anal. Chem. 48, 23R (1976).

[30] R. P. Buck, Anal. Chem. 50, 17R (1978).

[31] J. Koryta, Ion-Selective Electrodes, Cambridge University Press, Cambridge, London, New York, and Melbourne, 1975.

[32] J. Koryta and M. Březina, in Electroanalytical Chemistry (A. J. Bard, ed.), Marcel Dekker, New York, in press.

[33] W. E. Morf, D. Ammann, R. Bissig, E. Pretsch, and
W. Simon, in Progress in Macrocyclic Chemistry, Vol. 1
(R. M. Izatt and J. J. Christensen, eds.) Wiley-Inter-
science, New York, 1979.

[34] J. Schneider, Dissertation ETHZ 6239, Juris, Zürich, 1978.

[35] J. Šenkyr, D. Ammann, P. C. Meier, W. E. Morf, E. Pretsch,
and W. Simon, Anal. Chem. 51, 786 (1979).

[36] D. J. Cram, R. C. Helgeson, L. R. Sousa, J. M. Timko,
M. Newcomb, P. Moreau, F. deJong, G. W. Gokel, D. H. Hoff-
man, L. A. Domeier, S. C. Peacock, K. Madan, and
L. Kaplan, Pure Appl. Chem. 43, 327 (1975).

[37] D. Bedeković, Dissertation ETHZ No. 5777, Juris, Zürich,
1976.

[38] A. P. Thoma, Z. Cimerman, U. Fiedler, D. Bedeković,
M. Güggi, P. Jordan, K. May, E. Pretsch, V. Prelog, and
W. Simon, Chimia 29, 344 (1975).

[39] M. Güggi, U. Fiedler, E. Pretsch, and W. Simon, Anal.
Lett. 8 (12), 857 (1975).

[40] M. Güggi, M. Oehme, E. Pretsch, and W. Simon, Helv.
Chim. Acta 58, 2417 (1976).

[41] R. P. Scholer and W. Simon, Chimia (Switzerland) 24,
372 (1970).

[42] R. E. Cosgrove, C. A. Mask, and I. H. Krull, Anal. Lett.
3, 457 (1970).

[43] D. Ammann, M. Güggi, E. Pretsch, and W. Simon, Anal.
Lett. 8, 709 (1975).

[44] E. W. Baumann, Anal. Chem. 47, 959 (1975).

[45] R. J. Levins, Anal. Chem. 43, 1045 (1971); 44, 1544
(1972).

[46] A. M. Y. Jaber, G. J. Moody, and J. D. R. Thomas,
Analyst 101, 179 (1976).

[47] M. Güggi, E. Pretsch, and W. Simon, Anal. Chim. Acta 91,
107 (1977).

[48] J. Růžička, E. H. Hansen, and J. Chr. Tjell, Anal. Chim.
Acta 67, 155 (1973).

[49] T. Treasure, Intens. Care Med. 4, 83 (1978).

[50] A. P. Thoma, A. Viviani-Nauer, S. Arvanitis, W. E. Morf, and W. Simon, Anal. Chem. 49, 1567 (1977).

[51] S. M. Ciani, G. Eisenman, and G. Szabo, J. Membr. Biol. 1, 1 (1969).

[52] G. Eisenman, in Ion-Selective Electrodes (R. A. Durst, ed.), Nat. Bur. Stand. (U.S.), Spec. Publ. 314, Washington, D. C.,1969.

[53] S. M. Ciani, G. Eisenman, R. Laprade, and G. Szabo, in Membranes, Vol. 2 (G. Eisenman, ed.), Marcel Dekker, New York, 1973.

[54] J. H. Boles and R. P. Buck, Anal. Chem. 45, 2057 (1973).

[54a] S. B. Lewis and R. P. Buck, Anal. Lett. 9, 439 (1976).

[55] O. Kedem, M. Perry, and R. Bloch, IUPAC International Symposium on Selective Ion-Sensitive Electrodes, paper 44, Cardiff, 1973.

[56] W. E. Morf, G. Kahr, and W. Simon, Anal. Lett. 7, 9 (1974).

[57] W. E. Morf, D. Ammann, and W. Simon, Chimia (Switzerland) 28, 65 (1974).

[58] H. Seto, A. Jyo, and N. Ishibashi, Chem. Lett. (Jpn.) 483 (1975).

[59] H.-R. Wuhrmann, W. E. Morf, and W. Simon, Helv. Chim. Acta 56, 1011 (1973).

[60] W. E. Morf, P. Wuhrmann, and W. Simon, Anal. Chem. 48, 1031 (1976).

[61] O. Ryba and J. Petránek, J. Electroanal. Chem. 67, 321 (1976).

[62] R. Büchi, E. Pretsch, W. E. Morf, and W. Simon, Helv. Chim. Acta 59, 2407 (1976).

[63] T. Higuchi, C. R. Illian, and J. L. Tossounian, Anal. Chem. 42, 1674 (1970).

[64] R. Scholer, Dissertation ETH, Zürich, 1972.

[65] W. E. Morf, Anal. Chem. 49, 810 (1977).

[66] S. G. A. McLaughlin, G. Szabo, S. Ciani, and G. Eisenman, J. Membrane Biol. 9, 3 (1972).

[67] N. N. L. Kirsch and W. Simon, Helv. Chim. Acta 59, 357 (1976); N. N. L. Kirsch, Dissertation ETH, Zürich, 1977.

[68] E. Eyal and G. A. Rechnitz, Anal. Chem. 43, 1090 (1971).

[69] G. Szabo, G. Eisenman, R. Laprade, S. M. Ciani, and S. Krasne, in Membranes, Vol. 2 (G. Eisenman, ed.), Marcel Dekker, New York, 1973.

[70] W. E. Morf, D. Ammann, E. Pretsch, and W. Simon, Pure Appl. Chem. 36, 421 (1973).

[71] H. K. Frensdorff, J. Am. Chem. Soc. 93, 600 (1971).

[72] G. A. Rechnitz and E. Eyal, Anal. Chem. 44, 370 (1972).

[73] U. Fiedler, Anal. Chim. Acta 89, 111 (1977).

[74] W. Simon, W. E. Morf, E. Pretsch, and P. Wuhrmann, in Calcium Transport in Contraction and Secretion (E. Carafoli, ed.), North-Holland, Amsterdam, 1975.

[75] S. Lal and G. D. Christian, Anal. Lett. 3, 11 (1970).

[76] A. Hulanicki, private communication.

[77] E. Lindner, P. Wuhrmann, W. Simon, and E. Pungor, in Ion-Selective Electrodes (E. Pungor and I. Buzás, eds.), Akadémiai Kiadó, Budapest, 1977.

[78] M. Güggi, Dissertation ETH, Zürich, 1977.

[79] R. C. Thomas, W. Simon, and M. Oehme, Nature (London) 258, 754 (1975).

[80] M. Oehme, M. Kessler, and W. Simon, Chimia (Switzerland) 30, 204 (1976).

[81] U. Heinemann, H. D. Lux, and M. J. Gutnick, Exp. Brain Res. 27, 237 (1977).

[82] M. Oehme and W. Simon, Anal. Chim. Acta 86, 21 (1976).

[83] M. Oehme, Dissertation ETH, Zürich, 1977.

[84] W. Simon, D. Ammann, M. Oehme, and W. E. Morf, Ann. New York Acad. Sci. 307, 52 (1978).

[85] R. Schwyzer, Aung Tun-Kyi, M. Caviezel, and P. Moser, Helv. Chim. Acta 53, 15 (1970).

[86] H. Diebler, M. Eigen, G. Ilgenfritz, G. Maass, and
R. Winkler, Pure Appl. Chem. 20, 93 (1969).

[87] S. Krasne and G. Eisenman, section V of ref. [69].

[88] D. Ammann, R. Bissig, M. Güggi, E. Pretsch, W. Simon,
I. J. Borowitz, and L. Weiss, Helv. Chim. Acta 58, 1535
(1975).

[89] R. Büchi and E. Pretsch, Helv. Chim. Acta 58, 1573 (1975).

[90] K. Neupert-Laves and M. Dobler, Helv. Chim. Acta 60,
1861 (1977).

[91] R. Büchi, E. Pretsch, and W. Simon, Helv. Chim. Acta
59, 2327 (1976).

[92] A. Leo, C. Hansch, and D. Elkins, Chem. Rev. 71, 515
(1971).

[93] E. Pretsch, R. Büchi, D. Ammann, and W. Simon, in
Analytical Chemistry, Essays in Memory of Anders Ring-
bom (E. Wänninen, ed.), Pergamon Press, Oxford and
New York, 1977, p. 321.

[94] R. Bissig, U. Oesch, W. E. Morf, E. Pretsch, and
W. Simon, Helv. Chim. Acta 61, 1531 (1978).

[95] W. L. Duax, H. Hauptman, C. M. Weeks, and D. A. Norton,
Science 176, 911 (1972).

[96] E. Clementi, Lecture Notes in Chemistry, Vol. 2
(G. Berthier et al., eds.), Springer-Verlag, Berlin,
1976.

[97] E. Clementi, F. Cavallone, and R. Scordamaglia, J. Am.
Chem. Soc. 99, 5531 (1977).

[98] G. Corongiu, E. Clementi, E. Pretsch, and W. Simon,
J. Chem. Phys. 70, 1266 (1979).

CHAPTER 13

GLASS ELECTRODES

13.1. INTRODUCTION

In 1902 Bernstein [1] founded a modern membrane theory on
the bioelectrical phenomena of cells and tissues. His ingenious hypotheses gave much impetus to research on model
systems, based on various types of compact or porous membranes. Only a few years later, Cremer [2] experimented with
electrochemical cells using interposed thin glass layers. He
then observed an electrical potential difference which was
very sensitive to changes in the acidity of the aqueous solutions. The discovery of the pH-glass electrode was completed by
Haber and Klemensiewicz [3] who demonstrated more systematically the hydrogen-ion response of glass membrane electrode
assemblies. The commercial production of pH glasses, however,
was started only around 1930 when the first comparative studies of different glass compositions were terminated [4, 5].
MacInnes and Dole [5] recommended the composition 22% Na_2O-
6% CaO-72% SiO_2, which was brought into production by Corning
Glass Works under the code number 015. During the next decade,
the superiority of lithia glasses for pH measurements was
discovered [6], and further improvements were achieved by the
introduction of additional glass-modifying components, such
as BaO [7], Cs_2O, and La_2O_3 [8] (see also Table 13.1).

The significance of glass composition in respect to the
ion sensitivity, the electrical resistivity, and the chemical
durability of glass electrodes was soon recognized. Some of
the earliest studies [4, 10-12] already indicated the existence
of a glass electrode response to species other than hydrogen
ions. Thus the introduction of Al_2O_3 or B_2O_3 into a glass was

337

Table 13.1. Glass compositions recommended for pH- and cation-selective glass electrodes [9]

Ion to be determined	Glass composition (in mol%)	Selectivity	References
H^+	Li_2O, Cs_2O, La_2O_3, SiO_2, CaO, BaO	$K_{HNa}^{Pot} \approx 10^{-13}$	6-8
Li^+	15% Li_2O, 25% Al_2O_3, 60% SiO_2	$K_{LiNa}^{Pot} \approx 0.3$; $K_{LiK}^{Pot} < 10^{-3}$	
Na^+	11% Na_2O, 18% Al_2O_3, 71% SiO_2	$K_{NaK}^{Pot} \approx 10^{-3}$	15-17
K^+	27% Na_2O, 5% Al_2O_3, 68% SiO_2	$K_{KNa}^{Pot} \approx 10^{-1}$	15-17
Ag^+	11% Na_2O, 18% Al_2O_3, 71% SiO_2	$K_{NaAg}^{Pot} \approx 10^{3}$	

found to cause a considerable loss of H^+-specificity, respectively an increased sensitivity towards Na^+ and other alkali ions. These observations were confirmed and extended by Lengyel and Blum [13], Nicolsky and Tolmacheva [14], and Eisenman et al. [15]. Eisenman was primarily interested in the development of analytically useful glass electrodes for the measurement of Na^+ and K^+, respectively. Glass composition with optimized selectivities for Na^+ or K^+ ions were then realized by a systematic study of ternary glasses of the type $Na_2O-Al_2O_3-SiO_2$ [15] (see also Table 13.1).

A comprehensive introduction into the principles and the practice of glass electrodes is given in the book "Glass Electrodes for Hydrogen and Other Cations", edited by G. Eisenman [16], as well as in the references found therein. The fundamental contributions by Isard [17] and by Nicolsky and his colleagues [18] cover in great detail the historical developments and the chemical (structural) aspects of glass electrodes. Therefore, we shall focus here on the theoretical aspects and the underlying principles of these systems.

Since the nineteen thirties, a considerable number of different theories was devoted to the interpretation of glass membrane potentials (for a review, see also [16 - 19]). These approaches may be classified, according to their conceptual features, into the following categories:
a) Simple ion-exchange theories, assuming homogeneous properties and idealized behavior of the glass membrane [20-22].
b) Modifications correcting for the nonideality of the glass phase [15, 23-28], preferably by invoking an n-type description of ion activities [15, 29].
c) Solid-state approaches that account for multiple cation-exchange sites of different bonding strengths (heterogeneous-site glasses) [19, 30].

d) Theories related to the concepts for liquid membranes, treating anionic sites (vacancies) as discrete ligands for cations [31-33].

To elucidate the parallels as well as the discrepancies between specific models, a unified description of the glass membrane potential has been attempted [34]. Nearly all of the ion-exchange theories or solid-state approaches cited under a)-c) follow from this derivation and essentially correspond to special cases (see Section 13.2). A different extension of glass electrode theory can be obtained on the basis of a liquid-membrane approach (see d) and Section 13.5). Both, solid-state and liquid-membrane, approaches are realistic for glass electrodes and lead to rather universal formulae for the emf-response of these devices.

13.2. ION-EXCHANGE THEORIES AND n-TYPE DESCRIPTIONS OF GLASS MEMBRANE POTENTIALS

The ion-specific behavior of glass electrodes is largely determined by the cation-exchange equilibria established between the sites R_i^- of the glass phase (e. g., $(SiO_{3/2})O^-$ or $(AlO_{4/2})^-$) and the external solution. Here we focus on the selectivity between hydrogen ions H^+ and metal ions M^+, which is governed by equilibria of the type (13.1) [*]:

---

[*] The theoretical formalism evolved in this chapter can be applied to any pair of monovalent cations by simply inserting the corresponding symbols instead of H^+ and M^+. For simplicity, the subscript HM to the ion-exchange constant K_i has been omitted.

340

$$HR_i \text{(membrane)} + M^+ \text{(solution)} \underset{}{\overset{K_i}{\rightleftharpoons}} MR_i \text{(membrane)} + H^+ \text{(solution)} \qquad (13.1)$$

By following the suggestion of Rothmund and Kornfeld [29], which was referred to as n-type behavior of solid ion-exchangers [15, 16, 25-27] and was corroborated by empirical [15, 16, 24] and theoretical studies on glass membranes [23], we obtain the following expression for the law of mass action:

$$K_i = \frac{a_H' \, (N_{iM})^{n_{iM}}}{a_M' \, (N_{iH})^{n_{iH}}} \qquad (13.2)$$

where a_H' and a_M' are the activities of H^+ and M^+ in the external solution (sample), N_{iH} and N_{iM} the mole fractions of sites R_i^- in the glass surface occupied by ions H^+ and M^+, respectively. The coefficients n_{iH} and n_{iM} were associated with the interchange energy and the coordination numbers of the cations [23]; in the original work by Rothmund and Kornfeld [29] they were assumed to be identical which corresponds to a regular-solution approach to the glass phase [35, 36] (see also Chapter 10). Equation (13.2) implies that the activities of exchangeable ions from sites R_i^- can be generally formulated as

$$a_{iH} = \alpha_{iH} \, (N_{iH})^{n_{iH}} \quad ; \quad a_{iM} = \alpha_{iM} \, (N_{iM})^{n_{iM}} \qquad (13.3)$$

whereas the concentrations of the same species are obviously given by

$$c_{iH} = N_{iH} \, C \qquad ; \qquad c_{iM} = N_{iM} \, C \qquad (13.4)$$

$$N_{iH} + N_{iM} = N_i^{tot} \qquad (13.5)$$

The activity coefficients α_{iH} and α_{iM} characterize the bonding strength for the given cations in the ionogenic groups, C is the total concentration of all ion-exchange sites residing in the glass. Assuming equilibrium-distribution of free cations across the membrane/solution interface, the activities a_{iH} and a_{iM} are related to the external activities a_H' and a_M' as follows:

$$\frac{a_{iM}}{a_{iH}} = \frac{k_M \, a_M'}{k_H \, a_H'} \qquad (13.6)$$

where k_H and k_M are the ionic distribution coefficients characteristic of the "solvated" rather than the bound cations within the glass. In contrast to the ratio k_M/k_H, the ion-exchange constant K_i evidently includes terms for both ionic distribution and ion binding since we find from Eqs. (13.2), (13.3), and (13.6) that

$$K_i = \frac{k_M \, \alpha_{iH}}{k_H \, \alpha_{iM}} \qquad (13.7)$$

Combination of Eqs. (13.3) and (13.5)-(13.7) leads to the following relationship which, in principle, allows to determine all the individual ion activities established in the surface layer of the glass membrane:

$$\left(\frac{a_{iH}}{\alpha_{iH}}\right)^{1/n_{iH}} + \left(\frac{a_{iH}}{\alpha_{iH}} \cdot K_i \, \frac{a_M'}{a_H'}\right)^{1/n_{iM}} = N_i^{tot} \qquad (13.8)$$

In general cases, glass mixtures may contain N different sorts of ionogenic groups R_i^- each of which exhibits a different bonding strength for H^+ ions and other cations, M^+. This inhomogeneity of ion binding in the glass was first treated quantitatively in one version of Nicolsky's theories, published in 1953 [30]. It was assumed that the total activities a_H and a_M of ions H^+ and M^+ in the glass boundary are the sum of the partial activities a_{iH} and a_{iM}, respectively, as contributed by each ionogenic group:

$$a_H = \sum_{i=1}^{N} a_{iH} \quad ; \quad a_M = \sum_{i=1}^{N} a_{iM} \qquad (13.9)$$

These total ion activities are related to the interfacial electrical potential difference $\phi - \phi'$, according to Eq. (3.6a) or (9.3), as follows:

$$\phi - \phi = \frac{RT}{F} \ln \frac{k_H a_H'}{a_H} = \frac{RT}{F} \ln \frac{k_M a_M'}{a_M} \qquad (13.10)$$

where ϕ and a refer to the external solution (sample), ϕ and a refer to the membrane boundary (at x=0). The system of Eqs. (13.8) and (13.10) offers a general theoretical solution for the phase-boundary potential which is presumed to be the dominant contribution to the glass electrode potential. This description encompasses a series of apparently different approaches that were developed by several pioneers of the glass electrode. Some of the corresponding special cases are briefly discussed below.

First, we consider a glass phase that contains only one type of ionogenic site and which approximates ideal behavior, i. e. $n_H = n_M = 1$. In this case, Eq. (13.8) reduces to

$$\frac{a_H}{\alpha_H} \left[1 + K \frac{a'_M}{a'_H}\right] = N^{tot} = 1 \tag{13.11}$$

Insertion into Eq. (13.10) yields:

$$\phi - \phi' = \frac{RT}{F} \ln \frac{k_H}{\alpha_H} + \frac{RT}{F} \ln \left[a'_H + K a'_M\right] \tag{13.12}$$

If the phase-boundary potential difference between the glass electrode and the sample solution is the only variable contribution to the cell potential E:

$$E = \phi - \phi' + const \tag{13.13}$$

we immediately obtain

$$E = \left(\frac{RT}{F} \ln \frac{k_H}{\alpha_H} + const\right) + \frac{RT}{F} \ln \left[a'_H + K a'_M\right]$$

respectively:

$$E = E_H^o + \frac{RT}{F} \ln \left[a'_H + K_{HM}^{Pot} a'_M\right] \tag{13.14}$$

Equation (13.14) offers a rough description, for example, of the "alkali error" (interference by alkali ions) observed for pH-glass electrodes. Expressions of this type were introduced in 1931-37 by Lark-Horovitz [20], Dole [21], and Nicolsky [22] in their pioneering thermodynamic or statistical treatments of the glass electrode potential, and later found general acceptance in the daily routine of electrode applications [37]. However, the so-called Nicolsky equation, when applied to glass electrodes, often does not agree quantitatively with experimental results. The major discrepancy between this simple theory and experiment is observed in the region of the emf-response function intermediate between the pure H^+ function (for $a_H' \gg K_{HM}^{Pot} a_M'$) and the M^+ function (for $a_H' \ll K_{HM}^{Pot} a_M'$). One possibility to overcome the problem is to impose an n-type description of membrane activities (see below). A different correction for activity coefficients was proposed by Lengyel et al. [28].

In the work by Landqvist [23] and by Schwabe and Dahms [24], considerations were still restricted to glasses with one ionogenic group. However, the nonideality of these phases was taken into account, either theoretically [23] or empirically [24], by introducing individual coefficients n_H and n_M. Equation (13.8) then retains its general form (for i=1):

$$\left(\frac{a_H}{\alpha_H}\right)^{1/n_H} + \left(\frac{a_H}{\alpha_H}\right)^{1/n_M} \left(K \frac{a_M'}{a_H'}\right)^{1/n_M} = 1 \qquad (13.15)$$

A relationship between the term a_H/α_H and the electrical potential is obtained from Eqs. (13.10) and (13.13):

$$\frac{a_H}{\alpha_H} = \frac{k_H a_H'}{\alpha_H} \exp\left[-\frac{F}{RT}(\phi - \phi')\right]$$

$$= a_H' \exp\left[-\frac{F}{RT}(E - E_H^0)\right] \qquad (13.16)$$

This leads to an implicit solution for the observable emf E:

$$(a_H')^{1/n_H} \exp\left[-\frac{F}{n_H\,RT}(E - E_H^O)\right]$$

$$+ (K\,a_M')^{1/n_M} \exp\left[-\frac{F}{n_M\,RT}(E - E_H^O)\right] = 1 \qquad (13.17)$$

Landqvist [23] gave his basic result for the deviation of the glass-electrode potential from that expected ideally for the hydrogen half-cell:

$$\Delta E = E - (E_H^O + \frac{RT}{F}\ln a_H') \;.$$

Hence:

$$\exp\left(-\frac{F\Delta E}{n_H\,RT}\right) = 1 - \left[K\,\frac{a_M'}{a_H'}\exp\left(-\frac{F\Delta E}{RT}\right)\right]^{1/n_M} \qquad (13.18a)$$

An equivalent expression appears in an article by Schwabe and Dahms [24]. Accordingly, Eq. (13.17) may be rewritten as

$$\Delta pH + n_M \log (1 - 10^{-\Delta pH/n_H}) = \log\left(K\,\frac{a_M'}{a_H'}\right) \qquad (13.18b)$$

where

$$\Delta pH = \frac{F}{2.3\,RT}\,\Delta E\;.$$

It was shown that Eqs. (13.18a) and (13.18b) permit a very close fit to the experimental data obtained for different glass compositions [23, 24]. On the other hand, such transcendental equations are difficult to wield. A more convenient - and nevertheless successful - description of the emf-response of glass electrodes is based on the assumption that

$$n_H = n_M = n \tag{13.19}$$

In this case, Eq. (13.17) reduces simply to

$$E = E_H^\circ + \frac{n \, RT}{F} \ln \left[(a_H')^{1/n} + (K \, a_M')^{1/n} \right] \tag{13.20}$$

An analogous result was obtained previously for solid-state membrane electrodes where K is identical to the ratio of solubility products for the given species (see Chapter 10, Eq. (10.32)). If formation of a mixed solid phase is hindered it holds that $n \to 0$, whereas an ideal mixed phase or adsorption isotherm corresponds to n=1. For glass membranes it is very often found that n>1. Equation (13.20) then predicts a smoother transition from the H^+ function to the M^+ function of the electrode than the unmodified Nicolsky equation (13.14) does (see Figure 13.1).

Equation (13.20) was extended by Eisenman et al. [15, 16, 25-27] who also took into account the internal diffusion potential of the membrane. The theory is based on the Nernst-Planck equation (13.21a,b), describing the ion fluxes J_H and J_M in the membrane phase:

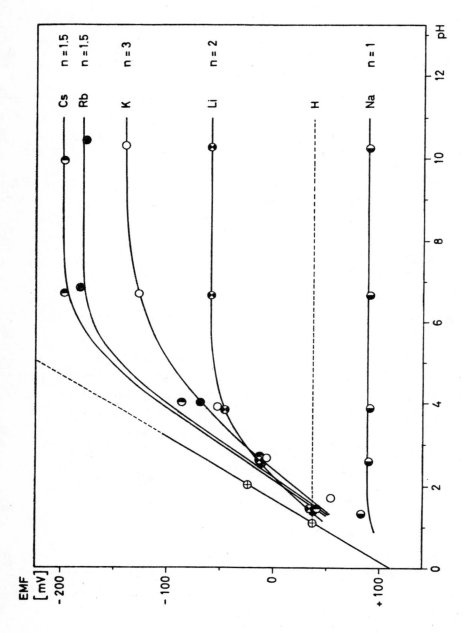

Figure 13.1. The pH-response of a sodium-selective glass electrode in the presence of different alkali metal ions (0.1M solutions) [38]. The solid lines are drawn according to Eq. (13.20) resp. (13.26) [38]. The experimental points for the larger alkali ions indicate some tendency towards a stepwise response function.

$$J_H = -u_H c_H \frac{d}{dx} [RT \ln a_H + F\phi] \qquad (13.21a)$$

$$J_M = -u_M c_M \frac{d}{dx} [RT \ln a_M + F\phi] \qquad (13.21b)$$

Assuming a constant ratio of cation mobilities, u_M/u_H, and recalling Eqs. (13.3), (13.4), and (13.19), one can write:

$$J_H = -u_H C \, n \, RT \left[\frac{d}{dx} (N_H) + N_H \frac{F}{n \, RT} \frac{d\phi}{dx} \right] \qquad (13.22a)$$

$$J_M = -u_H C \, n \, RT \left[\frac{d}{dx} \left(\frac{u_M}{u_H} N_M \right) + \frac{u_M}{u_H} N_M \frac{F}{n \, RT} \frac{d\phi}{dx} \right] \qquad (13.22b)$$

At zero-current conditions, it holds that $J_H + J_M = 0$. Hence:

$$\frac{d\phi}{dx} = - \frac{n \, RT}{F} \frac{d}{dx} \ln \left(N_H + \frac{u_M}{u_H} N_M \right)$$

which, upon integration over the membrane interior from x=0 to x=d, yields the diffusion potential:

$$E_D = \phi(d) - \phi(0) = \frac{n \, RT}{F} \ln \frac{N_H(0) + \frac{u_M}{u_H} N_M(0)}{N_H(d) + \frac{u_M}{u_H} N_M(d)} \qquad (13.23)$$

The two phase-boundary potentials are given by expressions of the type (13.10):

$$\phi(0) - \phi' = \frac{RT}{F} \ln \frac{k_H a_H'}{a_H(0)} \quad ; \quad \phi(d) - \phi'' = \frac{RT}{F} \ln \frac{k_H a_H''}{a_H(d)} \qquad (13.10a,b)$$

Correspondingly, the boundary potential difference assumes the form (see also Eq. (13.3)):

$$E_B = \phi'' - \phi(d) + \phi(0) - \phi'$$

$$= \frac{n\ RT}{F}\ \ln \frac{(a_H')^{1/n}/N_H(0)}{(a_H'')^{1/n}/N_H(d)} \qquad (13.24)$$

The total membrane potential E_M is finally obtained by simply adding Eqs. (13.23) and (13.24), making use of (13.2):

$$E_M = \frac{n\ RT}{F}\ \ln \frac{(a_H')^{1/n} + \frac{u_M}{u_H}\ (K\ a_M')^{1/n}}{(a_H'')^{1/n} + \frac{u_M}{u_H}\ (K\ a_M'')^{1/n}} \qquad (13.25)$$

For membrane electrodes having a constant internal solution, the emf-response function reduces to

$$E = E_H^o + \frac{n\ RT}{F}\ \ln \left[(a_H')^{1/n} + (K_{HM}^{Pot} a_M')^{1/n} \right] \qquad (13.26)$$

where $K_{HM}^{Pot} = (u_M/u_H)^n\ K$. As Eisenman's equation (13.26) turns out to be formally identical to the former result (13.20), addition of the diffusion potential according to (13.23) has no obvious effect on the shape of the emf-response curve[*],

---

[*] This is no longer true if a more sophisticated model is used to describe the diffusion potential (see Sections 13.4 and 13.5).

except for a more general definition of the potentiometric
selectivity factor. These findings are in favor of the aforementioned pure ion-exchange concepts of glass membrane electrodes, an approach which indeed shows good agreement with
experiment [15, 18, 23, 24, 30, 31]. Nevertheless, Eisenman's
equation has probably become the most widely used formula in
the field of glass electrodes. It is worthy of note that the
ideal form of (13.26) (for n=1, but including the mobility
ratio) was published as early as in 1931 [20]!

All the theories discussed so far account for situations
where the selectivity behavior of the glass electrode is
dictated by the properties of one ionogenic group. In typical pH-glasses, it is the strongly basic group $(SiO_{3/2})O^-$
that is selectivity-determining, whereas glasses selective
for sodium or other cations contain a relatively high concentration of weakly basic groups such as $(AlO_{4/2})^-$. The
glass compositions of commercially available electrodes have
been optimized and their response can approximately be described
on the basis of Eq. (13.20) resp. (13.26) (see also Figure
13.1). For more general cases, however, glasses must be considered to contain a variety of anionic sites (i = 1,2,...N)
of different bonding strengths. All these groupings contribute to the ion-exchange properties of the membranes. The
potentiometric behavior of typical heterogeneous-site glasses
is characterized by a stepwise response in mixed electrolytes
(varying pH at constant pM). A formal description of such behavior was initiated by Nicolsky [18, 30]. His solution
corresponds to Eqs. (13.8)-(13.10) of the present generalized
ion-exchange theory when the ideality assumption $n_{iH}=n_{iM}=1$
is used. Thus, one finds in analogy to (13.11):

$$\frac{a'_{iH}}{\alpha_{iH}} \left[1 + K_i \frac{a'_M}{a'_H} \right] = N_i^{tot} \qquad (13.27)$$

and hence:

$$a_H = \sum_i a_{iH} = \sum_i \frac{\beta_i \, a_H'}{a_H' + K_i \, a_M'} \qquad (13.28)$$

where $\qquad \beta_i = \alpha_{iH} \, N_i^{tot}$

Nicolsky's result for the emf-response function is then readily obtained from Eqs. (13.10), (13.13), and (13.28):

$$E = E_H^0 + \frac{RT}{F} \ln \sum_i \beta_i - \frac{RT}{F} \ln \sum_i \frac{\beta_i}{a_H' + K_i \, a_M'} \qquad (13.29)$$

This theory has been shown to permit a nearly perfect fit of several "normal" potential vs. pH curves [30]. The most important consequence, however, is its capability to produce a stepwise response to a_H' at constant a_M', consisting of regions with Nernstian or near-Nernstian slopes separated by shoulders [18, 30]. Formation of such step-curves requires the presence of at least two sorts of competing ion-exchange sites, having significantly different binding properties (e. g., β_1 and $\beta_2 > 0$ and $K_1 \ll K_2$ in Eq. (13.29)). Since the theoretical curves according to (13.29) did not compare well with all of the experimental data, Nicolsky and Shults [18, 31] later developed a second version of "generalized" theory, which led to somewhat different results (see Section 13.4).

The aim of the treatment above was to give a unified derivation of earlier approaches to the theory of glass electrodes. In view of general practical applications, it would be of prime interest to arrive at a closed formula incorporating Eisenman's familiar n-type description, Eq. (13.20) resp.

(13.26), and Nicolsky's heterogeneous-site theory, Eq. (13.29).
Such an all-encompassing formula could be found more intuitively,
but here it follows strictly from the previous Eqs. (13.8)-
(13.10). For simplicity, we make use of (13.13) and (13.19):

$$n_{iH} = n_{iM} = n_i \tag{13.19}$$

Then the following relations can be written, in analogy to
(13.8) and (13.28):

$$\left(\frac{a_{iH}}{\alpha_{iH}}\right)^{1/n_i} \left[1 + \left(K_i \frac{a_M'}{a_H'}\right)^{1/n_i}\right] = N_i^{tot} \tag{13.30}$$

and

$$a_H = \sum_i a_{iH} = \sum_i \frac{\beta_i \, a_H'}{[(a_H')^{1/n_i} + (K_i \, a_M')^{1/n_i}]^{n_i}} \tag{13.31}$$

where:

$$\beta_i = \alpha_{iH} \, (N_i^{tot})^{n_i}$$

This leads to the final result:

$$E = E_H^o + \frac{RT}{F} \ln \sum_i \beta_i - \frac{RT}{F} \ln \sum_i \frac{\beta_i}{[(a_H')^{1/n_i} + (K_i a_M')^{1/n_i}]^{n_i}}$$

$$\tag{13.32}$$

353

This generalized formula combines the advantages of Eisenman's equation (variable coefficients n) with those of Nicolsky's result (various sites i). For evident reasons, Eq. (13.32) is successful in reconstructing all the response functions that could be obtained from either of these theories. Beyond that, it affords a quantitative description of experimental data in cases where the mentioned theories fail and where other, basically different, models had to be constructed (see below).

13.3. POTENTIAL RESPONSES OF $Na_2O-Al_2O_3-SiO_2$ GLASSES

Figure 13.2 illustrates the pH response for a series of sodium aluminosilicate glasses, $Na_2O-Al_2O_3-SiO_2$, of varying alumina contents [18, 31]. The potentiometric behavior of these glasses can be interpreted on the basis of Eq. (13.32) if two terms are taken with $K_1 << K_2$. The term with i=1 corresponds to the silica sites and gives rise to the sodium error at high pH values, whereas the intermediate step in response arises from the term with i=2, corresponding to the alumina sites. For simplicity, the same values of $n_1=5$ and $n_2=1$ were used throughout. Hence, the following simplified form of Eq. (13.32) was applied for all calculations, involving only three adjustable selectivity-parameters:

$$E = E_H^O + \frac{RT}{F} \ln (\beta+1) - \frac{RT}{F} \ln \left[\frac{\beta}{[(a_H')^{0.2} + (K_1 a_M')^{0.2}]^5} \right.$$

$$\left. + \frac{1}{a_H' + K_2 a_M'} \right] \qquad (13.32a)$$

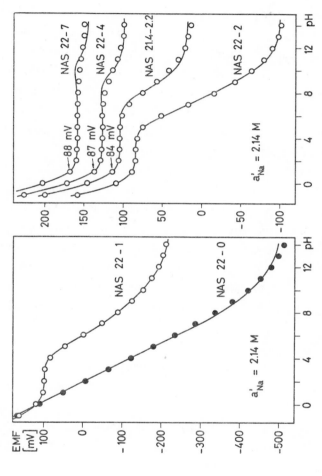

Figure 13.2. Computed pH-response of different sodium aluminosilicate glasses at constant sodium background ($c'_{Na} = 3M$, $a'_{Na} = 2.14M$). Circles: values obtained from Eq. (13.32a), using the parameters given in Table 13.2. Solid lines: values expected from Buck's theory (Eq. (13.34) and Table 13.2). Both sets of curves are in agreement with experimental data [18, 19, 31]. For convenience, some of the curves were shifted vertically, as indicated by the values obtained for pH=1. The glass numbers denote mol -% of Na and Al: NAS 22-0 is a 22% Na_2O -78% SiO_2 glass, NAS 22-1 corresponds to 22% Na_2O - 1% Al_2O_3 - 77% SiO_2, etc.

with

$$\beta = \beta_1/\beta_2$$

In Figure 13.2 computed emf-values according to Eq. (13.32a) are compared with curves given by Buck [19] on the basis of a more involved theory (see Table 13.2 and Section 13.4). The latter were shown to nearly coincide with the experimental data [18, 19, 31], except for the glasses with 4 and 7 mol % Al_2O_3 where the present theory seems preferable. For glasses in the high alumina regime, Buck's theory predicts spurious local maxima of the potential vs. pH curves [19] (see also Figure 13.2) which do not appear in experiment [18, 31]. Nevertheless, the agreement between the new Eq. (13.32), respectively (13.32a), and Eq. (13.34) suggested by Buck is excellent - in spite of the formal differences. For the five $Na_2O-Al_2O_3-SiO_2$ glasses in Figure 13.2, the mean deviation between the two approaches is less than 2 mV. An even better agreement between single curves, especially for the system 22% Na_2O- 78% SiO_2, could be achieved by optimizing the parameter n_1 for each glass composition, instead of inserting an average value of $n_1=5$.

These results clearly demonstrate the equivalence of Eq. (13.32a) and Buck's theory. The advantages of the present treatment are:
a) The compactness and clarity of the basic formula, Eq. (13.32), which is a logical extension of more familiar expressions.
b) The comparatively small number of parameters involved (Table 13.2).
c) The systematic variation of these parameters with varying glass composition.
Table 13.2 shows that the selectivity coefficient K_2, charac-

Table 13.2. Fundamental parameters obtained from data fit (Na₂O-Al₂O₃-SiO₂ glasses; varying pH at 3M sodium levels) [34]

mol % Al_2O_3	K_1 (ion-exchange on silica sites)	K_2 (ion-exchange on alumina sites)	k_M/k_H (ion-exchange for solvated ions)	$K_{H/M}$ (mobility and defect generation ratio)	β	E_H^o (mV)
Values from Eq. (13.32a)						
0	$4.7 \cdot 10^{-12}$	–			$\sim\infty$	110
1	$2.0 \cdot 10^{-10}$	0.40			$1.45 \cdot 10^{-4}$	103
2	$2.8 \cdot 10^{-9}$	0.29			$1.9 \cdot 10^{-5}$	97
2.2	$1.1 \cdot 10^{-8}$	0.12			$4 \cdot 10^{-6}$	110
4	$2.2 \cdot 10^{-8}$	0.10			$5.5 \cdot 10^{-7}$	117
7	$(2.2 \cdot 10^{-8})$ [a]	0.10			(10^{-7}) [a]	118
Values from Eq. (13.34), according to Buck [19]						
0	$2.5 \cdot 10^{-9}$	–		$7.8 \cdot 10^{-3}$	$\sim\infty$	~115
1	$6.7 \cdot 10^{-8}$	0.40		$1.9 \cdot 10^{-3}$	$6.7 \cdot 10^{-5}$	103
2	$9.7 \cdot 10^{-7}$	0.29		$2.4 \cdot 10^{-3}$	$6.7 \cdot 10^{-6}$	97
2.2	$3.9 \cdot 10^{-8}$	0.12		$3.5 \cdot 10^{-2}$	~0	110
4	$5.5 \cdot 10^{-10}$	0.10		$3.3 \cdot 10^{-1}$	~0	117
7	$2.2 \cdot 10^{-11}$	0.12		$5.7 \cdot 10^{-1}$	~0	114

(continued)

	Values from Eq. (13.35),		according to Nicolsky and Shults [18, 31]	
0	$4.1 \cdot 10^{-13}$	–	$4.6 \cdot 10^{-10}$	~ 8	110
1	$4.7 \cdot 10^{-13}$	0.47	0.47	10^{-1}	104
2	10^{-11}	0.1	0.1	10^{-4}	124
2.2	10^{-11}	0.1	0.1	10^{-7}	118
4	10^{-11}	0.1	0.1	10^{-9}	118
7	10^{-11}	0.1	0.1	10^{-10}	118

a Different combinations of K_1 and β values led to nearly the same results.
For simplicity, the same values K_1 and K_2 were used as for 4% Al_2O_3.

teristic of the alumina sites, remains roughly independent
of the membrane composition. In contrast, β is, by definition,
a direct measure of the heterogeneity, and the selectivity
coefficient K_1 for silica sites also shows a pronounced variation. The last effect is probably caused by the formation of
some mixed-type sites, the population of which should increase with decreasing β. In fact, the following relationship was established empirically:

$$K_1 = 4.7 \times 10^{-12} + 3.0 \times 10^{-14}/\beta$$

Estimated and observed values of K_1 agree within a factor of
≤ 2.5 for glasses with 0 - 4% Al_2O_3, although K_1 varies over
a range of 10^4. No such correlation is found for the parameter K_1 used by Buck (see Table 13.2).

Prior to the present or Buck's approach, only the "second
variant ion-exchange theory" of Nicolsky and Shults [18, 31]
was capable of fitting part of the curves in Figure 13.2.
Calculations by Nicolsky and Shults gave a surprisingly good
fit of data from 2 to 7% Al_2O_3 by varying only one parameter
(Section 13.4), but they failed in the crucial case with 1%
Al_2O_3, where maximal deviations from experiment exceeded
50 mV. Therefore, the use of the present extension of
Nicolsky's theory for ternary glasses is to be encouraged.

13.4. ALTERNATIVE APPROACHES TO HETEROGENEOUS-SITE GLASSES

The alternative solid-state approach by Buck [19] does not
use n-type nonideality corrections, but includes a diffusion
potential term. The assumption was made that only the fraction
of interstitial cations in the glass is mobile and contributes

to the diffusion potential. These defects are generated according to:

$$H_i(\text{lattice}) + \text{interstitial site} \rightleftharpoons H^+(\text{interstitial}) + v_i^{H-} \qquad (13.33a)$$

$$M_i(\text{lattice}) + \text{interstitial site} \rightleftharpoons M^+(\text{interstitial}) + v_i^{M-} \qquad (13.33b)$$

Although the distinction between vacancies v_i^{H-} and v_i^{M-} for H^+
and M^+ ions, respectively, may be questionable (memory effect
of lattice sites), the theory offers some interesting new
features. Unfortunately, Buck's derivation is cast in terms
of lumped parameters, which makes comparison with other theories difficult. A detailed examination reveals, however, that
his basic equation 29 [19] can be transformed into

$$E = E_H^o + \frac{RT}{F} \ln (\beta+1) - \frac{RT}{F} \ln \left[\frac{\beta}{a_H' + K_1 a_M'} + \frac{1}{a_H' + K_2 a_M'} \right]$$

$$+ \frac{RT}{F} \ln \frac{(a_H')^{1/2} + K_{H/M}(K_1 a_M')^{1/2}}{(a_H' + K_1 a_M')^{1/2}} \qquad (13.34)$$

Evidently, this result corresponds to Nicolsky's interfacial
potential, Eq. (13.29) for N=2, plus an additional term
describing the diffusion potential. The parameter β in
Eq. (13.34) stands for Buck's quantity T_4/α, and $K_{H/M}=T_3\alpha^{1/2}$
represents a new selectivity term, characterizing the ratio
of defect generation and mobility of interstitial cations.
The parameters are summarized in Table 13.2. For pure silicate
glasses containing only type 1 sites ($\beta\sim\infty$), Eq. (13.34) reduces to an expression different from Eisenman's equation

(13.20) or (13.26), and hence gives another meaning to n-type
behavior [19, 39]. However, n-type nonideality was first observed for purely interfacial phenomena on solid ion-exchangers [29] where diffusion potentials do not come into
play.

A liquid-state approach of heterogeneous-site glasses was
initiated by Nicolsky and Shults [18, 31]. The authors used
a formalism analogous to ordinary solution theory to describe
complexation between anionic sites or vacancies and solvated
cations in the glass phase. Accordingly, the ion-exchange
constant K_i for sites i includes the ratio of complex-formation
constants (term α_i in Nicolsky's work) and the ratio of
free-cation distribution coefficients, k_M/k_H in our terminology (Nicolsky: K_{HM}). The following result is obtained when
using the assumptions of nearly complete association, zero
diffusion potential, and ideal behavior (for consistency with
Eqs. (13.32a) and (13.34), the subscript 1 is used for silica
sites and 2 for alumina sites):

$$E = E_H^O + \tau \frac{RT}{F} \ln (\beta+1) + (1-\tau) \frac{RT}{F} \ln \left[a_H' + (k_M/k_H) \, a_M' \right]$$

$$- \tau \frac{RT}{F} \ln \left[\frac{\beta}{a_H' + K_1 a_M'} + \frac{1}{a_H' + K_2 a_M'} \right] \qquad (13.35)$$

with $\tau = 0.5$.

The numerical parameters of Eq. (13.35) for the system
$Na_2O-Al_2O_3-SiO_2$ are also included in Table 13.2. Constancy of
selectivity coefficients for glasses with 2 - 7% Al_2O_3 is
most remarkable, but values for 1% Al_2O_3 are based on a very

poor fit of experimental data, and an enormous decrease of k_M/k_H (by a factor of 10^9) is required for rationalizing the pH-response of pure silicate glasses. More recently, Shults and coworkers [32, 33] extended the theory of homogeneous-site glasses by accounting for the diffusion potential. For cases where either cations move in the glass by an interstitialcy mechanism ("solvated" ions), or cation transport is coupled with a countertransport of negative vacancies, a description analogous to the Sandblom-Eisenman-Walker theory [40] (Chapter 11) of liquid ion-exchange membranes is obtained. This demonstrates the parallels between the second variant theory of Nicolsky and Shults and the usual concepts of liquid membranes. An extension of these theories to heterogeneous-site ion exchangers, allowing for contributions from the diffusion potential (i. e., variable values of τ, see Section 13.5), offers more insight into the parameters of Eq. (13.35) and into the basic mechanisms for ion selectivity.

13.5. FURTHER DEVELOPMENT OF GLASS ELECTRODE THEORY (LIQUID-MEMBRANE CONCEPTS)

It is well known that pH- or cation-sensitive glasses, upon exposure to aqueous solutions or humid atmospheres, are subject to continuous corrosion. This slow process is accompanied by absorption of water and concomitant ion-exchange reactions in the glass surface, and it gives rise to the formation of surface layers of differing compositions. The existence of such hydrated layers is essential to the functioning of glass membranes as ion sensors, that is, as reversible cation-exchangers.

In the past decade, the structural aspects (e. g., the ion concentration profiles) and the ion-transport behavior of the hydrated glass layers were investigated to considerable detail. Notable work in this field was done by Baucke and Bach [41-43], Boksay, Csákvári et al.[44-47], Wikby [48-50], and others [51-54]. An excellent report on the cation-exchange properties of dry, non-hydrated silicate membranes, on the other hand, was contributed by Garfinkel [36]. For glass membranes conditioned in aqueous solutions of relatively low pH, two major regions of the surface domain were distinguished [41-43, 45]. The outer part of the surface was termed "leached" or "protonated" layer because nearly all the alkali ions of the original host network are replaced here by hydrogen ions. The adjacent "transition" layer shows a state of continuous change from the intact glass to the leached-layer structure. This mixture region is the origin of membrane-internal diffusion potentials. The low proton mobility (local interdiffusion coefficients as low as 10^{-18} cm^2 s^{-1} were reported [41]) protects the glass from being leached in depth during measurable periods, but leads to a pronounced maximum in electrical resistivity for this region [41, 47, 49]. For the migration of cations within the hydrated glass layer, interstitialcy mechanisms (see also Section 13.4) as well as vacancy mechanisms were called upon [19, 32, 33, 46-48, 55]. Stephanova and Shults [32] were the first to offer a glass electrode theory accounting for both transport mechanisms. A simplified rederivation and extension of their theory is given below.

The transport of interstitial ("solvated") cations M^+ in the glass proceeds according to the following mechanism:

$$M^+(x) + \text{interstitial site } (x+\Delta x) \xrightarrow{\tilde{k}_M/\Delta x}$$

$$M^+ (x+\Delta x) + \text{interstitial site } (x) \qquad (13.36)$$

Hence, the transport rate according to the theory of reaction rates is given by

$$J_M = \frac{\tilde{k}_M}{\Delta x} c_M(x) \, c_{I.S.}(x+\Delta x) - \frac{\tilde{k}_M \exp(F\Delta\phi/RT)}{\Delta x} c_M(x+\Delta x) \, c_{I.S.}(x)$$

$$\cong - \tilde{k}_M \, c_M \, c_{I.S.} \left[\frac{d \ln c_M}{dx} + \frac{F}{RT} \frac{d\phi}{dx} - \frac{d \ln c_{I.S.}}{dx} \right] \qquad (13.37)$$

where c_M and $c_{I.S.}$ are the concentrations of free ions and of interstitial sites, respectively, and $\tilde{k}_M$ is a rate constant characteristic of the interstitial transport process. Since $c_{I.S.}$ is nearly invariant with space and time, Equation (13.37) can be transformed into a flux equation of the Nernst-Planck type:

$$J_M = - u_M c_M \frac{d\tilde{\mu}_M}{dx} = - u_M RT \frac{dc_M}{dx} - u_M c_M \, F \frac{d\phi}{dx} \qquad (13.38a)$$

with

$$u_M = \frac{\tilde{k}_M \, c_{I.S.}}{R\,T}$$

An analogous expression is applicable to the electrodiffusion of other free cations, e. g. for H^+:

$$J_H = - u_H c_H \frac{d\tilde{\mu}_H}{dx} = - u_H RT \frac{dc_H}{dx} - u_H c_H \, F \frac{d\phi}{dx} \qquad (13.38b)$$

In contrast, a vacancy mechanism has to be invoked to allow for transport processes that directly involve the fraction of lattice- ("bound") cations, i. e.:

$$MR\ (x) + R^-\ (x+\Delta x) \xrightarrow{\ \tilde{k}_{MR}/\Delta x\ } MR\ (x+\Delta x) + R^-\ (x) \qquad (13.39)$$

The flux of complexed cations according to this mechanism is formulated as follows:

$$J_{MR} = \frac{\tilde{k}_{MR}}{\Delta x}\ c_{MR}(x)\ c_R(x+\Delta x)\ -\ \frac{\tilde{k}_{MR}\ \exp(F\Delta\phi/RT)}{\Delta x}\ c_{MR}(x+\Delta x)\ c_R(x)$$

$$\cong\ -\ \tilde{k}_{MR}\ c_{MR}\ c_R \left[\frac{d\ \ln\ c_{MR}}{dx} + \frac{F}{RT}\frac{d\phi}{dx} - \frac{d\ \ln\ c_R}{dx}\right] \qquad (13.40)$$

respectively:

$$J_{MR} = -\ u_{MR}c_{MR}\frac{c_R}{c_R^{tot}}\frac{d\tilde{\mu}_M}{dx}$$

$$= -\ u_{MR}K_{MR}\frac{c_R^2}{c_R^{tot}}\left[RT\frac{dc_M}{dx} + c_M\ F\ \frac{d\phi}{dx}\right] \qquad (13.41a)$$

where

$$u_{MR} = \frac{\tilde{k}_{MR}\ c_R^{tot}}{RT}\quad ;\quad K_{MR} = \frac{c_{MR}}{c_M\ c_R}$$

Analogously, one can write

$$J_{HR} = - u_{HR} c_{HR} \frac{c_R}{c_R^{tot}} \frac{d\tilde{u}_H}{dx}$$

$$= - u_{HR} K_{HR} \frac{c_R^2}{c_R^{tot}} \left[RT \frac{dc_H}{dx} + c_H F \frac{d\phi}{dx} \right] \qquad (13.41b)$$

where u_{HR} is the mobility and K_{HR} the association constant for hydrogen ions in lattice positions of the glass. The flux equations (13.41a,b) are comparable but not identical to the familiar Nernst-Planck formalism. The validity of such relations was implicitly postulated by Stephanova and Shults [32] in their description of the cation transport numbers t[*]:

$$t_M + t_{MR} = 1 - (t_H + t_{HR})$$

$$= \frac{u_M c_M + u_{MR} c_{MR} c_R / c_R^{tot}}{u_H c_H + u_M c_M + (u_{HR} c_{HR} + u_{MR} c_{MR}) \ c_R / c_R^{tot}} \qquad (13.42)$$

These transport numbers can be used to formulate the membrane-internal potential gradient under zero-current conditions, that is for $J_H + J_M + J_{HR} + J_{MR} = 0$:

$$\frac{d\phi}{dx} = - \frac{RT}{F} \left[(t_H + t_{HR}) \frac{d \ln c_H}{dx} + (t_M + t_{MR}) \frac{d \ln c_M}{dx} \right] \qquad (13.43)$$

---

[*] These authors used the mobility symbols $u_M^{(s)}$ and $\bar{u}_M^{(v)}$ instead of the present u_M and u_{MR}, whereas the basic parameter $u_M^{(v)} \hat{=} u_{MR}/c_R^{tot}$ obviously does not have the same dimension.

A significant reduction of Eq. (13.42) can be obtained when imposing the assumptions of electroneutrality, of nearly complete association between sites and counterions, and of identical mobilities for all interstitial cations and for all vacancies, respectively. Thus:

$$c_R = c_H + c_M \tag{13.44}$$

$$c_R^{tot} \cong c_{HR} + c_{MR} = (K_{HR}c_H + K_{MR}c_M) \, c_R \tag{13.45}$$

$$u_H = u_M \quad ; \quad u_{HR} = u_{MR} = u_R \tag{13.46}$$

which leads to
$$\tag{13.47}$$

$$t_M + t_{MR} = 1 - (t_H + t_{HR}) = \frac{u_M}{u_M + u_R} \cdot \frac{c_M}{c_H + c_M} + \frac{u_R}{u_M + u_R} \cdot \frac{K_{MR}c_M}{K_{HR}c_H + K_{MR}c_M}$$

After insertion of Eq. (13.47), integration of (13.43) is now easily accomplished. The diffusion potential, as established predominantly within the hydrated surface layer of a glass membrane electrode (constant internal solution and constant bulk of the membrane), is then determined as

$$E_D = const + \frac{u_M}{u_M + u_R} \, \frac{RT}{F} \, \ln \, [c_H(0) + c_M(0)]$$

$$+ \frac{u_R}{u_M + u_R} \, \frac{RT}{F} \, \ln \, [K_{HR}c_H(0) + K_{MR}c_M(0)] \tag{13.48}$$

Combination with Eq. (13.10) yields the final result for the emf:

$$E = E_H^O + (1-\tau) \; \frac{RT}{F} \; \ln \; [a_H' + K_{HM}^{(1)} a_M']$$

$$+ \; \tau \; \frac{RT}{F} \; \ln \; [a_H' + K_{HM}^{(2)} a_M'] \qquad (13.49)$$

This expression is formally identical to the result given by Sandblom, Eisenman, and Walker [40] for liquid ion-exchange membranes (see Chapter 11), and the selectivity parameters involved have basically the same meaning:

$$K_{HM}^{(1)} = \frac{k_M}{k_H} \quad ; \quad K_{HM}^{(2)} = \frac{K_{MR} k_M}{K_{HR} k_H} \quad ; \quad \tau = \frac{u_R}{u_M + u_R} \qquad (13.50)$$

Accordingly, the potentiometric selectivity of ion-exchange membranes generally reflects the ion affinity of the charged ligand sites or vacancies only if a transport of these species becomes operative, that is for $u_R \gg u_M$. In the other limit, $u_M \gg u_R$, the selectivity depends merely on the distribution ratio of the free, "solvated" counterions. This important point was made explicitly for glass electrodes only recently [33]. Stephanova and Shults [32] succeeded in deriving an even more general result for the potential of homogeneous-site glass membranes, allowing for an individual choice of cation mobilities. Based on Eqs. (13.42) - (13.45) an explicit solution was obtained for systems with $K_{HR} \gg K_{MR}$, which has exactly the same form as Eq. (13.49) but involves slightly modified parameters:

$$K_{HM}^{(1)} = \frac{u_M + u_{HR}}{u_H + u_{HR}} \frac{k_M}{k_H} \quad ; \quad K_{HM}^{(2)} = \frac{u_M + u_{MR}}{u_M + u_{HR}} \frac{K_{MR} k_M}{K_{HR} k_H} \quad ; \quad \tau = \frac{u_{HR}}{u_M + u_{HR}} \qquad (13.51)$$

These selectivity terms prove to be nearly identical to the parameters of the Sandblom-Eisenman-Walker theory.

So far, no attempt has been made to extend Eq. (13.49) to solid or liquid ion-exchangers containing multiple cation-specific ligands R_i^-. Such a generalization indeed turns out to be nearly impracticable when a rigorous solution is sought for. However, a reasonable first-order approximation can be deduced straightforwardly along the lines of Eqs. (13.42) – (13.49). To this end, the same mobilities u_M and u_R have to be used for all uncomplexed cations and for all vacant ligands in the membrane phase, respectively. The total transport number for cations M^+ will then be given, in analogy to Eq. (13.47), as follows:

$$t_M + \sum_i t_{MR_i} = \frac{u_M}{u_M + u_R} \frac{c_M}{c_H + c_M} + \frac{u_R}{u_M + u_R} \sum_i \frac{c_{R_i}}{\sum c_{R_i}} \frac{K_{MR_i} c_M}{K_{HR_i} c_H + K_{MR_i} c_M} \qquad (13.52)$$

with $\quad c_{R_i} = \dfrac{c_{R_i}^{tot}}{K_{HR_i} c_H + K_{MR_i} c_M}$

Such expressions can be inserted into Eq. (13.43) to yield

$$\frac{d\phi}{dx} = -\frac{RT}{F} \left[\frac{u_M}{u_M + u_R} \frac{d}{dx} \ln [c_H + c_M] - \frac{u_R}{u_M + u_R} \frac{d}{dx} \ln [\sum_i c_{R_i}] \right] \qquad (13.53)$$

$$= -\frac{RT}{F} \left[\frac{u_M}{u_M + u_R} \frac{d}{dx} \ln [c_H + c_M] - \frac{u_R}{u_M + u_R} \frac{d}{dx} \ln \left[\sum_i \frac{c_{R_i}^{tot}}{K_{HR_i} c_H + K_{MR_i} c_M} \right] \right]$$

369

Upon integration and combination with equations of the type (13.10), the result for the emf assumes the form:

$$E = E_H^O + \tau \, \frac{RT}{F} \, \ln \sum_i \beta_i + (1-\tau) \, \frac{RT}{F} \, \ln \left[a_H' + \frac{k_M}{k_H} \, a_M' \right]$$

$$- \tau \, \frac{RT}{F} \, \ln \sum_i \frac{\beta_i}{a_H' + K_i \, a_M'} \tag{13.54}$$

where

$$K_i = \frac{K_{MR_i} \, k_M}{K_{HR_i} \, k_H}$$

$$\beta_i = \frac{c_{R_i}^{tot}}{\sum c_{R_i}^{tot}} \, \frac{1}{K_{HR_i}} = \frac{N_i^{tot}}{K_{HR_i}}$$

This new formula is consistent with a liquid-membrane approach and may be applied to both heterogeneous-site glasses and liquid membranes containing different negatively charged ligands. In contrast, the earlier universal relationship, Eq. (13.32), was based on solid-state principles and is therefore valid for glass or solid-state membranes. Both approaches are equally suited for a general and nearly exact characterization of glass electrodes; the choice between the two descriptions depends on which structural model of glasses is believed to be more realistic. The present liquid-membrane concept takes account of the diffusion potential, but corrections for the nonideality of the membrane phase could not be phrased in simple terms. The universal character of Eq. (13.54) is clearly demonstrated by the fact that it in-

corporates all versions of ion-exchange theories developed
so far by Nicolsky, Shults, and coworkers [18, 22, 30-32].
The "first variant theory" corresponds to the situation where
$\tau = 1$, which means that charge transfer in the glass be performed exclusively by the fraction of bound cations via a
vacancy mechanism. The selectivity-determining ion-binding
parameters α entering in the former Eq. (13.7) are here replaced consistently by the reciprocals of association constants. The "second variant theory" is identical to the limiting case realized for $\tau = 0.5$, which implies that the diffusion potential becomes negligible. Here the overall ion-selectivity reflects the energy levels for both complexed and free
cations in the glass phase. For $\tau = 0$, Equation (13.54) simply
reduces to an expression of the classical Nicolsky type
(13.14), the selectivity being dictated by the ion-exchange
equilibrium for the purely solvated cations, however. Finally,
the behavior of homogeneous-site glasses is, of course, still
characterized by the simplified theory of Stephanova and
Shults, Eq. (13.49).

It can be concluded that the new theory of glass membrane
potentials is of like import as the alternative description
presented previously in Section 13.2. Both approaches evidently bridge the gap that heretofore existed between the
earlier, more specific treatments. It has been established
that glass electrodes share the features with crystal membrane sensors, on the one hand, and liquid-membrane electrodes,
on the other. For that reason, these more recent ion-sensor
systems have already been treated in the Chapters 10 and 11.

REFERENCES

[1]   J. Bernstein, Pflüger's Arch. Ges. Physiol. 92, 521
      (1902).

[2]   M. Cremer, Z. Biol. 47, 562 (1906).

[3]   F. Haber and Z. Klemensiewicz, Z. Physik. Chem. 67, 385
      (1909).

[4]   W. S. Hughes, J. Chem. Soc. 1928, 491.

[5]   D. A. MacInnes and M. Dole, J. Am. Chem. Soc. 52, 29
      (1930).

[6]   S. I. Sokolof and A. H. Passynsky, Z. Physik. Chem.
      A160, 366 (1932).

[7]   H. H. Cary and W. P. Baxter, U.S. Pat. 2462843 (1945).

[8]   G. A. Perley, Anal. Chem. 21, 391 (1949).

[9a] G. A. Rechnitz, Chem. Eng. News 45(25), 146 (1967).

[9b] W. Simon, H.-R. Wuhrmann, M. Vašák, L. A. R. Pioda,
      R. Dohner, and Z. Štefanac, Angew. Chemie 82, 433
      (1970); Angew. Chem. Intern. Ed. 9, 445 (1970).

[10] W. S. Hughes, J. Am. Chem. Soc. 44, 2860 (1922).

[11] K. Horovitz, Z. Physik 15, 369 (1923); Z. Physik. Chem.
      115, 424 (1925).

[12] H. Schiller, Ann. Physik 74, 105 (1924).

[13] B. Lengyel and E. Blum, Trans. Faraday Soc. 30, 461
      (1934).

[14] B. P. Nicolsky and T. A. Tolmacheva, Zh. Fiz. Khim. 10,
      504, 513 (1937).

[15] G. Eisenman, D. O. Rudin, and J. U. Casby, Science 126,
      831 (1957).

[16] G. Eisenman, ed., Glass Electrodes for Hydrogen and
      Other Cations, Marcel Dekker, New York, 1967.

[17] J. O. Isard, 'The dependence of glass-electrode proper-
      ties on composition', chapter 3 of ref. [16].

[18] B. P. Nicolsky, M. M. Shults, A. A. Belijustin, and A. A. Lev, 'Recent developments in the ion-exchange theory of the glass electrode and its application in the chemistry of glass', chapter 6 of ref. [16].

[19] R. P. Buck, Anal. Chem. 45, 654 (1973).

[20] K. Lark-Horovitz, Naturwiss. 19, 397 (1931); Nature 127, 440 (1931).

[21] M. Dole, J. Chem. Phys. 2, 862 (1934).

[22] B. P. Nicolsky, Zh. Fiz. Khim. 10, 495 (1937).

[23] N. Landqvist, Acta Chim. Scand. 9, 595 (1955).

[24] K. Schwabe and H. Dahms, Z. Elektrochem. 65, 518 (1961).

[25] G. Eisenman, Biophys. J. 2, Pt. 2, 259 (1962).

[26] G. Karreman and G. Eisenman, Bull. Math. Biophys. 24, 413 (1962).

[27] F. Conti and G. Eisenman, Biophys. J. 5, 247 (1965).

[28] B. Lengyel, B. Csákvári, and Z. Boksay, Acta Chim. Hung. 25, 225 (1960).

[29] V. Rothmund and G. Kornfeld, Z. Anorg. Allgem. Chem. 103, 129 (1918).

[30] B. P. Nicolsky, Zh. Fiz. Khim. 27, 724 (1953).

[31] B. P. Nicolsky and M. M. Shults, Zh. Fiz. Khim. 34, 1327 (1962); Vestn. Leningrad.Univ., No. 4, 1963, 73.

[32] O. K. Stephanova and M. M. Shults, Vestn. Leningrad. Univ., No. 4, 1972, 80.

[33] B. P. Nicolsky, M. M. Shults, and A. A. Belijustin, Wiss. Z. TH Leuna-Merseburg 18, 573 (1976).

[34] W. E. Morf, Talanta 26(8), 719 (1979).

[35] R. M. Garrels and C. L. Christ, Solution, Minerals, and Equilibria, Harper and Row, New York, 1965.

[36] H. Garfinkel, in Membranes, Vol. 1 (G. Eisenman, ed.), Marcel Dekker, New York, 1972.

[37] IUPAC Recommendations for Nomenclature of Ion-Selective Electrodes, Pure Appl. Chem. 48, 127 (1976).

[38] Z. Štefanac and W. Simon, <u>Anal. Lett.</u> <u>1</u>, 1 (1967).

[39] R. P. Buck, J. H. Boles, R. D. Porter, and J. A. Margo-
      lis, <u>Anal. Chem.</u> <u>46</u>, 255 (1974).

[40] J. P. Sandblom, G. Eisenman, and J. L. Walker, Jr.,
      <u>J. Phys. Chem.</u> <u>71</u>, 3862 (1967).

[41] F. G. K. Baucke, <u>J. Non-Cryst. Solids</u> <u>14</u>, 13 (1974).

[42] H. Bach and F. G. K. Baucke, <u>Phys. Chem. Glasses</u> <u>15</u>,
      123 (1974).

[43] F. G. K. Baucke, in <u>Ion and Enzyme Electrodes in Biology
      and Medicine</u> (M. Kessler et al., eds.), Urban & Schwarzen-
      berg, Munich, 1976, p. 77.

[44] Z. Boksay, G. Bouquet, and S. Dobos, <u>Phys. Chem. Glasses</u>
      <u>8</u>, 140 (1967); <u>9</u>, 69 (1968).

[45] B. Csákvári, Z. Boksay, G. Bouquet, and I. Ivanovskaya,
      in <u>Stekloobraznoe Sostoyanie</u>, Tr. 5 - Vses. Soveshch.,
      Leningrad, 1969, Nauka, Leningrad, 1971, p. 310.

[46] Z. Boksay and B. Csákvári, <u>Acta Chim. Hung.</u> <u>67</u>, 157
      (1971).

[47] Z. Boksay, in <u>Ion-Selective Electrodes</u> (E. Pungor, ed.),
      Akad. Kiadó, Budapest, 1978, p. 245.

[48] A. Wikby and G. Johansson, <u>J. Electroanal. Chem.</u> <u>23</u>, 23
      (1969).

[49] A. Wikby, <u>J. Electroanal. Chem.</u> <u>33</u>, 145 (1971); <u>38</u>, 429
      (1972); <u>39</u>, 103 (1972).

[50] A. Wikby, <u>Phys. Chem. Glasses</u> <u>15</u>, 37 (1974).

[51] R. P. Buck, <u>J. Electroanal. Chem.</u> <u>18</u>, 363, 381 (1968).

[52] R. P. Buck and I. Krull, <u>J. Electroanal. Chem.</u> <u>18</u>, 387
      (1968).

[53] M. J. D. Brand and G. A. Rechnitz, <u>Anal. Chem.</u> <u>41</u>, 1788
      (1969); <u>42</u>, 304 (1970).

[54] A. A. Belijustin, M. M. Shults, et al., <u>Dokl. Akad. Nauk
      (U.S.S.R.)</u> <u>240</u>(6), 1376 (1978); <u>241</u>(1), 155 (1978);
      <u>Fiz. Khim. Stekla (Phys. Chem. Glass, U.S.S.R.)</u> <u>4</u>(4),
      465, 473 (1978).

[55] M. M. Shults and O. K. Stephanova, <u>Vestn. Leningrad.
      Univ.</u>, No. 4, <u>1971</u>, 22.

CHAPTER 14

DYNAMIC RESPONSE BEHAVIOR OF ION-SELECTIVE ELECTRODES

One of the critical limiting factors in the use of ion-selective electrodes, especially in routine analysis, is their so-called response time. Extensive studies of the dynamic behavior of ion sensors were initiated mainly by research groups working on the development of continuous monitoring systems [1 - 6]. Most of these authors were also engaged in the elaboration of appropriate theories [3, 5, 7 - 10] some of which were reviewed recently [10 - 14].

The current theories have proved more or less successful in describing the usual time course of the electrode potential after a step-change in the sample activity ($a_i^o \rightarrow a_i$). Exceptions aside [15, 16] they predict a monotonic and asymptotic transition of the emf from the initial value at $t \leqslant 0$:

$$E(0) = E_i^o + s \log a_i^o \qquad (14.1)$$

to the final value at $t \rightarrow \infty$:

$$E(\infty) = E_i^o + s \log a_i \qquad (14.2)$$

The formal descriptions of the intermediate values $E(t)$ may be widely dissimilar, depending on the model assumptions made in respect to the rate-determining process. Theoretical and experimental studies on ion-selective electrodes and related membrane systems suggest that there are generally several

25*   375

time-dependent processes and associated time constants τ_n. The following semi-empirical sum formula was proposed by Shatkay [14] for the potential vs. time response to an activity step on one side of the membrane:

$$E(t) \approx E(0) + \sum_n \Delta E_n \left(1 - e^{-t/\tau_n}\right) \qquad (14.3)$$

where ΔE_n is the contribution by the n-th process to the final emf change:

$$E(\infty) - E(0) = \Delta E = \sum_n \Delta E_n \qquad (14.4)$$

Equation (14.3) is valid only in the linear regime where ΔE is proportional to the activity step $a_i - a_i^o$:

$$|\Delta E| << |s/2.303| \cong |RT/z_i F|$$

$$\Delta E \approx \frac{s}{2.303} \frac{a_i - a_i^o}{a_i} \qquad (14.5)$$

Furthermore, some of the slowest processes, those involving diffusion of ions within liquid membranes or dissolution of solid membrane materials to establish interfacial equilibrium, can hardly be represented by a single exponential-type term; a series of exponentials adding to a different time dependence is more appropriate (see below). Nevertheless, Eq. (14.3) nicely shows that the earliest stage of transient response usually does not bear much distinctive information because it is largely given by the sum of equilibration steps. Only a few processes, ideally but the slowest one, dominate in the last stage. The possible time constants and basic processes determining the "response time" of different ion-selective electrodes are examined and catalogued below.

14.1. ELECTRICAL RELAXATION PROCESSES

In a series of reviews, Buck [11 - 13] gave a profound description of the electrical relaxation processes contributing to the time response of ion-selective electrodes. The treatment leaned on the impedance theory of Macdonald [17 - 19]. Accordingly, the RC time constants associated with the following steps should be important for ion-selective electrodes.

a) Charging of the external space-charge surface regions coupled to the ac bulk-electrode resistance.

The corresponding time constant is given by

$$\tau_\infty = R_\infty C_g \tag{14.6}$$

C_g is the geometric capacitance of the membrane, and R_∞ is the high-frequency resistance, depending on the concentrations and mobilities of ionic charges in the electroneutral system. This time constant is independent of the membrane thickness and area. Values between 0.3 ms and 0.2 s have been measured for solid-state and glass electrode membranes [20].

b) Slow surface rates coupled to the relaxed capacitance of each interface.

For slow exchange kinetics at the surfaces of thick membranes, another time constant may be observed:

$$\tau_k = R_\Theta C_{o,b} = M\ R_\Theta C_g \tag{14.7}$$

R_Θ is the linearized surface resistance, being inversely proportional to the total exchange current density, and $M = d/2L_D$ is half the number of Debye lengths (L_D) across the membrane. It should be noted that the time constant τ_k does not appear when surface processes are rapid and reversible, which is consistent with high exchange current densities.

c) Concentration polarization of charge carriers within the membrane.

The longest time constant is associated with the Warburg finite-diffusion process. This requires the presence of at least two charge-carrying species within the membrane phase. Typical examples for Warburg behavior are liquid, mobile-site membranes, especially when interfering ions are present on one side (see also Section 14.4). Diffusion processes in the membrane interior then lead to a $t^{-1/2}$ dependence of $E(t)$ over a range of relatively short times, whereas the long-time term is of the form (14.3) but with a single time constant:

$$\tau_w = \frac{M^2}{2.53}\ \tau_\infty = \frac{d^2}{2\ D} \tag{14.8}$$

τ_w is a measure for the time that is required to adjust the steady-state concentration profiles throughout the membrane. A more detailed discussion is given in Section 14.4.

The equivalent-circuit description of the measuring cell becomes more complicated when electrode membranes of extremely high internal resistances are involved. The corresponding RC time constants should become limiting for certain microelectrodes which have a relatively low concentration of charge-carrying species and a membrane area of only around 10^{-8} cm^2 [10, 21].

Although different processes contribute to the dynamic behavior of ion-selective electrodes, the overall response is commonly dominated by the slowest step [5, 10 - 14], while the other (faster) relaxation steps may give rise to an apparent response delay at short times. Hence, it is often convenient to use the following first-order approximation, based on Eq. (14.3):

$$E(t) \approx E(\infty) - \Delta E_1\, e^{-t/\tau_1}, \text{ for } t > \tau_1 \qquad (14.9)$$

Several theoretical models that corroborate such expressions, as well as some contrasting cases characterized by different time dependences are summarized below. Some of these approaches are equivalent or related to the former cases a) - c) obtained from impedance theory.

14.2. KINETICS OF INTERFACIAL REACTIONS

The early theories by Rechnitz and Hameka [7] and by Johansson and Norberg [22] were devoted to the dynamic response of glass electrodes. They used an energy-barrier concept for the membrane/solution interface together with a capacitor model for the adjoining membrane section (see also Section 14.1.b and [12, 14]). This approach leads to exponential time-relationships analogous to Eq. (14.9). The time constant was found to depend on the external activities a_i and a_i^o [22]. A similar behavior is expected from Eq. (14.7) where the exchange current density entering in the term R_Θ is also activity-dependent.

A somewhat different description was obtained by Tóth and Pungor [9] who assumed rate control by an ion transfer reaction of first-order kinetics. The reaction rate may generally be formulated as

$$\frac{d}{dt} \Delta a_i(t) = -k[\Delta a_i(t)]^n \tag{14.10}$$

where n is the reaction order, k is the rate constant, and $\Delta a_i(t)$ is the difference between the actual activity a_i', as sensed by the electrode, and the final equilibrium value a_i:

$$\Delta a_i(t) = |a_i'(t) - a_i| \tag{14.11}$$

By solving the differential equation (14.10) for n=1, one gets the following result for the measured activity [3, 8]:

$$a_i' = a_i - (a_i - a_i^o) \, e^{-kt} \tag{14.12}$$

Hence

$$E(t) = E(\infty) + s \, \log \frac{a_i'}{a_i} \tag{14.13}$$

$$E(t) = E(\infty) + s \, \log \left[1 - (1 - \frac{a_i^o}{a_i}) \, e^{-kt} \right] \tag{14.14}$$

380

For sufficiently small activity steps, this expression can be linearized according to Eq. (14.5) and approximates the "classical" form (14.9). Here the time constant is evidently given by the reciprocal of the reaction rate constant k.

Buffle and Parthasarathy [23, 24] studied the dissolution of crystalline membrane materials (e. g., AgCl in diluted chloride solutions) that is a prerequisite for reaching inter-facial equilibrium. They found ample evidence for a second-order reaction, i. e., n=2 in Eq. (14.10). This leads to the solution

$$a_i' = a_i - (a_i - a_i^o) \frac{1}{1 + |a_i - a_i^o| \, k \, t} \tag{14.15}$$

For reversible electrodes, a_i' is the measured activity of the primary ions (e. g., of silver ions in the case of AgCl membranes). From this the emf is obtained in the linearized form

$$E(t) \cong E(\infty) - \Delta E \frac{1}{\kappa t + 1} = E(0) + \Delta E \frac{\kappa t}{\kappa t + 1} \tag{14.16a}$$

$$\kappa = |a_i - a_i^o| \, k$$

respectively

$$E(t) \cong E(\infty) - \frac{1}{A \, t + B} \tag{14.16b}$$

$$A \cong \frac{F}{RT} k \, a_i \quad ; \quad B \cong \frac{F}{RT} \frac{a_i}{|a_i - a_i^o|}$$

381

The first expression is Müller's equation [25] which was introduced in discussions of ion-selective electrodes by Mertens et al. [26]; the second formulation was proposed by Buffle and Parthasarathy [23, 24]. Potential vs. time functions of this type were observed for solid-state fluoride- [23, 24, 26] and chloride-sensitive electrodes [24]. Among other aspects, the fundamental dependence of the rate parameter A on the primary ion activity a_i (at equilibrium) was verified experimentally [24]. The fastest response was indeed obtained in concentrated fluoride solutions and in diluted chloride solutions, providing for high values of a_F and $a_{Ag} = L_{AgCl}/a_{Cl}$, respectively.

14.3. DIFFUSION THROUGH A STAGNANT LAYER

In the theories by Markovic and Osburn [5] and by Morf et al. [8, 10] (see also 14.1.c) it was recognized that diffusion processes are crucial for the response time of many membrane electrode systems. One such process is the diffusion of sample ions through the unstirred film adhering to the ion-sensing membrane surface (Nernstian diffusion layer, see Figure 14.1). If this is the rate-determining step, one gets the approximation [5, 8, 10, 14]

$$E(t) = E(\infty) + s \log \left[1 - (1 - \frac{a_i^o}{a_i}) \frac{4}{\pi} e^{-t/\tau'} \right]$$

$$= E(\infty) + s \log \left[1 - (1 - \frac{a_i^o}{a_i}) e^{-t/\tau'} + 0.24 \right] \qquad (14.17)$$

which gives nearly perfect results for $t \gtrsim 0.5\tau'$ [10]. The time constant τ' depends on the thickness δ of the diffusion layer and on the ionic diffusion coefficient D':

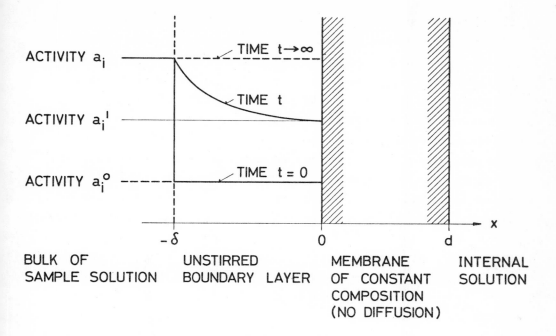

Figure 14.1. Dynamic model proposed for nondiffusive membrane
electrodes [8].
Relaxation of concentration polarization in the Nernstian boundary layer at the electrode surface is assumed to be rate-
controlling. The time course of the activity profile after a
step change $a_i^o \to a_i$ at t=0 and x=-δ is shown schematically.

$$\tau' = \frac{4\delta^2}{\pi^2 D'} \approx \frac{\delta^2}{2 D'}$$
(14.18)

Equation (14.17) seems to be appropriate for fast-responding
membrane electrodes based on charged sites as long as no inter-

fering species are present (Section 14.4). Expressions of the same type may be applied to describe 1) the sluggish response behavior of glass electrodes, as arising from mass transport through thick surface films of protonated and hydrolized glass [15, 27, 28], 2) the dynamic characteristics of gas-sensing electrodes where the gas-permeable membrane represents the external diffusion layer (see Chapter 15), and 3) the time-response of enzyme electrodes in the analytically useful range (Chapter 15). According to Eqs. (14.17) and (14.18), one of the parameters affecting the dynamic response characteristics of membrane electrodes is the thickness of the aqueous diffusion layer which can be drastically reduced by stirring; besides, it depends on shape and condition of the electrode surface and on the composition of the sample solution [29]. The other major factor is the direction of the activity change in the sample solution. It is clearly evident from Figure 14.2 and Table 14.1 that the response time must be expected to increase considerably when changing from a high activity a_i^o to a low activity a_i, as compared to a change in the opposite direction. Moreover, the response time values should be independent of absolute activity levels. These theoretical predictions are in agreement with usual findings (Table 14.2 and [2 - 4, 9]). Discrepancies may be expected for electrodes giving a slow response which is due to other than diffusional limitations.

14.4. DIFFUSION WITHIN THE ION-SENSING MEMBRANE

A certain increase of the response time has to be endured if ionic diffusion within the membrane phase cannot be excluded (see Fig. 14.3). The reason is that, usually, the membrane-internal steady-state is attained rather slowly as compared to the outside equilibration. Hence the response time, which mainly reflects the slowest equilibration process,

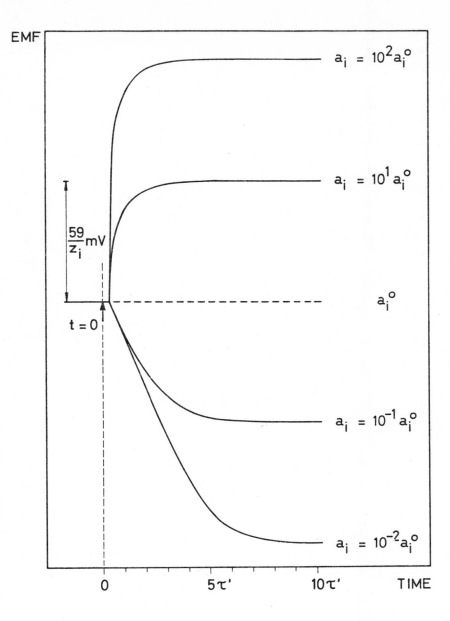

Figure 14.2. Theoretical EMF-response vs. time profiles for ion-exchange membrane electrodes, calculated according to Eq. (14.17).

Table 14.1. Theoretical response time values for membrane electrodes with diffusion control

Response time parameter		Values [s] calculated from Eqs. (14.17) and (14.22)				
		$\tau = 0$		$\tau = 1ms$		
		$\tau' = 0.1s$	$\tau' = 1s$	$\tau' = 0$	$\tau' = 0.1s$	$\tau' = 1s$
$t_{1/2}$	for 10-fold activity increase	0.05	0.52	<0.01	0.07	0.57
t_{95}		0.24	2.35	0.07	0.30	2.53
$t_{99.5}$		0.46	4.61	6.18	6.18	7.21
$t_{1/2}$	for 10-fold activity decrease	0.17	1.67	0.02	0.20	1.77
t_{95}		0.45	4.54	5.44	5.44	6.79
$t_{99.5}$		0.69	6.90	604	604	604

Table 14.2. Experimental response time values for a Ca^{2+} ion-exchanger membrane electrode [4]

Activity change	t_{95} [s]	τ' [s] calculated from Eq. (14.17)
$10^{-4} \rightarrow 10^{-3}M$	2.3 ± 0.2	0.98 ± 0.08
$10^{-3} \rightarrow 10^{-2}M$	2.2 ± 0.2	0.93 ± 0.08
$10^{-2} \rightarrow 10^{-1}M$	2.2 ± 0.1	0.93 ± 0.04
$10^{-4} \rightarrow 10^{-3}M$	2.3 ± 0.2	0.98 ± 0.08

is then to a large degree determined by the dynamic behavior of the membrane itself. As pointed out in Section 14.1.c, time-dependent variations of the membrane composition are of general importance if two or more sorts of sample ions are participating in the interfacial equilibrium. In the case of neutral carrier membranes, for example, these are the cations (primary ions) and the anions (interfering ions) offered by the sample solution. A theoretical model developed by Morf et al. [8], which was later refined [10], led to the following approximation for the time response of thick carrier membrane electrodes:

$$E(t) = E(\infty) + s \log \left[1 - (1 - \frac{a_i^o}{a_i}) \frac{1}{\sqrt{t/\tau} + 1} \right] \qquad (14.19)$$

This result is strikingly different from the exponential time-relationships found before. The new time constant,

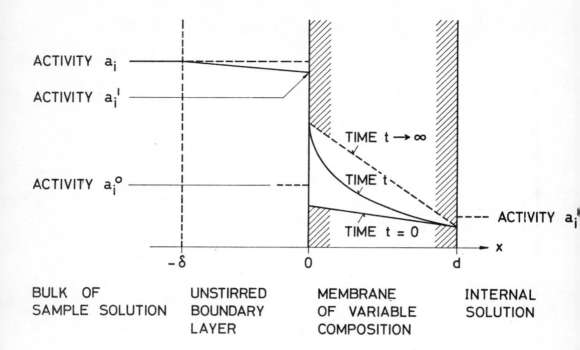

ACTIVITY a_i

ACTIVITY a_i^l

ACTIVITY a_i^o

TIME $t \to \infty$

TIME t

TIME $t = 0$

--- ACTIVITY a_i^l

x

$-\delta$ 0   d

BULK OF
SAMPLE SOLUTION

UNSTIRRED
BOUNDARY
LAYER

MEMBRANE
OF VARIABLE
COMPOSITION

INTERNAL
SOLUTION

Figure 14.3. Diffusion model suited for neutral carrier membrane electrodes [8, 10]. The establishment of steady-state concentration profiles for $t > 0$ is assumed to be rate-controlling.

$$\tau = \frac{DK^2\delta^2}{\pi D'^2} \approx \tau' \frac{D}{D'} K^2, \tag{14.20}$$

is not only affected by the parameters δ and D' referring to the aqueous boundary layer but also depends on the salt diffusion coefficient D and on a salt distribution parameter K for the carrier membrane [8, 10]. Expressions of this type can directly be deduced from the flux equations

$$J_i(t) \cong D' \frac{a_i - a_i'}{\delta}$$ (14.21a)

$$J_i(t) \cong D \frac{c_i' - c_i^o}{\sqrt{\pi\,Dt}} = DK \frac{a_i' - a_i^o}{\sqrt{\pi\,Dt}}$$ (14.21b)

where Eq. (14.21a) is the Nernstian approximation for the ion flux across the unstirred solution film, and Eq. (14.21b) is the well known description for the diffusion into an infinite layer after a concentration step change $c_i^o \to c_i'$ at the boundary. In a more rigorous theoretical analysis of the diffusion problem illustrated in Figure 14.3 [10] the steady-state assumption (14.21a) is no longer upheld. Thus the following combined expression gives the best fit of the exact theory for $t > \tau'$:

$$E(t) = E_i^o + s \log \left[a_i - (a_i - a_i^o) \left(\frac{1}{\sqrt{t/\tau}} + \frac{4}{\pi}\,e^{-t/\tau'} \right) \right]$$ (14.22)

This means that the specific influence of membrane-internal diffusion becomes dominant only in the later period where the outside equilibration is nearly completed.

As a consequence of the square-root time dependence, the practical response time of diffusion-type membranes is usually somewhat increased (see Table 14.1). On the other hand, the qualitative influence of the direction of the sample-activity change on the rate of response is clearly the same as shown before in Figure 14.2. One may generally expect a considerably

26 W. E. Morf   389

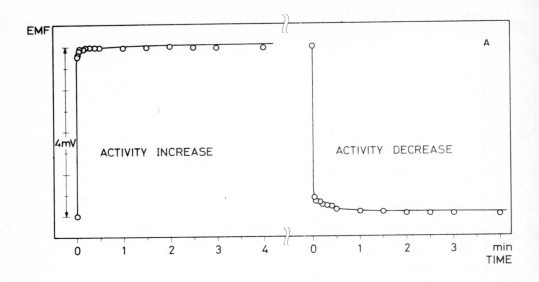

<u>Figure 14.4.</u> EMF-response vs. time profiles of valinomycin-
based PVC membrane electrodes after a step change in KCl activity corresponding to ± 4mV change in final emf [8].
(A) Effect of the direction of sample activity change.
(B) Effect of the membrane composition (see also Table 14.3).
(C) Effect of the stirring rate.
Points are experimental; curves are calculated from Eq. (14.19)
with τ = 0.02 s (nonpolar membrane, fast stirring, activity
increase or decrease), τ = 0.45 s (nonpolar membrane, slow
stirring), and τ = 6 s (polar membrane, fast stirring),
respectively.

390

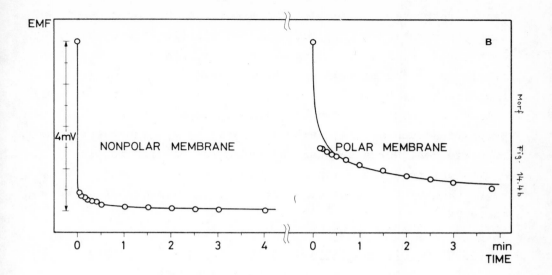

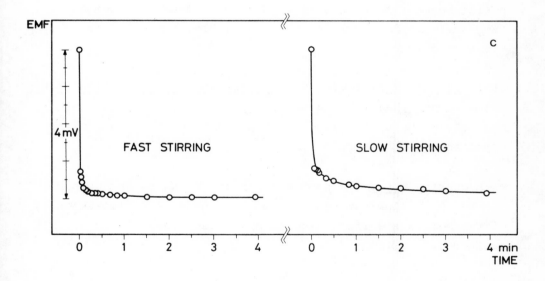

Figure 14.4. (continued)

faster response to a sample if the membrane has been conditioned beforehand with a more diluted solution or, ideally, with a very similar solution where $a_i^o \approx a_i$. In contrast to the traits found for membranes without internal diffusion ($\tau = 0$), the dynamic response characteristics of carrier membranes and related systems are shown to be highly dependent on membrane properties such as the extraction capacity and the resistance to diffusion. For a substantial reduction of τ, which is equivalent to a reduction of the response time, the following requirements have to be considered [8, 10] (Figure 14.4 and Table 14.3).

a) Reduction of salt extraction into the carrier membrane (reduction of K). The membrane components used, i. e. membrane solvent and matrix, should be as nonpolar as possible. The sample solution should contain no highly extractable lipophilic anions. Membranes with incorporated anionic components, such as tetraphenylborate, are preferable in view of an efficient coion exclusion.

b) Reduction of diffusion within the membrane (reduction of D). The membrane phase should be highly viscous, which can readily be realized by specifying a high percentage of the polymeric component.

c) Reduction of δ. The sample solution has to be thoroughly stirred or a flow-through cell must be used. A minimization of the membrane surface (use of microelectrodes) is to be preferred.

In addition, a moderate reduction of the membrane thickness may be advantageous [8, 10].

The response of liquid ion-exchange membranes, based on charged sites of a fixed concentration X, is usually rather

Table 14.3. Experimental time constants for different K$^+$-selective electrodes in KCl solutions

| Membrane composition[a] | | | | Time constants[b] | |
Valinomycin	Tetraphenyl-borate	Solvent (residual percentage)	Matrix	τ	τ'
3.3 %	-	66.7% o-NPOE	PVC	6 s	- [8]
3.3 %	-	66.7% DPP	PVC	0.02 s	- [8]
2.2 %	-	70.2% DBP	PVC	0.02 s	- [6]
2.4 %	0.66 %	70.5% DBP	PVC	KCl : 0.0035 s KSCN: 0.01 s	- - [6]
4.76%	-	-	SR	~0	0.225s [6]

a Abbreviations: o-NPOE: o-nitrophenyl octyl ether, DPP: dipentyl phthalate, DBP: dibutyl phthalate, PVC: polyvinylchloride, SR: silicon rubber.

b All measurements were performed at high stirring frequencies [8] or high flow rates [6].

fast and determined by Eq. (14.17). However, considerable
slowing effects of the type (14.22) are induced by interfering
ions that encroach upon the membrane composition. Then ionic
diffusion or interdiffusion in the membrane phase can no longer
be neglected. For a system with primary and interfering ions
of the same charge, concentration polarization at the interface may be approximated by

$$a_i' \approx a_i - \frac{D\delta}{D'\sqrt{\pi Dt}} (c_i - c_i^o) - (a_i - a_i^o) e^{-t/\tau'} \tag{14.23}$$

where:

$$\frac{c_i}{c_i + c_j} = \frac{a_i}{a_i + K_{ij}a_j} \; ; \; \frac{c_i^o}{c_i^o + c_j^o} = \frac{a_i^o}{a_i^o + K_{ij}a_j^o}$$

An analogous expression holds for a_j'. If the cell potential
follows a simple Nicolsky-type behavior at any time $t >> 0$, the
approximation to the final steady-state must occur according
to the function

$$E(t) = E(0) + s \log \left[A - (A-1) e^{-t/\tau'} - B \frac{1}{\sqrt{t/\tau}} \right] \tag{14.24}$$

with

$$A = \frac{a_i + K_{ij}a_j}{a_i^o + K_{ij}a_j^o} \tag{14.25}$$

$$B = \frac{1 - K_{ij}}{a_i^o + K_{ij}a_j^o} \left(\frac{a_i}{a_i + K_{ij}a_j} - \frac{a_i^o}{a_i^o + K_{ij}a_j^o} \right) \tag{14.26}$$

394

$$\tau = \frac{D \, (X/z)^2 \delta^2}{\pi D'^2} \qquad\qquad (14.27)$$

Equations (14.24) - (14.27) are excellently suited to rationalize the observed response of liquid membrane electrodes. A typical example is given in Figure 14.5 where the experimental emf vs. time profiles of the Orion Ca^{2+}-electrode in the presence of Mg^{2+} ions [4] are fitted by computed values. The agreement between theory and experiment is surprising, so much the more since the same values of K_{CaMg}, τ', and τ were used for all calculations.

One interesting feature of Eq. (14.24) is its capacity to simulate transient responses or potential overshoots. Such phenomena were observed for the mentioned electrode system when magnesium ions were added to samples having a constant background of calcium ions [30]. Indeed, the condition for Eq. (14.24) to predict anomalous response to increasing activities is $B<<0$, respectively $a_j/a_j^o >> a_i/a_i^o$ for $K_{ij}<1$. Hence transient emf-excursions are usually found for those kinds of interfering ions which on the other hand induce sluggish response in solutions of the primary ion ($B>>0$ resp. $a_i/a_i^o >> a_j/a_j^o$, see Figure 14.5). Alternative theories on the origin of transient response phenomena were presented earlier. Stover and Buck [16] demonstrated by computer simulation of liquid ion-exchange membranes that internal equilibration processes may give rise to non-monotonic variations of the diffusion potential. The requirement to produce an overshoot in potential-time response was low mobility of the entering ions, relative to the mobility of the ions already in the membrane [16]. Exactly the same argument was used intuitively by Rechnitz and Kugler [1] to explain the observed transient response of glass electrodes to sudden changes in the activity of interferents. Belijustin et al. [31] reported on similar

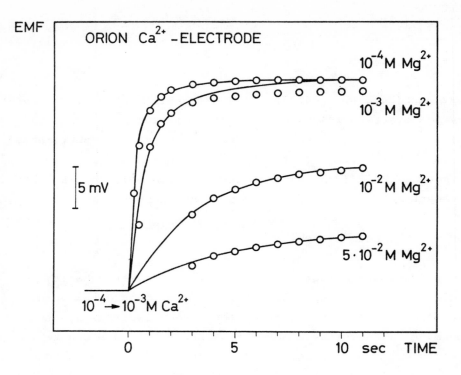

Figure 14.5. EMF-response vs. time profiles of a calcium-selective liquid membrane electrode (Orion 92-20) after an activity step $10^{-4}M \rightarrow 10^{-3}M$ Ca^{2+} in the presence of various activities of Mg^{2+}.

The experimental curves are taken from fig. 2 in Ref. 4 (s = 25.5 mV). The points were computed from Eqs. (14.24) – (14.26) with K_{CaMg} = 0.011, τ' = 0.8s, and τ = $6.25 \cdot 10^{-6}$ M^2s.

results. Finally, transient response or sluggish response of diffusive membrane electrodes was ascribed to possible inhomogeneities of the membrane phase [15]. It was shown that the initial emf-excursion reflects the selectivity behavior of the membrane surface, while the long-time response is dominated by the bulk membrane properties.

The preceding short review on available theories makes it clear that the definition of a meaningful and universally acceptable response-time parameter remains problematic [32]. In fact, the speed and mode of electrode response depend not only on the membrane type used but also on the composition of the sample solution and on various other parameters. It is even conceivable that different points on an electrode surface exhibit different time courses of potential-generating processes [3]. Therefore, the ingenious flow-analyzer technique proposed by Růžička et al. [33], which actually measures the potential $E(t)$ after a constant and relatively short interval t instead of the final value $E(\infty)$, may be ill-suited for analytical work of high reproducibility. In cases where waiting for the final steady-state is inacceptable, it should rather be attempted to extrapolate the values $E(\infty)$ from the available response curve sections $E(t_1)$ .... $E(t_2)$. Figure 14.6 illustrates the application of such procedures [10] which may find use as practical aids.

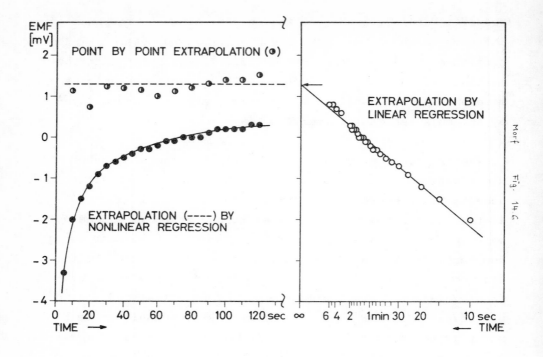

Morf Fig. 14.6

Figure 14.6. Determination of the steady-state EMF of a valinomycin-based membrane electrode (slowly stirred sample) using
different extrapolation procedures [10]. Nonlinear regression:
curve fit by Eq. (14.19). Linear regression: curve fit based
on the linearized version $E(t) = E(\infty) - C\,t^{-\frac{1}{2}}$. Point by point
extrapolation: early stage estimation inserting $E(\infty) =$
$3.414\,E(t) - 2.414\,E(t/2)$.

REFERENCES

[1] G. A. Rechnitz and G. C. Kugler, Anal. Chem. 39, 1682 (1967).

[2] K. Tóth, I. Gavallér, and E. Pungor, Anal. Chim. Acta 57, 131 (1971).

[3] E. Lindner, K. Tóth, and E. Pungor, Anal. Chem. 48, 1051 (1976).

[4] B. Fleet, T. H. Ryan, and M. J. D. Brand, Anal. Chem. 46, 12 (1974).

[5] P. L. Markovic and J. O. Osburn, AIChE J. 19, 504 (1973).

[6] E. Lindner, K. Tóth, E. Pungor, W. E. Morf, and W. Simon, Anal. Chem. 50, 1627 (1978).

[7] G. A. Rechnitz and H. F. Hameka, Fresenius' Z. Anal. Chem. 214, 252 (1965).

[8] W. E. Morf, E. Lindner, and W. Simon, Anal. Chem. 47, 1596 (1975).

[9] K. Tóth and E. Pungor, Anal. Chim. Acta 64, 417 (1973).

[10] W. E. Morf and W. Simon, in Ion-Selective Electrodes in Analytical Chemistry (H. Freiser, ed.), Plenum, New York, 1978, p. 211.

[11] R. P. Buck, Crit. Rev. Anal. Chem. 5, 323 (1975).

[12] R. P. Buck, in Ion-Selective Electrodes in Analytical Chemistry (H. Freiser, ed.), Plenum, New York, 1978, p. 1.

[13] F. S. Stover, T. R. Brumleve, and R. P. Buck, Anal. Chim. Acta, in press (1979).

[14] A. Shatkay, Anal. Chem. 48, 1039 (1976).

[15] W. E. Morf, Anal. Lett. 10(2), 87 (1977).

[16] F. S. Stover and R. P. Buck, Biophys. J. 16, 753 (1976).

[17] J. R. Macdonald, J. Chem. Phys. 60, 343 (1974); 61, 3977 (1974).

[18] J. R. Macdonald, J. Applied Phys. 45, 73 (1974).

[19] J. R. Macdonald, J. Electroanal. Chem. 53, 1 (1974).

[20] J. R. Sandifer and R. P. Buck, J. Electroanal. Chem. 56, 385 (1974).

[21] W. E. Morf, M. Oehme, and W. Simon, in Ionic Actions on Vascular Smooth Muscle (E. Betz, ed.), Springer-Verlag, Berlin, Heidelberg, New York, 1976, p. 1; M. Oehme, Dissertation ETH, Zürich, 1977.

[22] G. Johansson and K. Norberg, J. Electroanal. Chem. 18, 239 (1968).

[23] J. Buffle and N. Parthasarathy, Anal. Chim. Acta 93, 111 (1977).

[24] N. Parthasarathy, J. Buffle, and W. Haerdi, Anal. Chim. Acta 93, 121 (1977).

[25] R. H. Müller, Anal. Chem. 41, 113A (1969).

[26] J. Mertens, P. Van den Winkel, and D. L. Massart, Anal. Chem. 48, 2 (1976).

[27] R. P. Buck, J. Electroanal. Chem. 18, 363 (1968).

[28] B. Karlberg, J. Electroanal. Chem. 42, 115 (1973); 49, 1 (1974).

[29] W. Jaenicke and M. Haase, Z. Elektrochem. 63, 521 (1959).

[30] G. A. Rechnitz, in Ion-Selective Electrodes (R. A. Durst, ed.), National Bureau of Standards Spec. Publ. 314, Washington, 1969.

[31] A. A. Belijustin, I. V. Valova, and I. S. Ivanovskaja, in Ion-Selective Electrodes (E. Pungor, ed.), Akadémiai Kiadó, Budapest, 1978, p. 235.

[32] IUPAC Recommendations for Nomenclature of Ion-Selective Electrodes, Pure Appl. Chem. 48, 127 (1976).

[33] J. Růžička, E. H. Hansen, and E. A. Zagatto, Anal. Chim. Acta 88, 1 (1977).

Chapter 15

Special Arrangements: Gas-Sensing Electrodes and Enzyme Electrodes

A new field of analytical applications was opened with the design of compound electrodes in which an ion-selective electrode is combined with a specific chemical reaction or a separation step. In gas-sensing electrodes a conventional ion sensor is contacted with a thin film of reagent solution. A gas-permeable membrane is interposed between this electrochemical cell and the sample under test. The gas to be determined diffuses through the separation layer until an equilibrium is established in the internal electrolyte film. The dissolved gas and the reagent of the internal electrolyte constitute a buffering system, the activity of the buffered ion being sensed by the ion-selective electrode.

Enzyme electrodes usually consist of an "active" membrane, containing an immobilized enzyme, coupled to an ion-selective electrode or a gas sensor. Such systems operate by converting a substrate to a species that can be sensed potentiometrically.

Gas-sensing probes and enzyme electrodes represent rather ingenious applications of conventional electrochemical sensors. Their intriguing feature is that they combine a highly specific reaction with a selective detection of the products formed. Such compound electrodes have become a valuable tool in analytical chemistry and they certainly have a promising future.

401

15.1. GAS-SENSING ELECTRODES

In 1956, Clark [1] had the pioneering idea to couple a gas-permeable membrane with an electrochemical sensor. He developed an oxygen electrode, consisting of a membrane and a platinum electrode, separated by a thin layer of indifferent electrolyte. An amperometric method was used to detect the oxygen diffusing through the membrane. The principle of the Clark electrode is still applied in modern oxygen sensors [2 - 6]. A similar construction was introduced by Severinghaus [7] for the measurement of carbon dioxide. However, in his CO_2-sensor the gas-permeable membrane was combined with a complete ion-selective electrode cell of the type

$$\text{reference electrode} \mid NaHCO_3(aq) \mid \text{pH-electrode} \qquad (15.1)$$

Carbon dioxide, entering through the membrane into the internal electrolyte, participates in the equilibrium

$$CO_2 + H_2O \overset{K_{CO_2}}{\rightleftharpoons} HCO_3^- + H^+ \qquad (15.2)$$

Since the bicarbonate activity is kept constant ($a'_{HCO_3} = a$), the sensed pH is directly related to the amount of CO_2 dissolved in the internal solution:

$$a'_H = \frac{K_{CO_2}}{a} \cdot a'_{CO_2} \qquad (15.3)$$

When equilibrium is established throughout the membrane, the activity of internal CO_2 has approximated that of the gas in the sample, a_{CO_2}. Hence the response of the gas-sensing probe is Nernstian and is a direct measure of the CO_2 level in the sample.

The original idea of Severinghaus was not followed up until the early seventies when an ammonia-sensing electrode and subsequently a series of other gas sensors were developed (for a review, see [8]). In the NH_3-electrode, the following electrochemical cell is applied [8][*]:

AgCl reference electrode | 0.01M NH_4Cl | pH-electrode   (15.4)

Hence the detection of ammonia is based on the equilibrium

$$NH_3 + H^+ \xrightleftharpoons{1/K_{NH_4}} NH_4^+ \qquad (15.5)$$

$$a_H' = K_{NH_4} \, 10^{-2}M/a_{NH_3}' \qquad (15.6)$$

Again a Nernstian response is obtained, but here the slope is -59 mV instead of the +59 mV found above for the acid species. The calibration curve of an ammonia-sensing electrode using the cell (15.4) is shown in Figure 15.1. The experimental curve can be nicely rationalized from theory. Evidently, the linear range of the response function lies between $10^{-1}M$ and $10^{-5}M$ NH_3. Deviations from linearity occurring at higher levels are largely a consequence of the interference of NH_3 with the AgCl reference electrode, as treated theoretically in Chapter 10. The lower detection limit of the gas-sensing electrode is due to the fact that the basic equilibrium (15.5) obtains even in absence of external gas. Accordingly, a small amount of NH_3 is generated which only slowly diffuses out of the internal electrolyte (see below).

------

[*] The reference electrode directly measures the chloride level of the electrolyte; no liquid junction is required.

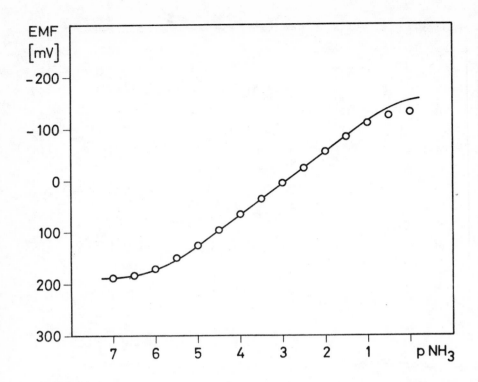

Figure 15.1. Calibration curve for an ammonia-sensing elec-
trode based on cell (15.4).
Solid line: experimental working curve of the electrode [8].
Circles: calculated values [9]; comment is found in the text.

A critical factor in the use of gas-sensing electrodes is their response time. An adequate theoretical analysis was presented by Ross et al. [8]. Their results show that the time course of equilibration between the internal and the external solution is given by a function (written for the NH_3-electrode)

$$a'_{NH_3} \approx a_{NH_3} - (a_{NH_3} - a^o_{NH_3})\, e^{-t/\tau} \qquad (15.7)$$

with

$$\tau = \frac{V\, d}{A\, D\, k} \qquad (15.8)$$

The activity a'_{NH_3} sensed at the time t obviously depends on both the initial activity $a^o_{NH_3}$ and the final (sample) activity a_{NH_3}, as well as on a time constant τ. The latter parameter includes contributions by the volume V of the internal electrolyte, the thickness d and the active area A of the gas-permeable membrane, as well as the diffusion coefficient D and the partition coefficient k characteristic of the neutral species in the membrane. As a rule of thumb, a 99% response time of about 2 min (tenfold activity increase) was reported for ammonia- and related gas-sensing electrodes [8]. This would correspond to a time constant on the order of 30 s. Indeed a value of τ = 31.5 s was obtained from the data fit in Figure 15.1 where the practical limit of detection was defined as the activity level indicated by the electrode for a_{NH_3} = 0, $a^o_{NH_3}$ = 10^{-2}M, and t = 5 min [8].

Equation (15.8) reveals that different factors may contribute to a reduction of the time constant τ, which means a reduction of the response time. First, the geometric parameters d and V/A should be kept as small as possible; the latter parameter may be identified with the thickness of the internal

solution film [8]. Second, the values of D and k must be sufficiently high. The best choice in this respect is a membrane
consisting essentially of an air layer. Such systems were
realized by using microporous membranes that are not wetted
by contact with aqueous solutions [8] or by creating a complete
air gap between the internal solution film and the sample [10,
11]. Homogeneous, nonporous plastic membranes were found to be
less suited because diffusion coefficients in such condensed
phases are typically 10^4 times smaller than in air. Efforts
aimed at a fundamental understanding of gas-sensing electrodes
[8] have finally led to significant improvements and very elegant new constructions. At present, useful sensors for NH_3,
CO_2, SO_2, NO_x, H_2S, HCN, and other gaseous compounds are
commercially available (for a review, see [8, 12, 13]). These
systems offer a high specificity for the species to be determined.

15.2. ENZYME ELECTRODES

The first enzyme-coupled electrochemical sensor was introduced in 1962 by Clark and Lyons [14]. They designed a glucose
sensor based on immobilized glucose oxidase; the hydrogen
peroxide generated by the enzymatic reaction was determined
amperometrically. Guilbault and coworkers [15 - 18] followed
up this idea by constructing a urea-sensitive electrode,
using an acrylamide gel membrane with trapped urease, coupled
to an ion-selective electrode. The original version [15, 16]
consisted of a coated cation-selective glass electrode,
responding to the ionic product of the enzymatic reaction

$$CO(NH_2)_2 + 2 H_3O^+ \xrightarrow{\text{urease}} 2 NH_4^+ + CO_2 + H_2O \qquad (15.9)$$

In a more recent version, the glass electrode has been replaced by a NH_4^+-selective electrode using the carrier nonactin [17, 18]. An alternative construction [19] uses a gas-sensing electrode for monitoring the CO_2 formed by reaction (15.9).

During the past decade, the field of enzyme-coupled electrochemical sensors expanded rapidly. There are several comprehensive reviews relating enzymes to electrodes [12, 13, 20 - 25]. The scope of this section is to simply give a short introduction into the principles of potentiometric enzyme electrodes.

The specific reaction of a substrate S with an enzyme E, leading to a ionic product I, is usually modeled kinetically by a Michaelis-Menten mechanism:

$$S + E \underset{k_{-1}}{\overset{k_1}{\rightleftharpoons}} SE \overset{k_2}{\longrightarrow} E + n \ I + \ldots \qquad (15.10)$$

At steady-state, the rate of the enzymatic reaction is then determined as

$$-\left(\frac{\partial a_S'}{\partial t}\right)_e = \frac{k_2 \ a_E \ a_S'}{K_M + a_S'} \qquad (15.11)$$

$$\left(\frac{\partial a_I'}{\partial t}\right)_e = - \ n \ \left(\frac{\partial a_S'}{\partial t}\right)_e \qquad (15.12)$$

where a_S' denotes the activity of substrate available in the reaction system, a_E is the total enzyme activity, and K_M is the Michaelis constant (mol/l) characteristic of the enzyme-substrate reaction:

$$K_M = \frac{k_{-1} + k_2}{k_1}$$

In the context of enzyme electrodes, the activities a_S' and a_I' refer to the enzyme layer contacting the ion sensor. Exchange of substrate and product species between the ion-selective electrode and the external solution (a_S, a_I) occurs by diffusion. The rate of diffusional uptake into the enzyme membrane, respectively release from the membrane, may be formulated by the following approximation which is correct for limiting cases [9]:

$$\left(\frac{\partial a_S'}{\partial t}\right)_d \cong k_d \ (a_S - a_S') \tag{15.13}$$

$$\left(\frac{\partial a_I'}{\partial t}\right)_d \cong k_d \ (a_I - a_I') \tag{15.14}$$

with $k_d \cong \frac{2\,D}{\delta^2}$. D is the mean diffusion coefficient and δ the thickness of the diffusion layer (defined by the enzyme membrane and additional electrode coatings). The assumption of a steady-state requires that

$$\frac{\partial a_S'}{\partial t} = \left(\frac{\partial a_S'}{\partial t}\right)_e + \left(\frac{\partial a_S'}{\partial t}\right)_d = 0 \tag{15.15}$$

$$\frac{\partial a_I'}{\partial t} = \left(\frac{\partial a_I'}{\partial t}\right)_e + \left(\frac{\partial a_I'}{\partial t}\right)_d = 0 \tag{15.16}$$

Combination of Eqs. (15.11) - (15.16) leads to the following solution:

$$a_S' + \frac{k_2 a_E}{k_d} \frac{a_S'}{K_M + a_S'} = a_S \qquad (15.17)$$

$$a_I' = a_I + n (a_S - a_S') \qquad (15.18)$$

The first expression constitutes a quadratic equation which can easily be solved to yield a_S'. Subsequently, a_I' is determined from the second expression. The electrode response is finally given by

$$E = E_I^o + s \log [a_I' + \sum_J K_{IJ}^{Pot} a_J'^{z_I/z_J}]$$

$$= E_I^o + s \log [n (a_S - a_S') + a_0] \qquad (15.19)$$

where $a_0 \cong a_I + \sum K_{IJ}^{Pot} a_J^{z_I/z_J}$ is the total activity of the ionic background in the sample, as sensed by the ion-selective electrode in the absence of substrate.

Figure 15.2 illustrates calibration plots of urea-sensitive electrodes. Equations (15.17) - (15.19) were used to fit the experimental data [16] obtained for different enzyme activities. The same kinetic parameters were used for all calculations which, nevertheless, show good agreement with experiment. Three major regions, corresponding to different limiting cases, are discerned for each response curve in Figure 15.2. These are discussed in the following.

For high substrate activities, i. e. a_S and $a_S' >> K_M$,

409

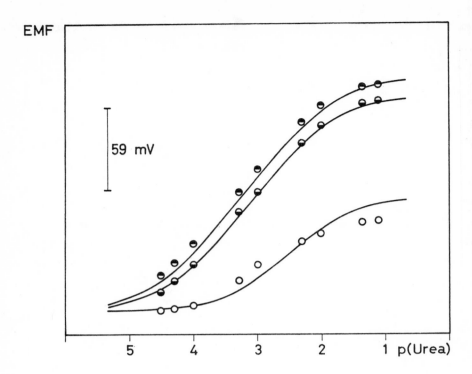

Figure 15.2. Calibration curves for different urea-sensitive electrodes based on the enzyme urease.

Points: experimental results obtained for 233, 140, and 8 mg urease per ml of membrane gel, respectively [16].

Solid lines: calculated response from Eqs. (15.17) – (15.19) with $K_M = 0.015$ mol l^{-1}, $k_2 a_E / k_d K_M = 1.786 \times 10^{-4} \times a_E$ (in g l^{-1}), and $a_0 = 2 \times 10^{-6}$ mol l^{-1}. (●) $a_E = 233$ g l^{-1}; (◖) $a_E = 140$ g l^{-1}; (o) $a_E = 8$ g l^{-1}. Note: the same curves are obtained when using approximation (15.21) with $K_{IS}^{Pot} = nk_2 a_E / (k_2 a_E + k_d K_M + k_d a_S)$.

410

Eq. (15.11) predicts a pseudo first-order reaction, the reaction rate depending only on the enzyme activity. Accordingly, the emf-response of the enzyme electrode becomes independent of the substrate activity (see Figure 15.2):

$$E = E_I^o + s \log \left[n \frac{k_2 a_E}{k_d} + a_0 \right] \tag{15.20}$$

This generally imposes an upper limit ($a_S \sim K_M$) on the detection of a substrate by an enzyme-coupled electrochemical sensor. However, this working range may be attractive for the study of inhibitors [26, 27] that reversibly affect the activity of the enzyme.

At lower activities, a_S and $a_S' << K_M$, a second-order reaction results, which fact is observable in Figure 15.2 by a linear response of the enzyme electrode to the substrate activity. Equations (15.17) - (15.19) here lead to the simplified relationship

$$E = E_I^o + s \log \left[K_{IS}^{Pot} a_S + a_0 \right] \tag{15.21}$$

$$K_{IS}^{Pot} = n \frac{k_2 a_E}{k_2 a_E + k_d K_M} \leqslant n$$

It may be recognized that a high enzyme activity is required for the electrode to show optimized selectivity for substrate relative to interfering species. An additional improvement in this respect could be achieved by a moderate reduction of the diffusion rate constant k_d, involving a reduction of diffusion coefficients D in the electrode coating and/or an increase of the diffusion layer thickness δ. However, one then has to

endure a certain slowing of the electrode response since diffusional relaxation is characterized by a time constant [28]

$$\tau' = \frac{4 \, \delta^2}{\pi^2 \, D} \approx 1/k_d \tag{15.22}$$

The fact of diffusion control of reactants and products through
coated electrodes was treated theoretically by Racine and Mindt
[29], Tran – Minh and Broun [30], Blaedel [37], and more recently
by Carr [28] who presented a rigorous analysis of the response
vs. time behavior of enzyme electrodes.

For values of $K_{IS}^{Pot} \, a_S << a_0$, enzyme electrodes become insen-
sitive to the substrate activity. The existence of such a
lower limit of detection is clearly demonstrated by Eq. (15.21)
and Figure 15.2. Two points are essential for an extension of
the linear response range towards lower levels of a_S. First,
a high enzyme activity must be applied in order to increase
the fraction of substrate reacting in the active layer of the
electrode (increase of K_{IS}^{Pot}). This leads to an upward shift
of the calibration curves, as is indicated in Figure 15.2.
Second, the electrochemical sensor should be highly specific
for the product of the enzymatic reaction, and the intrinsic
detection limit of the sensor must be low, which ensures a
low background activity a_0. Unfortunately, some of the attractive ion-selective electrodes, especially the ammonium ion
sensors, are severely limited in this respect. The lack of
specificity of the detection step can, of course, neutralize
the specificity offered by the enzymatic reaction. The parti-
cular problem of interfering ions may be eliminated by the
use of gas-sensing probes (e. g., NH_3 electrodes [31 - 33])
which allow a more specific detection of the reaction products.

At present, there still seem to be no dipping-type enzyme

electrodes commercially available - in spite of the anticipated
simplicity of operation of such devices. Instead, the manufacturers have chosen to separate the enzyme and sensor functions (for a review, see [25]). This has the advantage that
the conditions for both the reaction and the detection step
can be optimized individually. The enzymes can be immobilized
more stably in reactors [25, 27, 34 - 36], the lifetime of the
catalysts thereby being increased considerably. Such enzyme
reactors have become an important tool in various branches of
chemistry. For analytical purposes, they may be coupled to
potentiometric sensing units or to amperometric, spectrophoto-
metric, or thermal detectors [25].

There is no doubt that the technology of enzyme electrodes
will be much improved in future. Considerable efforts are
concentrated on new developments and more in-depth investi-
gations in this field. Since there exist several thousands of
enzymes and specific substrates, the marriage of enzyme-cata-
lyzed reactions and electrochemical sensors will certainly
bring forth a rapid increase in attractive analytical applications.

REFERENCES

[1] L. C. Clark, Trans. Am. Soc. Artificial Internal Organs
     2, 41 (1956).

[2] P. Eberhard, K. Hammacher, and W. Mindt, Biomed. Technik
     18, 216 (1973).

[3] H. Degn, I. Balslev, and R. Brook, Measurement of Oxygen,
     Elsevier, Amsterdam, 1976.

[4] N. Lakshminarayanaiah, Membrane Electrodes, Academic
     Press, New York, 1976.

[5] J. W. Severinghaus, J. Peabody, A. Thunstrom, P. Eber-
     hard, and E. Zappia, eds., 'Methodologic aspects of
     transcutaneous blood gas analysis', Acta Anaesthes.
     Scand. Suppl. 68, 1ff (1978).

[6] M. L. Hitchman, Measurement of Dissolved Oxygen, Wiley,
     New York, 1978.

[7] J. W. Severinghaus and A. F. Bradley, J. Appl. Physiol.
     13, 515 (1958).

[8] J. W. Ross, J. H. Riseman, and J. A. Krueger, Pure
     Appl. Chem. 36, 473 (1973).

[9] W. E. Morf, unpublished results.

[10] J. Růžička and E. H. Hansen, Anal. Chim. Acta 69, 129
     (1974); 72, 215 (1974).

[11] U. Fiedler, E. H. Hansen, and J. Růžička, Anal. Chim.
     Acta 74, 423 (1975).

[12] R. P. Buck, Anal. Chem. 48, 23R (1976).

[13] R. P. Buck, Anal. Chem. 50, 17R (1978).

[14] L. C. Clark and C. Lyons, Ann. N. Y. Acad. Sci. 102,
     29 (1962).

[15] G. G. Guilbault and J. G. Montalvo, Anal. Lett. 2, 283
     (1969).

[16] G. G. Guilbault and J. G. Montalvo, J. Am. Chem. Soc.
     92, 2533 (1970).

[17] G. G. Guilbault and G. Nagy, Anal. Chem. 45, 417 (1973).

[18] G. G. Guilbault and G. Nagy, Anal. Lett. 6, 301 (1973).

[19] G. G. Guilbault and F. R. Shu, Anal. Chem. 44, 2161 (1972)

[20] G. D. Christian, in Advance in Biomedical Engineering and Medical Physics, Vol. IV (S. N. Levine, ed.), Inter-science, New York, 1971, p. 95.

[21] C. B. Wingard, ed., Biotechnol. and Bioeng. Symp. 3, Wiley, New York, 1972.

[22] M. Kessler, L. C. Clark, Jr., D. W. Lübbers, I. A. Silver, and W. Simon, eds., Ion and Enzyme Electrodes in Biology and Medicine, Urban and Schwarzenberg, Munich, 1976.

[23] G. G. Guilbault, Handbook of Enzymatic Methods of Analysis, M. Dekker, New York, 1976.

[24] G. G. Guilbault, 'Enzyme electrodes in analytical chemistry', in Comprehensive Analytical Chemistry, Vol. Vlll (G. Svehla, ed.), Pergamon Press, Amsterdam, 1977.

[25] D. N. Gray, M. H. Keyes, and B. Watson, Anal. Chem. 49, 1067A (1977).

[26] C. Tran-Minh and J. Beaux, Anal. Chem. 51, 91 (1979).

[27] L. Ögren and G. Johansson, Anal. Chim. Acta 96, 1 (1978).

[28] P. W. Carr, Anal. Chem. 49, 799 (1977).

[29] P. Racine and W. Mindt, Experientia Suppl. 18, 525 (1971).

[30] C. Tran-Minh, Thesis, Rouen, 1971; C. Tran-Minh and G. Broun, Anal. Chem. 47, 1359 (1975).

[31] R. A. Llenado and G. A. Rechnitz, Anal. Chem. 46, 1109 (1974).

[32] H. Thompson and G. A. Rechnitz, Anal. Chem. 46, 246 (1974).

[33] D. S. Papastathopoulos and G. A. Rechnitz, Anal. Chim. Acta 79, 17 (1975).

[34] G. Johansson and L. Ögren, Anal. Chim. Acta 84, 23 (1976).

[35] G. Johansson, K. Edström, and L. Ögren, Anal. Chim. Acta 85, 55 (1976).

[36] G. Johansson and L. Ögren, in <u>Ion-Selective Electrodes</u>
      (E. Pungor and I. Buzás, eds.), Akadémiai Kiadó, Buda-
      pest, 1977, p. 93.

[37] W. J. Blaedel, T. R. Kissel, and R. C. Boguslaski,
      <u>Anal. Chem.</u> <u>44</u>, 2030 (1972).

AUTHOR INDEX

Alfenaar, M., [57] 16, *24*
Ammann, D., [24, 25, 27] 1, [43] 7, [27] 8, [43] 17, [43] 20, *22, 23,* [7] 28, [7, 15] 30, [15] 31, *34,* [11] 42, *43,* [33] 58, *63,* [5] 66, [5] 73, *74,* [8] 79, *86,* [16] 106, *112,* [6, 9] 113, [6, 9] 114, [9] 134, [9] 145, [6, 9] 148, *150,* [38] 167, *209,* [57–59] 213, [57] 218, [73] 255, [59, 73] 257, *262, 263,* [8] 264, [8, 19, 23, 24, 27, 33, 35] 268, [33] 269, [8, 24, 43] 270, [27] 271, [8, 43] 272, [43] 273, [57] 277, [33] 288, [70] 289, [70] 291, [70] 292, [70] 293, [27, 57] 296, [57] 297, [57] 303, [57] 305, [35] 308, [57] 309, [57] 312, [57] 314, [23, 33] 315, [84] 316, [8, 33, 43, 84, 88] 322, [33] 234, [33] 325, [88, 93] 326, [88] 327, [8, 24] 328, [27] 330, *331–336*
Arvanitis, S., [7] 28, [7] 30, *34,* [16] 106, *112,* [10] 113, [10] 127, [10] 145, [10] 148, *150,* [58] 213, *262,* [50] 275, [50] 278, [50] 279, [50] 280, [50] 281, *334*

Bach, H., [42] 363, *374*
Bäck, S., [67] 223, [67] 229, *262*
Bagg, J., [45] 170, *209*
Bailey, P. L., [42] 1, *23,* [24] 165, [24] 172, *208*
Balslev, I., [3] 402, *414*
Bates, R. G., [56–59] 16, [59] 17, [59] 19, *24,* [14] 73, *74*
Baucke, F. G. K., [40] 165, [40] 173, *209,* [41–43] 363, *374*
Baum, G., [40] 212, [40, 48, 62] 214, [40, 48] 216, *261, 262*
Baumann, E. W., [44] 270, [44] 272, *333*
Baxter, W. P., [7] 337, [7] 338, *372*
Bean, C. P., [22] 211, *260*
Beaux, J., [26] 411, *415*
Bedeković, D., [37, 38] 268, *333*
Behn, U., [11] 98, *112*
Belijustin, A. A., [18] 339, [33] 340, [18] 351, [18] 352, [18] 354, [18] 355, [18] 356, [18] 358, [18] 359, [18] 361, [33] 362, [33, 54] 363, [33] 368, [18] 371, *373, 374,* [31] 395, *400*
Bergveld, P., [47] 9, *23*
Bernstein, J., [1] 337, *372*
Bertrand, C., [47] 216, *261*
Beutner, R., [24] 211, *260*
Bissig, R., [25] 1, *22,* [6, 9] 113, [6, 9] 114, [9] 134, [9] 145, [6, 9] 148, *150,* [8] 264, [8, 24, 33] 268, [33] 269, [8, 24] 270, [8] 272, [33] 288, [33] 315, [8, 33, 88] 322, [33] 324, [33] 325, [88, 94] 326, [88] 327, [8, 24] 328, *331–333, 336*
Blaedel, W. J., [37] 412, *416*
Bloch, R., [3] 113, *150,* [37] 212, [37] 217, *261,* [11] 264, [11] 270, [55] 277, [55] 281, [55] 296, [55] 297, [55] 309, *331, 334*
Blum, E., [13] 339, *372*
Boguslaski, R. C., [37] 412, *416*
Boksay, Z., [28] 339, [28] 345, [44–47] 363, *373, 374*
Boles, J. H., [38] 145, *152,* [54] 277, [54] 281, [54] 296, [54] 297, [54] 300, [54] 303, [54] 306, [54] 307, *334,* [39] 361, *374*
Bonhoeffer, K. F., [25] 211, *260*
Borowitz, I. J., [88] 322, [88] 326, [88] 327, *336*
Bound, G. P., [39] 167, [39] 205, [39] 206, *209*
Bouquet, G., [44, 45] 363, *374*
Bradley, A. F., [7] 402, *414*
Brand, M. J. D., [53] 363, *374,* [4] 375, [4] 384, [4] 387, [4] 395, [4] 396, *399*
Březina, M., [32] 268, [32] 270, *332*
Brook, R., [3] 402, *414*
Broun, G., [30] 412, *415*
Brumleve, T. R., [13] 375, [13] 377, [13] 379, *399*
Büchi, R., [62] 277, [62] 284, [62] 296, [62] 297, [62] 298, [62] 299, [62] 303, [62] 307, [89, 91] 322, [93] 326, *334, 336*
Buck, R. P., [5–8] 1, [50] 9, *21, 24* [12, 16] 30, *34,* [12] 41, *43,* [23] 48, [31] 55, [23] 58, *63,* [10] 79, *86,* [38] 145, *152,* [12] 166,

Buck, R. P., (cont.)
[12, 18–22, 36] 167, [36] 168, [44] 169, [19, 21] 172, [36] 173, [19, 21] 181, [19, 21] 183, [36] 190, [36] 191, *207–208*, [69] 240, *262*, [28–30] 268, [29, 54] 277, [54] 281, [28–30, 54, 54a] 296, [29, 54] 297, [54] 300, [54, 54a] 303, [54] 306, [54] 307, *332, 334*, [19] 339, [19] 355, [19] 356, [19] 357, [19] 359, [19] 360, [19, 39] 361, [19, 51, 52] 363, *373, 374*, [11–13, 16] 375, [11–13, 20] 377, [11–13] 379, [27] 384, [16] 395, *399, 400*, [12, 13] 406, [12, 13] 407, *414*
Buffle, J., [23, 24] 381, [23, 24] 382, *400*
Butler, J. A. V., [2] 37, *43*, [31] 123, *153*
Butler, J. N., [11] 165, [11] 167, *207*
Buzás, I., [18, 19] 1, *22*

Cammann, K., [16] 1, *21*, [7, 8] 39, *43*, [23] 167, *208*
Carmack, G. P., [66] 223, [66] 229, *262*
Carr, P. W., [28] 412, *415*
Cary, H. H., [7] 337, [7] 338, *372*
Casby, J. U., [1] 78, *86*, [15] 338, [15] 339, [15] 341, [15] 347, [15] 351, *372*
Cavallone, F., [97] 330, *336*
Caviezel, M., [85] 315, *335*
Choy, E. M., [5] 28, *34*
Christ, C. L., [35] 341, *373*
Christian, G. D., [53] 217, *261*, [75] 296, [75] 306, *335*, [20] 407, *415*
Ciani, S. M., [19] 48, [19] 51, *62*, [6] 78, *86*, [3] 87, *112*, [17, 18] 113, [17, 23, 24] 115, [23] 117, [23] 119, [23] 126, [17, 23, 24] 128, [24] 129, [23, 24] 131, [17, 18, 24] 133, [23, 24] 134, [23, 24] 136, [17, 18, 23, 24] 138, [17, 23, 24] 141, [18, 24] 142, [24] 144, [17, 18, 24] 148, [24] 149, *151*, [51, 53] 274, [66] 287, [66] 288, [51, 53, 69] 289, [69] 291, [51] 300, *334, 335*
Cimerman, Z., [7] 28, [7] 30, *34*, [16] 106, *112*, [6] 113, [6] 114, [6] 148, *150*, [58] 213, *262*, [8] 264, [8, 38] 268, [8] 270, [8] 272, [8] 322, [8] 328, *331, 333*
Clark, L. C., [4] 1, *21*, [1] 402, [14] 406, [22] 407, *414, 415*
Claus, R., [27] 165, *208*
Clementi, E., [96–98] 330, *336*
Coetzee, C. J., [38] 212, [38] 213, [38] 214, [38] 215, [38] 223, [38] 229, *261*
Conti, F., [16] 48, [16] 51, *62*, [2] 78, *86*, [27] 339, [27] 341, [27] 347, *373*
Corongiu, G., [98] 330, *336*
Cosgrove, R. E., [12] 264, [12, 42] 270, *331, 333*
Covington, A. K., [9–11] 1, *21*, [13] 167, *207*
Cram, D. J., [36] 268, *333*

Cremer, M., [2] 337, *372*
Csákvári, B., [28] 339, [28] 345, [45, 46] 363, *373, 374*
Curran, P. F., [20] 211, *260*
Cussler, E. L., [4, 5] 28, *34*

Dahms, H., [24] 339, [24] 341, [24] 345, [24] 346, [24] 347, [24] 351, *373*
Date, K., [28] 167, *208*
Davies, C. W., [52] 14, *24*
Davies, J. E. W., [43] 215, *261*
Dawson, D. G., [23] 211, *260*
Degn, H., [3] 402, *414*
de Jong, F., [36] 268, *333*
Denesi, P. R., [45] 215, [45] 223, [45] 229, [45] 248, *261*
Diebler, H., [86] 315, *336*
Dietrich, B., [18] 268, [18] 315, *332*
Dobler, M., [90] 322, [90] 324, *336*
Dobos, S., [44] 363, *374*
Dohner, R., [22] 1, *22*, [9b] 338, *372*
Dole, M., [5] 337, [21] 339, [21] 344, *372, 373*
Domeier, L. A., [36] 268, *331*
Donnan, F. G., [15, 16] 42, *43*
Duax, W. L., [95] 325, *336*
Durst, R. A., [36, 37] 1, [37] 10, *23*, [9] 165, [9] 167, [48] 181, [48] 182, [50] 183, *207, 209*

Eberhard, P., [2, 5] 402, *414*
Edström, K., [35] 413, *415*
Eigen, M., [86] 315, *336*
Eisenmann, G., [1–3] 1, [1, 2] 6, *21*, [13] 30, [13] 32, *34*, [16–20] 48, [16–20] 51, *62, 63*, [1–7] 78, *86*, [3] 87, *112*, [16–18] 113, [17, 24, 25] 115, [17, 24] 128, [24] 129, [24] 131, [17, 18, 24] 133, [24] 134, [24, 25] 136, [17, 18, 24, 25] 138, [17, 24] 141, [18, 24, 25] 142, [24] 144, [16–18, 24, 25] 148, [24] 149, *151*, [55] 190, *209*, [21] 211, [29, 39] 212, [39] 216, [29] 236, [29] 239, *260, 261*, [51–53] 274, [66] 287, [66] 288, [51–53, 69] 289, [69] 291, [51] 300, [87] 315, [87] 320, *334–336*, [15, 16] 338, [15, 16, 25–27] 339, [15, 16, 25–27] 341, [15, 16, 25–27] 347, [15] 351, [40] 362, [40] 368, *372–374*
Elkins, D., [92] 324, *336*
Erdey-Gruz, T., [3] 37, *43*, [32] 123, *152*
Erne, D., [7] 28, [7] 30, *34*, [16] 106, *112*, [58] 211, *262*
Evans, D. F., [5] 28, *34*
Evans, D. H., [39] 167, [39] 205, [39] 206, *209*
Eyal, E., [41] 148, *152*, [68] 289, [68, 72] 291, *335*
Eyring, H., [4–6] 37, *43*, [29, 30] 121, [29, 30] 123, *151*

Fiedler, U., [6] 113, [6] 114, [6] 148, *150*, [8, 9] 264, [8, 38] 268, [8, 9, 39] 270, [8, 39] 272, [73] 294, [8] 322, [8, 39] 328, *331, 333, 335*, [11] 406, *414*

Fleet, B., [39] 167, [39] 205, [39, 61] 206, *209, 210,* [4] 375, [4] 384, [4] 387, [4] 395, [4] 396, *399*

Frant, M. S., [8] 165, [8] 167, *207*, [6] 264, [6] 270, [6] 306, *331*

Freiser, H., [31] 1, *22*, [7] 113, [7] 148, *150*, [38] 212, [38] 213, [38] 214, [38] 215, [38, 66] 223, [38, 66] 229, *261, 262*

Frensdorff, H. K., [71] 291, *335*

Fujita, A., [2] 211, *259*

Furmansky, M., [11] 264, [11] 270, *331*

Gaboriaud, R., [27] 48, *63*, [6] 65, *74*

Garfinkel, H., [56] 190, *210*, [36] 341, [36] 363, *373*

Garrels, R. M., [55] 15, *24*, [35] 341, *373*

Gavach, C., [47, 49] 216, *261*

Gavallér, I., [2] 375, [2] 384, *399*

Gokel, G. W., [36] 268, *333*

Goldman, D. E., [9] 48, [9] 55, *62*, [5] 88, *112*, [28]117, [28] 131, *151*

Goossen, J., [57] 193, *210*

Gordon, S., [11] 264, [11] 270, *331*

Gray, D. N., [25] 407, [25] 413, *415*

Grekovich, A. L., [46] 216, *261*

Grove-Rasmussen, K. V., [10] 68, *74*

Grubb, W. T., [13] 264, [13] 270, *331*

Guggenheim, E. A., [51] 14, [56] 16, *24*, [1] 35, [16] 42, *43*, [1] 47, [1] 48, *62*

Güggi, M., [25] 1, *22*, [6] 113, [6] 114, [6] 148, *150*, [8] 264, [8, 24, 38] 268, [8, 24, 39, 40, 43, 47] 270, [8, 39, 40, 43, 47] 272, [43] 273, [78] 308, [8, 43, 88] 322, [88] 326, [88] 327, [8, 24, 39] 328, [40] 330, *331–333, 335, 336*

Guilbault, G. G., [15–18] 406, [17–19, 23, 24] 407, [16] 409, [16] 410, *414, 415*

Gulens, J., [52] 183, *209*

Gutnick, M. J., [81] 309, *335*

Haase, M., [35] 167, [35] 206, *208*, [29] 384, *400*

Haber, F., [3] 337, *372*

Hackleman, D. E., [50] 9, *24*

Haerdi, W., [24] 381, [24] 382, *400*

Hall, J. E., [33] 131, [33] 134, *152*

Hameka, H. F., [7] 375, [7] 379, *399*

Hammacher, K., [2] 402, *414*

Hansch, C., [92] 324, *336*

Hansen, E. H., [31] 167, *208* [56] 213, [56] 218, [56] 236, *261*, [48] 273, *333*, [33] 397, *400*, [10, 11] 406, *414*

Hartman, K., [59] 213, [59] 257, *262*

Hauptman, H., [95] 325, *336*

Havas, J., [41] 167, [41] 173, *209*

Haydon, D. A., [21] 115, [21] 123, [21] 128, [35] 133, [21] 134, *151, 152*

Heinemann, U., [81] 309, *335*

Helfferich, F., [14, 15] 48, *62*, [13, 14] 211, *259*

Helgeson, R. C., [36] 268, *333*

Henderson, P., [7] 48, [7] 60, *62*, [3] 64, *74*

Hennig, I., [27] 167, *208*

Higashiyama, K., [28–30] 167, *208*

Higuchi, T., [51] 216, *261*, [63] 280, *333*

Hirata, H., [28–30] 167, *208*

Hitchman, M. L., [6] 402, *414*

Hladky, S. B., [21 22] 115, [22] 117, [21, 22] 123, [21, 22] 128, [22] 131, [21, 22] 134, *151*

Hocking, C. S., [26] 211, *260*

Hodgkin, A. L., [6] 89, [6] 91, *112*

Hoffman, D. H., [36] 268, *333*

Hollós-Rokosinyi, E., [4] 165, [4] 167, *207*

Horovitz, K., [11] 337, *372*

Hughes, W. S., [4, 10] 337, *372*

Hulanicki, A., [58] 193, [58] 195, [58] 197, [58] 206, *210*, [44] 214, [44] 215, [44] 223, [44] 229, [44] 247, [44] 248, [44] 252, [44] 253, [44] 255, [44] 257, *261*, [76] 296, *335*

Ikeda, B., [52] 183, *209*

Ilgenfritz, G., [86] 315, *336*

Illian, C. R., [51] 216, *261*, [63] 280, *334*

Isard, J. O., [17] 338, [17] 339, *372*

Ishibashi, N., [42] 215, [72] 247, [72] 248, [72] 251, *261, 262*, [58] 277, [58] 296, [58] 297, [58] 303, *334*

Ivanov, V. T., [1] 264, *331*

Ivanovskaja, I. S., [45] 363, *374*, [31] 395, *400*

Jaber, A. M. Y., [46] 270, *333*

Jaenicke, W., [35] 167, [35] 206, *208*, [29] 384, *400*

James, H. J., [66] 223, [66] 229, *262*

Janáček, K., [8] 90, *112*

Janata, J., [48] 9, *23*

Jensen, O. J., [25] 167, [25] 181, [25] 182, [25] 204, *208*

Johansson, G., [48] 363, *374*, [22] 379, *400*, [27] 411, [27, 34–36] 413, *415, 416*

Johnson, C. C., [48] 9, *23*

Johnson, F. H., [5] 37, *43*, [29] 121, [29] 125, *152*

Johnson, K. R., [6] 48, *62*

Jordan, P., [38] 268, *333*

Jyo, A., [72] 247, [72] 248, [72] 251, *262*, [58] 277, [58] 296, [58] 297, [58] 303, *334*

419

Kahlweit, M., [25–28] 211, *260*
Kahr, G., [15, 42] 167, [42] 168, [15] 170, [15, 42] 171, [15, 42] 172, [15] 174, [42] 176, [42] 178, [42] 179, [42] 180, [42] 181, [42] 185, [15] 186, [15] 188, [42] 189, [42] 191, [42] 199, [42] 201, [15] 204, [42] 205, [42] 206, *207, 209,* [56] 277, [56] 296, [56] 297. [56] 303, [56] 309. [56] 312, [56] 313, [56] 322, *333*
Kane, P. O., [12] 73, *74*
Kaplan, L., [36] 268, *333*
Karlberg, B., [28] 384, *400*
Karreman, G., [26] 339, [26] 341, [26] 347, *373*
Katchalsky, A., [18–20] 211, *259, 260*
Katz, B., [6] 89, [6] 91, *112*
Kedem, O., [18, 19] 211, *259,* [11] 262, [11] 268, [55] 275, [55] 281, [55] 296, [55] 297, [55] 309, *331, 334*
Kelley, R. G., [49] 9, *23*
Kessler, M., [4, 25] 1, *21, 22,* [24] 268, [24] 270, [80] 309, [24] 328, *332, 335,* [22] 407, *415*
Keyes, M. H., [25] 407, [25] 413, *415*
Kirkwood, J. G., [16] 211, *259*
Kirsch, N. N. L., [67] 287, [67] 288, [67] 291, [67] 322, *335*
Kissel, T. R., [37] 412, *416*
Kitchener, J. A., [60] 213, *262*
Klasens, H. A., [57] 193, *210*
Klemensiewicz, Z., [3] 337, *372*
Kohara, H., [42] 215, *261*
Kolthoff, I. M., [2] 165, [2] 167, *207*
Kornfeld, G., [54] 190, *209,* [29] 339, [29] 341, [29] 361, *373*
Koryta, J., [12–15] 1, *21,* [14] 41, *43,* [5] 113, [5] 148, *150,* [16, 17] 167, [17] 183, [16] 199, [16] 206, *207,* [61] 214, *262,* [31, 32] 268, [32] 270, *332*
Kotyk, A., [8] 90, *112*
Krasne, S., [18] 113, [24, 25] 115, [24] 128, [24] 129, [24] 131, [18, 24] 133, [24] 134, [24, 25] 136, [18, 24, 25] 138, [24] 141, [18, 24, 25] 142, [24] 144, [18, 24, 25] 148, [24] 149, *151,* [69] 289, [69] 291, [87] 315, [87] 320, *335, 336*
Krueger, J. A., [39] 1, *23,* [8] 403, [8] 404, [8] 405, [8] 406, *414*
Krull, I. H., [12] 264, [12, 42] 270, *331, 333,* [52] 363, *374*
Kugler, G. C., [1] 375, [1] 395, *399*
Kunin, R., [60] 213, *262*
Kurey, M. J., [62] 214, *262*

Lakshminarayanaiah, N., [28–30] 1, *22,* [3] 28, [3] 30, *34,* [3] 47, [3] 48, [3] 55, *62,* [16] 82, 86, [1] 87, *112,* [4] 402, *414*

Lal, S., [53] 217, *261,* [75] 296, [75] 306, *335*
Lamm, C. G., [31] 167, *208,* [55] 217, *261*
Landqvist, N., [23] 339, [23] 341, [23] 345, [23] 346, [23] 347, [23] 351, *373*
Laprade, R., [3] 87, *112,* !17, 18] 113, [17] 115, [17] 128, [17, 18] 133, [17, 18] 138, [17] 141, [18] 142, [17, 18] 148, *151,* [53] 274, [53, 69] 289, [69] 291, *334, 335*
Lark–Horovitz, K., [29] 51, *63,* [20] 339, [20] 344, [20] 351, *373*
Läuger, P., [2] 87, *112,* [19, 20] 113, [19] 116, [19] 117, [20] 120, [19] 121, [19, 20, 34] 131, [19, 20, 34] 133, [19, 36] 135, [34] 138, [34] 139, *151, 152*
LeBlanc, O. H., [13] 264, [13] 270, *331*
Lehn, J. – M., [16–18] 268, [16–18] 315, [16] 320, [16] 324, *332*
Lengyel, B., [13, 28] 339, [28] 345, *372, 373*
Leo, A., [92] 324, *336*
Lev, A. A., [40] 147, *152*
Levins, R. J., [45] 270, [45] 272, *333*
Lewandowski, R., [44] 214, [44] 215, [44] 223, [44] 229, [44] 247, [44] 248, [44] 252, [44] 253, [44] 255, [44] 257, *261*
Lewenstam, A., [58] 193, [58] 195, [58] 197, [58] 206, *210*
Lewis, G. N., [32] 55, *63,* [9] 68, *74*
Lewis, S. B., [54a] 296, [54a] 303, *334*
Liberti, A., [32] 167, *208*
Linderholm, H., [28] 48, *63,* [12] 98, [12] 99, *112*
Lindner, E., [77] 303, *335,* [3, 6, 8] 375, [3, 8] 380, [8] 382, [8] 383, [3] 384, [8] 387, [8] 389, [8] 390, [8] 392, [6, 9] 393, [3] 397, *399*
Llenado, R. A., [31] 412, *415*
Loebel, E., [11] 264, [11] 270, *331*
Lübbers, D. W., [4] 1, *21,* [22] 407, *415*
Luterotti, S., [59] 213, [59] 257, *262*
Lux, H. D., [81] 309, *335*
Lynn, M., [40] 212, [40] 214, [40] 216, *261*
Lyons, C., [14] 406, *414*

Maass, G., [86] 315, *336*
Macdonald, J. R., [17–19] 377, *399, 400*
MacInnes, D. A., [54] 15, *24,* [30] 52, *63,* [8] 68, [8] 69, *74,* [5] 337, *372*
Madan, K., [36] 268, *333*
Malev, V. V., [40] 147, *152*
Mantella, L., [45] 215, [45] 223, [45] 229, [45] 248, *261*
Margolis, J. A., [39] 361, *374*
Markovic, P. L., [5] 375, [5] 379, [5] 382, *399*
Martell, A. E., [47] 174, *209*
Mascini, M., [32] 167, *208*
Mask, C. A., [12] 264, [12, 42] 270, *331, 333*
Massart, D. L., [26] 382, *400*

Materova, E. A., [46] 216, *261*
May, K., [38] 268, *333*
McLaughlin, S. G. A., [66] 287, [66] 288, *334*
Mead, C. A., [33] 131, [33] 134, *152*
Meares, P., [23] 211, *260*
Meier, P. C., [23] 1, [43] 7, [43] 17, [43] 20, *22,
  23,* [5] 66, [5] 73, *74,* [15] 113, [15] 148,
  151, [59] 213, [73] 255, [59, 73] 257,
  262, 263, [22, 27, 35] 268, [22] 269, [27]
  271, [22] 277, [22] 284, [22] 289, [22]
  291, [22] 295, [27] 296, [35] 308, [22]
  315, [22] 317, [22] 318, [22] 320, [27] 330,
  332, 333
Mertens, J., [26] 382, *400*
Meyer, K. H., [10, 11] 29, [10, 11] 30, *34,* [12]
  48, *62,* [13] 82, *86,* [8, 9] 211, *259*
Michaelis, L., [1, 2] 211, *259*
Mindt, W., [2] 402, [29] 413, *414, 415*
Mohan, M. S., [14] 73, *74*
Montalvo, J. G., [15, 16] 406, [16] 409, [16]
  410, *414*
Moody, G. J., [32–35] 1, [32] 8, *22, 23,* [4]
  113, *150,* [14] 167, *207,* [52] 213, [43]
  215, [52] 217, *261,* [46] 270, *333*
Moore, C., [12] 114, *151,* [2] 263, *331*
Moreau, P., [36] 267, *332*
Morf, W. E., [23–27] 1, [43] 7, [27] 8, [43] 17,
  [43] 20, *22, 23,* [7] 28, [7, 14, 15] 30, [14,
  15] 32, *34,* [9, 10] 41, [10, 11] 42, *43,*
  [24–26] 48, [25] 52, [25] 53, [25] 54, [25]
  55, [26, 33, 58, *63,* [2] 64, [5] 66, [2] 68,
  [2] 70, [5] 73, *74,* [8, 9] 79, [17] 83, [17]
  84, *86,* [9, 10] 92, [9, 10] 93, [9, 10] 94, [9,
  10] 103, [14] 104, [14, 15] 105, [16] 106,
  [9, 10, 17] 107, [9, 17] 108, [17] 110, [9,
  17] 111, *112,* [6, 8–11, 15] 113, [6, 9] 114,
  [26, 27] 115, [26] 117, [26] 123, [10, 26]
  127, [26] 128, [26] 131, [9] 134, [26] 143,
  [8–11, 26, 27, 39] 145, [26] 146, [6, 8–11,
  15, 26, 39] 148, *150–152,* [38, 42] 167,
  [42, 43] 168, [42] 171, [42] 172, [42] 176,
  [42] 178, [42] 179, [42] 180, [42] 181, [42]
  185, [42] 189, [43] 190, [42, 43] 191, [42]
  199, [42] 201, [42] 205, [42] 206, *209,* [31,
  32] 212, [57, 58] 213, [57] 218, [70] 244,
  [70] 245, [70] 246, [73] 255, [73] 257,
  260, 262, 263, [7, 8] 264, [7, 8, 20–22,
  24–27, 33, 35] 268, [21, 22, 33] 269, [7, 8,
  24] 270, [25–27] 271, [8] 272, [50] 275,
  [22, 25, 26, 56, 57, 59, 60, 62] 277, [50,
  60] 278, [50] 279, [50] 280, [50, 60] 281,
  [25, 26, 60] 282, [65] 283, [22, 25, 26, 59,
  60, 62] 284, [25, 26, 65] 287, [25, 26, 33]
  288, [22, 59, 60, 70] 289, [26] 290, [22,
  59, 70] 291, [70] 292, [70] 291, [74] 294,
  [22] 295, [25–27, 56, 57, 60, 62] 296, [26,
  56, 57, 60, 62] 296, [62] 297, [62] 298,
  [56, 57, 60, 62] 303, [57] 305, [26, 62]
  307, [35] 308, [26, 56, 57] 309, [26] 311,
  [26, 56, 57] 312, [26, 56] 313, [57] 314,
  [20–22, 33] 315, [84] 316, [20–22] 317,
  [20–22] 318, [20–22] 320, [8, 20, 21, 33,
  56, 84] 322, [33] 324, [33] 325, [94] 326,
  [8, 24] 328, [27] 330, *331–336,* [34] 340,
  [34] 356, *372,* [6, 8, 10, 15] 374, [10, 21]
  378, [8] 379, [8, 10] 382, [8] 383, [15]
  384, [8, 10] 387, [8, 10] 389, [8] 390, [8,
  10] 392, [6, 8] 393, [10, 15] 397, [10] 398,
  399, 400, [9] 404, [9] 408, *414*
Moser, P., [85] 315, *335*
Moss, S. D., [48] 9, *23*
Mueller, P., [1, 14] 113, *150, 151*
Müller, R. H., [25] 382, *400*
Myers, R. J., [60] 213, *262*

Nagy, G., [17, 18] 406, [17, 18] 407, *414, 415*
Naszódi, L., [46] 8, *23*
Nernst, W., [4] 48, [4] 51, *62,* [60] 194, *210*
Neumcke, B., [2] 87, *112,* [19] 113, [19] 116,
  [19] 117, [19] 121, [19] 131, [19] 133, [19]
  135, *151*
Neupert-Laves, K., [90] 322, [90] 324, *336*
Newcomb, M., [36] 268, *333*
Nicolaisen, B., [25] 167, [25] 181, [25] 182,
  [25] 204, *208*
Nicolsky, B. P., [44] 6, *23,* [14, 18, 22, 30]
  339, [31, 33] 340, [30] 343, [22] 344, [18,
  30, 31] 351, [18, 30, 31] 352, [18, 31] 354,
  [18, 31] 355, [18, 31] 356, [18, 31] 358,
  [18, 31] 359, [18, 31] 361, [33] 362, [33]
  363, [33] 368, [18, 22, 30, 31] 371, *372,*
  373
Norberg, K., [22] 379, *400*
Norton, D. A., [95] 325, *336*

Oehme, M., [27] 1, [27] 8, *22,* [13] 72, *74,* [6]
  113, [6] 114, [6] 148, *150,* [57, 59] 213,
  [57] 218, [59] 257, *262,* [8] 264, [8] 268,
  [8, 40,] 270, [8, 40] 272, [79, 80, 82, 83]
  309, [84] 316, [8, 84] 322, [8] 328, [40]
  330, *331, 333, 335,* [21] 379, *400*
Oesch, U., [74] 256, *263,* [94] 326, *336*
Ögren, L., [27] 411, [27, 34–36] 413, *415, 416*
Oke, R. B., [52] 213, [52] 217, *261*
Orme, F., [30] 212, [30] 236, [30] 237, [30]
  243, *260*
Osburn, J. O., [5] 375, [5] 379, [5] 382, *399*
Osipov, V. V., [40] 147, *152*
Osswald, H., [6] 113, [6] 114, [6] 148, *150,*
  [59] 213, [59] 257, *262,* [8] 264, [8] 268,
  [8] 270, [8] 272, [8] 322, [8] 328, *331*
Ovchinnikov, Yu. A., [1] 264, *331*

Papastathopoulos, D. S., [33] 412, *415*
Parlin, R. B., [6] 36, *43,* [30] 121, [30] 123, *151*
Parthasarathy, N., [23, 24] 381, [23, 24] 382, *400*
Passynsky, A. H., [6] 337, [6] 338, *372*
Peabody, J., [5] 402, *414*
Peacock, S. C., [36] 268, *333*
Pedersen, C. J., [14] 268, *332*
Perley, G. A., [8] 337, [8] 338, *372*
Perry, M., [55] 277, [55] 281, [55] 296, [55] 297, [55] 309, *334*
Petránek, J., [15] 268, [15] 270, [15] 272, [61] 277, [61] 284, [61] 296, [61] 297, [61] 303, *332, 334*
Pick, J., [10] 264, [10] 270, *331*
Pickard, W. F., [21] 48, *63,* [4] 65, *74*
Pioda, L. A. R., [22] 1, *22,* [5] 264, [5] 270, *331,* [9b] 338, *372*
Planck, M., [5] 48, [5] 52, [5] 53, *62,* [1] 64, *74*
Pleijel, H., [8] 48, [8] 58, *62*
Polissar, M. J., [5] 36, *43,* [29] 121, [29] 123, *151*
Porter, R. D., [39] 361, *374*
Prelog, V., [38] 268, *333*
Pressman, B. C., [12] 113, *150,* [2] 264, *331*
Pretsch, E., [24, 25] 1, *22,* [15] 30 [15] 32, *34,* [11] 42, *43,* [33] 57, *62,* [8] 79, *36,* [10] 92, [10] 93, [10] 94, [10] 103, [10] 107, *112,* [6, 9] 113, [6, 9] 114, [27] 115, [9] 134, [9, 27] 145, [6, 9] 148, *150, 151,* [38] 167, *209,* [8] 264, [8, 19, 23, 24, 33, 9, 35, 38] 268, [33] 269, [8, 24, 39, 40, 43, 47] 270, [8, 39, 40, 43, 47] 272, [43] 273, [62] 277, [62] 284, [33] 288, [70] 289, [70] 291, [70] 292, [70] 293, [74] 294, [62] 296, [62] 297, [62] 298, [62] 299, [62] 303, [62] 307, [35] 308, [23, 33] 315, [8, 33, 43, 88, 89, 91] 322, [33] 324, [33] 325, [88, 93, 94] 326, [88] 327, [8, 24, 39] 328, [40, 98] 330, *331–336*
Pron'kina, T. I., [46] 216, *261*
Pungor, E., [17–21] 1, *22,* [4–7] 165, [4–7, 37] 167, [7] 172, [37] 173, [5–7, 37] 188, [5, 7, 37] 197, [37] 205, [5, 37] 206, *207, 208,* [10] 264, [10] 270, [77] 303, *331, 335,* [2, 3, 6, 9] 375, [3, 9] 380, [2, 3, 9] 384, [6] 393, [3] 397, *399*

Racine, P., [29] 412, *415*
Rais, J., [64] 220, [64] 221, [64] 223, [64] 227, [64] 248, *262*
Rechnitz, G. A., [40, 41] 1, *23,* [7] 39, *43,* [41] 148, *152,* [45] 170, *209,* [71] 247, *262,* [68] 289, [68, 72] 291, *335,* [9a] 338, [53] 363, *372, 374,* [1, 7] 375, [7] 379, [1, 30] 395, *399, 400,* [31–33] 412, *415*
Reese, C. E., [4] 37, *43*
Riseman, J. H., [39] 1, *23,* [8] 403, [8] 404, [8] 405, [8] 406, *414*
Robinson, R. A., [53] 14, [53, 59] 16, [59] 17, [59] 19, *24*
Ross, J. W., [38, 39] 1, *23,* [8, 10] 165, [8, 10] 167, [10] 186, [10] 191, *207,* [34, 35] 212, [34, 35] 213, [35] 214, [35] 215, [34, 35] 217, [35] 227, [35] 232, [35] 236, [35] 257, *260,* [6] 264, [6] 270, [6] 306, *331,* [8] 403, [8] 404, [8] 405, [8] 406, *414*
Rothmund, V., [54] 190, *209,* [29] 339, [29] 341, [29] 361, *373*
Rudin, D. O., [1] 78, *86,* [1, 14] 113, *150, 151,* [15] 338, [15] 339, [15] 341, [15] 347, [15] 351, *372*
Růžička, J., [31] 167, *208,* [56] 213, [50] 216, [55] 217, [56] 218, [56] 236, *261, 262,* [9] 264, [9] 270, [48] 273, *331, 333,* [33] 397, *400,* [10, 11] 406, *414*
Ryan, T. H., [4] 375, [4] 384, [4] 387, [4] 395, [4] 396, *399*
Ryba, O., [15] 268, [15] 270, [15] 272, [61] 277, [61] 284, [61] 296, [61] 297, [61] 303, *332, 334*

Salvemini, F., [45] 215, [45] 223, [45] 229, [45] 248, *261*
Sandblom, J. P., [18] 48, [18] 51, *62,* [5] 78, *86,* [29, 30] 212, [29, 30] 236, [30] 237, [29] 239, [30] 243, *260,* [40] 362, [40] 368, *374*
Sanders, H. L., [2] 165, [2] 167, *207*
Sandifer, J. R., [12] 42, *43,* [23] 48, [23] 58, *63,* [10] 79, *86,* [20] 377, *400*
Sargent, L. W., [32] 55, *63,* [9] 68, *74*
Saroff, H. A., [3] 113, *150,* [37] 212, [37] 217, *261*
Sauvage, J. P., [17, 18] 268, [17, 18] 315, *332*
Schiller, H., [12] 337, *372*
Schlögl, R., [1, 2] 27, [1, 2] 30, *34,* [13] 42, *43,* [13, 14] 48, [13] 50, [13] 58, *62,* [11] 72, *74,* [14, 15] 82, *86,* [4] 88, [4] 94, [4] 95, [4, 13] 99, [4, 13] 100, *112,* [10–13] 211, *259*
Schneider, J., [34] 268, *333*
Scholer, R. P., [41] 212, [41] 214, [41] 216, [41] 221, [41] 222, [41] 248, *261,* [41] 270, [64] 280, [64] 296, *333, 334*
Schwabe, K., [24] 339, [24] 341, [24] 345, [24] 346, [24] 347, [24] 351, *373*
Schwyzer, R., [85] 315, *335*
Scibona, G., [45] 215, [45] 223, [45] 229, [45] 248, *261*

Scordamaglia, R., [97] 330, *336*

Scuppa, B., [45] 215, [45] 223, [45] 229, [45] 248, *261*

Šenkyr, J., [35] 268, [35] 308, *333*

Seta, P., [49] 216, *261*

Seto, H., [58] 277, [58] 296, [58] 297, [58] 303, *334*

Severinghaus, J. W., [5, 7] 402, *414*

Shatkay, A., [2, 3] 113, *150,* [36, 37] 212, [36, 37] 217, *260, 261,* [14] 375, [14] 376, [14], 379, [14] 382, *399*

Shean, G. M., [33] 212, [33] 213, [33] 215, *260*

Shkrob, A. M., [1] 264, *331*

Shu, F. R., [19] 407, *415*

Shults, M. M., [22] 48, *63,* [18] 339, [31–33] 340, [18, 31] 351, [18, 31] 352, [18, 31] 354, [18, 31] 355, [18, 31] 356, [18, 31] 358, [18, 31] 359, [18, 31] 361, [32, 33] 362, [32, 33, 54, 55] 363, [32] 366 [32, 33] 368, [18, 31, 32] 371, *373, 374*

Siemroth, J., [27] 167, *208*

Sievers, J. F., [10, 11] 29, [10, 11] 30, *34,* [12] 48, *62,* [13] 82, *86* [8, 9] 211, *259*

Sillén, L. R., [47] 174, *209*

Silver, I. A., [4] 1, *21,* [22] 407, *415*

Simon, W., [4, 22–27] 1, [43] 7, [27] 8, [43] 17, [43] 20, *21–23,* [7] 28, [7, 14, 15] 30, [14, 15] 32, *34,* [10] 40, [10, 11] 42, *43,* [24, 26] 48, [26, 33] 58, *63,* [5] 66, [5] 73, *74,* [8, 9] 79, *86,* [9, 10] 92, [9, 10] 93, [9, 10] 94, [9, 10] 103, [14] 104, [14, 15] 105, [16] 106, [9, 10, 17] 107, [9, 17] 108, [17] 110, [9, 17] 111, *112,* [6, 8–11, 13, 15] 113, [6, 9] 114, [26, 27] 115, [26] 117, [26] 123, [10, 26] 127, [26] 128, [26] 131, [9] 134, [26] 143, [8–11, 26, 27, 39] 145, [26] 146, [6, 8–11, 15, 26, 39] 148, [42] 149, *151–152,* [38, 42] 167, [42, 43] 168, [42] 171, [42] 172, [42] 176, [42] 178, [42] 179, [42] 180, [42] 181, [49] 182, [42] 185, [42] 189, [43] 190, [42, 43] 191, [42] 199, [42] 201, [42] 205, [42] 206, *209,* [32, 41] 212, [57–59] 213, [41] 214, [41] 216, [57] 218, [41] 221, [41] 222, [41] 248, [73] 255, [74] 256, [59, 73] 257, *260–263,* [3, 5, 7, 8, 10] 264, [7, 8, 19–27, 33, 35, 38] 268, [21, 22, 33] 269, [5, 7, 8, 10, 24, 39–41, 43, 47] 270, [25–27] 271, [8, 39, 40, 43, 47] 272, [43] 273, [50] 275, [22, 25, 26, 56, 57, 59, 60, 62] 277, [50, 60] 278, [50] 279, [50] 280, [50, 60] 281, [25, 26, 60] 282, [22, 25, 26, 59, 60, 62] 284, [25, 26, 67] 285, [25, 26, 33, 67] 288, [22, 59, 60, 70] 289, [26] 290, [22, 59, 67, 70] 291, [70] 292, [70] 293, [74] 294, [22] 295, [25–27, 56, 57, 60, 62] 296, [26, 56,

57, 60, 62] 297, [62] 298, [62] 299, [56, 57, 60, 62, 77] 303, [57] 305, [26, 62] 306, [35] 307, [26, 56, 57, 79, 80, 82] 308, [26] 310, [26, 56, 57] 312, [26, 56] 313, [57] 314, [20–23, 33] 315, [84] 316, [20–22] 317, [20–22] 318, [20–22] 320, [8, 20, 21, 33, 43, 56, 67, 84, 88, 91] 322, [33] 324, [33] 325, [88, 93, 94] 326, [88] 327, [8, 24, 39] 328, [27, 40, 98] 330, *331–336,* [9b] 338, [38] 348, *372, 374,* [6, 8, 10] 375, [10, 21] 379, [8] 380, [8, 10] 382, [8] 383, [8, 10] 387, [8, 10] 389, [8] 390, [8, 10] 392, [6, 8] 393, [10] 397, [*10*] 389, *399, 400,* [22] 407, *415*

Sokolof, S. I., [6] 337, [6] 338, *372*

Sollner, K., [8] 28, [8] 30, *34,* [3, 4] 211, [33] 212, [33] 213, [33] 215, *259, 260*

Sousa, L. R., [36] 268, *333*

Spiegler, K. S., [17] 211, *259*

Srinivasan, K., [71] 247, *262*

Stankova, V., [5] 264, [5] 270, *331* ₃

Staples, B. R., [59] 16, [59] 17, [59] 19, *24*

Stark, G., [20] 113, [20] 120, [20] 131, [20] 133, *151*

Staverman, A. J., [2] 47, *62,* [15] 211, *259*

Štefanac, Z., [22] 1, *22,* [13] 113, *150,* [3] 264, *331,* [9b] 338, [38] 348, *372, 374*

Stephanova, O. K., [22] 48, *63,* [32] 340, [32] 362, [32, 55] 363, [32] 366, [32] 368, [32] 371, *372, 374*

Stokes, R. H., [53] 14, [53] 16, *24*

Stover, F. S., [69] 240, *262,* [13, 16] 375, [13] 377, [13] 379, [16] 395, *399*

Strehlow, H., [25–27] 211, *260*

Stucky, G. L., [54] 217, *261*

Szabo, G., [19] 48, [19] 51, *62,* [6] 78, *86* [3] 87, *112,* [17, 18] 113, [17] 115, [17]128, [33] 131, [17, 18] 133, [33] 134, [17, 18] 138, [17] 141, [18] 142, [17, 18] 148, *151, 152,* [51, 53] 274, [66] 287, [66] 288, [51, 53, 69] 289, [69] 291, [51] 300, *334, 335*

Szepesváry, P., [46] 8, *23*

Tendeloo, H. J. C., [3] 165, *207*

Teorell, T., [9] 27, [9] 29, *34,* [10, 11] 48, [10, 11] 56, [11] 57, *62,* [12] 81, *86,* [5–7] 211, *259*

Thain, J. F., [23] 211, *260*

Thoma, A. P., [15] 105, *112,* [10] 113, [10] 127, [10] 145, [10] 148, *150,* [38] 268, [50] 275, [50] 278, [50] 279, [50] 280, [50] 281, *333, 334*

Thomas, J. D. R., [32–35] 1, [32] 8, *22, 23,* [4] 113, *150,* [14] 167, [51] 183, *207, 209,* [52] 213, [43] 215, [52] 217, *261,* [46] 270, *333*

Thomas, R. C., [79] 309, *335*
Thompson, H., [32] 412, *415*
Thunstrom, A., [5] 402, *414*
Timko, J. M., [36] 268, *333*
Ti Tien, H., [1] 113, *150*
Tjell, J. C., [56] 213, [50] 216, [55] 217, [56] 218, [56] 236, *261, 262,* [48] 273, *333*
Tolmacheva, T. A., [14] 339, *372*
Tossounian, J. L., [51] 216, *261,* [63] 280, *334*
Tóth, K., [17, 21] 1, *22,* [5–7] 165, [5–7, 34, 37] 167, [7] 172, [37] 173, [5–7, 37] 188, [5, 7, 37] 197, [37] 205, [5, 37] 206, *207, 208,* [10] 264, [10] 270, *331,* [2, 3, 6, 9] 375, [3, 9] 380, [2, 3, 9] 384, [6] 393, [3] 397, *399*
Tran-Minh, C., [26] 411, [30] 412, *415*
Treasure, T., [49] 271, *333*
Trümpler, G., [1] 165, *207*
Tun-Kyi, Aung, [85] 315, *335*

Ussing, H. H., [7] 90, *112*

Valova, I. V., [31] 395, *400*
Van den Winkel, P., [26] 382, *400*
Vašák, M., [22] 1, *22,* [10] 264, [10] 270, *331,* [9b] 338, *372*
Vesely, J., [25, 26, 33] 167, [25] 181, [25] 182, [25] 204, *208*
Viviani-Nauer, A., [10] 113, [10] 127, [10] 145, [10] 148, *150,* [50] 275, [50] 278, [50] 279, [50] 279, [50] 280, [50] 281, *334*
Volmer, M., [3] 37, *43,* [32] 123, *152*
von Rechenberg, A. V., [65] 223, [65] 225, [65] 226, [65] 229, *262*

von Storp, H., [39] 167, [39] 205, [39, 61] 206, *207, 210*
Vuilleumier, P., [37] 140, *152*

Walker, J. L., [18] 48, [18] 51, *62,* [5] 78, *86,* [29] 212, [63] 214, [29] 236, [29] 239, *260, 262,* [40] 362, [40] 368, *374*
Watson, B., [25] 407, [25] 413, *415*
Weeks, C. M., [95] 325, *336*
Weiss, L., [88] 322, [88] 326, [88] 327, *336*
Wescott, W. C., [1] 113, *150*
Wikby, A., [48–50] 363, *374*
Wingard, C. B., [21] 407, *415*
Winkler, R., [86] 315, *336*
Wipf, H.–K., [6] 28, *34*
Wise, W. M., [40] 212, [40, 62] 214, [40] 216, *261, 262*
Wuhrmann, H.–R., [22] 1, *22,* [14] 30, [14] 32, *34* [24] 48, *63,* [43] 168, [43]190,[43] 191, *209,* [59] 277, [59] 284, [59] 289, [59] 291, *334,* [9b] 338, *372*
Wuhrmann, P., [9, 10] 92, [9, 10] 93, [9, 10] 94, [9, 10] 103, [9, 10] 107, [9] 108, [9] 111, *112,* [26, 27] 115, [26] 117, [26] 123, [26] 127, [26] 128, [26] 131, [26] 143, [26, 27] 145, [26] 146, [26] 148, *151,* [60] 277, [60] 278, [60] 281, [60] 282, [60] 284, [60] 289 [74] 294, [60] 296, [60] 297, [60, 77] 301, *334, 335*

Yeh, Y. L., [8] 68, [8] 69, *74*

Zagatto, E. A., [33] 397, *400*
Zappia, E., [5] 402, *414*
Züst, Ch. U., [7] 264, [7] 268, [7] 270, *331*
Zwolinski, B. J., [4] 37, *43*

424

SUBJECT INDEX

Acetylcholine sensor, 214, 216
Activation barrier, *see* Barrier
Active or coupled transport, 28, 90
Activity coefficients, aqueous solution, 13–20
   membrane phase, 32, 48, 88, 95, 99, 190, 341
Activity ratio measurement, 73
Air gap membrane, 27, 406
Aliquat 336S, 215, 252–254, 257
Alkali error, 344, 354
All-solid-state electrode, 2, 169
Alumina group, 351, 354
   ion-exchange on, 357
Aluminosilicate glasses, 337–339
   response behavior, 354–359
Ammonia, interference with silver chloride electrodes, 403
Ammonia sensor, 403–405
Ammonium electrode, 270
   selectivity, 149, 272
Ammonium ions, organic, used as membrane components, 213–216, 225, 227, 253, 254, 257
   quaternary, electrodes for, 214, 216
   selectivity, 222–224
Amperometric detection, 402, 406
Anion-selective electrodes, liquid-membrane, 213–216, 225, 227, 252–257
   solid-state, 165–166, 175–183, 186–206
Anions, lipophilic,
   interference in anion-exchanger membranes, 255
   in cation-exchanger membranes, 280
   in neutral carrier membranes, 111, 296–315
      elimination or reduction of, 303–305, 309–314
   transport across bilayer membranes, 135
Antibiotics, structures of, 114, 265
Association between ionic sites and counterions, glass membranes, 361, 367–371
   liquid ion-exchangers, 223–236, 241–246, 249
   neutral carrier membranes, 307–309

Back-diffusion of carriers, 125, 134, 146, 278
Barium electrodes, 270
   selectivity, 272, 329
Barrier, interfacial, 123, 130, 379
   multi-, 118, 121–123, 129, 137
   rectangular, 120
   sharp, 120, 131–134
   trapezoidal, 115–120, 128–130, 136
Behn's transport equation, 98
Bilayer lipid membranes, comparison with bulk membranes, 143–149
   ion transport across, 113–149
Bis(p-chlorophenyl)propanedion, membrane based on, 106
Blood serum, applicability of sensors in, 214, 255, 271, 292, 330
Boltzmann distribution, 46, 310
Born equation, 320
Boundary potential, definition, 30
   general formulation, 35–42
   general result for reversible membrane electrodes, 40
   of glass membranes, 343, 350
   of liquid ion-exchanger membranes, 220, 231
   of neutral carrier membranes, 282, 300
   summary, 153
Bromide electrodes, liquid-membrane, 216, 254
   solid-state, 197
Buck's theory of glass electrodes, 356–361
Buffered samples, measurement in, 175–183, 218, 273, 401
Bulk membranes, comparison with bilyaer lipid membranes, 143–149
   ion transport across, 91–111
Butler-Volmer equation, 38

Calcium electrodes, based on liquid ion-exchangers, 213, 217
   calibration curves, 273
   pH-interference, 237
   response *vs.* time behavior, 387, 396
   selectivity, 218
   titration curves, 273

Calcium electrodes, (cont.)
based on neutral carriers, 270, 329
    anion interference, 111, 298, 313
    calibration curve, 273, 275
    comparison of theoretical and measured
        potential response curves, 292
    improvement of response characteristics
        by incorporation of tetraphenylborate,
        313–316
    precision, 7
    selectivity, 272, 329
    study of selectivity-determining parameters,
        295, 323, 327, 329
    titration curve, 273
Calcium fluoride electrode, 165
Calcium transport, across bilayer lipid mem-
    branes, 140
    across bulk membranes, 106, 108, 275
Carbon dioxide sensor, 402
Cation-selective electrodes, glass, 337–339
    liquid ion-exchanger, 213–218
    neutral carrier, 268–273
    solid-state, 165–167, 182–186
Cavity radius, 320
Cell, organization of, with liquid junction, 2, 3,
    170
    without liquid junction, 13, 71, 402, 403
Cell potential (emf), 2–8
Charge transfer reactions, 37–39, 123–125
Charged carriers, 28, 106, 229
Chemical potential, 35
Chemisorption, 172
Chiral recognition, 105, 268
Chloride electrodes, based on liquid ion-ex-
    changers, 214, 215
        calibration plots for interfering ions, 253
        precision, 7
        selectivity, 257
    based on silver chloride,
        application as reference electrodes, 403
        calibration curve, 176
        detection limit, 175
        response vs. time behavior, 381
        selectivity, 189, 197
Ciani–Eisenmann–Krasne   bilayer   model,
    134–143
Clark electrode, 402
Coions and counterions, definition, 56
Complex formation, and potentiometric selec-
    tivity, 289–291
    between mercuric and halide ions, 183–186
    between metal ions and neutral carriers,
        123, 286
    between silver ions and different ligands,
        198–206
    mean degree of, 186, 288
    values of stability constants, 174, 290

Complexing   agents,   membrane-active,   see
    Charged carriers and
    Neutral carriers
Concentration- or activity change, direction of,
    effect on response time, 384–387, 390
Concentration polarization, 247, 378, 383, 394
Concentration profiles, time course of, 383, 388
    in valinomycin membranes, 279
Constant field approximation, 55, 88, 92, 117,
    122, 145
Consumption   of   free   carriers,   effects   of,
    296–315
Coordinating   ligand   sites,   number   of,   269,
    318–320, 328
Copper sulfide electrode, 167
Countertransport, 106
Crown compounds, enantiomer-selective, 105,
    268
    potassium-selective, 268, 272
Current density, electrical, 38, 46, 131–135
Current-voltage characteristics, of bilayer lipid
    membranes, 133–135, 138–140
    of bulk membranes, 93, 146
    correlations between theoretical and mea-
        sured curves, 135, 139, 140
Cyanide electrode, 201–206

Debye-Hückel, convention, 16–19, 32
    theory, results from 14, 16, 65, 99
Decomplexation reaction, kinetic limitation by,
    134, 142
    rate of, 124, 134
Defect   mechanisms,   in   glass   membranes,
    360–365
    in solid-state membranes, 166
Detection limit, definition, 6
    of enzyme electrodes, 410–412
    of gas-sensing probes, 403–405
    of silver compound membrane electrodes,
        171–183
Dialkyl-   and   diarylphosphates,   membranes
    based on, 213, 217, 218
Diaphragm, 4, 64
Dielectric constant of membrane solvent, 289,
    294–295, 320–322
Diffusion, definition, 29
    into membrane, time dependence, 256, 378,
        384–397
    mean free path of, 255
    through boundary layer, 194–196, 205,
        249, 382–384, 388
Diffusion coefficient, definition, 45, 88
Diffusion   layer,   thickness   of,   194,   249,
    382–384, 408, 411

Diffusion model, for enzyme electrodes, 408
  for gas-sensing electrodes, 405
  for liquid ion-exchanger membrane electrodes, 248
  for membranes of constant composition, 383
  for neutral carrier membrane electrodes, 388
  for solid-state membrane electrodes, 195
Diffusion potential, definition, 30
  general formulation, 44–50
  general results, 47, 49
  practical solutions, 50–61
    Goldman, 55
    Henderson, 60
    ideally homogeneous membranes, 56
    ideally selective membranes, 75
    Lewis and Sargent, 55
    Nernst, extension, 51
    permselective membranes, 50, 56
    Planck, 52
    Schlögl, 58
    Teorell, 56, 109
  special results for
    glass membranes, 349, 360, 367, 369
    liquid ion-exchanger membranes, 219, 229, 233, 240, 242, 244, 245
    neutral carrier membranes, 281, 284, 300
  summary, 154
Dimensions of cation/carrier complexes, effect on selectivity, 289, 320–323
Dinactin, 114, 265
Dioxaoctanedioic diamides, structure-selectivity relationships, 322–328
Dipole moment of ligand sites, 318, 326
Dissociated ion-exchangers, 219–223, 240–241, 249
Dissolution processes at solid membrane/solution interfaces, 171, 362, 381
Distribution of ions at equilibrium,
  across membranes, 310
  between two phases, 36–38, 124
  within a phase, 46
Distribution coefficient, kinetic definition, 38
  thermodynamic definition, 36, 223
  overall, 282, 285
Divalent-ion sensors, 213, 217
Donnan exclusion (coion exclusion), failure of, 81–83
Donnan potential, 41, 77, 83, 154, 167
Drain current, 9
Driving forces, 28–30, 46
Dynamic response behavior, 375–398
  enzyme electrodes, 384, 412
  gas sensors, 405

glass electrodes, 379, 384
ion-exchanger membrane electrodes, 382–387, 394–396
microelectrodes, 378
neutral carrier membrane electrodes, 387–393, 398
solid-state membrane electrodes, 381–382

Eisenman equation, 6, 147, 156, 350
Electrical conductivity, 29; see also Membrane conductance and Resistance
Electrical current, 28, 29; see also Current density
Electrical potential, local, 30, 35
Electrical transference number, see Transference number
Electrodes, commercially available, 10
Electrodialysis, 29, 102–111, 274–280
Electrodialytic separation, 29
Electrodiffusion, 29, 88, 94
Electroneutrality, deviations from, 113, 277, 287
Electroosmosis, 48
Electrostatic model, 288, 317–322
EMF response, basic aspects, 2–8
  experimental curves,
    aluminosilicate glasses, 355
    ammonia sensor, 404
    anion-selective liquid-membrane electrodes, 225, 253
    calcium electrodes,
      liquid ion-exchanger, 237, 273
      neutral carrier, 111, 273, 292, 298, 313
    potassium electrodes, 299, 304
    silver bromide electrode, 201
    silver chloride electrode, 176, 177, 201
    silver iodide electrode, 179, 185, 201
    silver sulfide electrode, 180, 182, 201
    slopes of, 275
    sodium electrodes, 305, 314, 348
    urea electrode, 410
  general formulation, 75–85
  special results for
    enzyme electrodes, 409–412
    gas sensors, 402–405
    glass electrodes, 340–371
    ideally selective electrodes, 40, 76
    liquid ion-exchanger membrane electrodes, 219–258
    neutral carrier membrane electrodes, 281–315
    permselective membranes, 77–83
    silver compound membrane electrodes, 170–206
    solid-state membrane electrodes, 76, 169

427

EMF response, (cont.)
  summary, 156–159
  vs. time profiles, 385, 390, 396, 398; *see also* Dynamic response behavior and Response time
Enantiomer-selective electrodes, 268
  selectivity, 105
Energy barrier, *see* Barrier
Enzymatic reaction, 406
Enzyme electrodes, 406–413
Enzyme reactors, 413
Equilibrium assumption, 30, 36–39, 92, 124
Equilibrium domain, 138, 141, 145, 149
Equitransferent solution, 67–70
Equivalent-circuit description, 378
Equivalent ionic conductivity, definition, 45, 55
  values for aqueous solutions, 67
Exchange-current density, 38, 166, 378
Extraction, and potential response, 298, 299
  and selectivity, 222, 227
    reactions for neutral carrier membranes, 270, 296
Eyring formalism, 37, 87, 120–124, 137

FCCP (a proton carrier), 106
Fixed-site membranes, *see* Permselective membranes
Flow-analyzer technique, 397
Flow-through cell, advantage of, 256, 392
Flow velocity, 44
Fluoride electrode, 165–167
  dynamic response behavior, 382
  selectivity, 165
Flux equations,
  for charge transfer across an interface, 37, 123
  for electrodialysis across a membrane,
    Behn, 98
    bilayer membrane, cation-carrier, 138, 139
    Ciani-Eisenman-Krasne, 136
    Goldman-Hodgkin-Katz, 89
    Läuger-Stark, 132
    Schlögl, 94, 97–101
    Ussing, 90
  for electrodiffusion within a phase,
    Eyring, 121
    glass membrane, 349, 364–366
    Goldman, 89, 120, 146
    Nernst-Planck, 44–46, 88, 116
    Nernstian approximation, 194, 249, 389
    symmetrical bulk membrane, 93, 146
    time-dependent, 389
  general formulations, 44–46, 94, 116, 121, 123, 130
  summary, 159–161
Friction coefficient, 45

Gas-permeable membranes, 406
Gas-sensitive electrodes, 401–406
  application in enzyme electrodes, 407, 412
  dynamic response behavior, 405
  geometric parameters, 405
Geometric capacitance, 377
Glass electrodes, 337–371
  application in compound electrodes, 402, 403, 406
  dynamic response behavior, 379, 384
  glass compositions, 337–339
  selectivities, 338
  theories, classification of, 339
    liquid-state approaches, 361–371
    solid-state approaches, 340–361
    unified description, 352–359, 369–371
Glucose sensor, 406
Goldman equation, diffusion potential, 55
  ion flux, 89, 96, 120, 146
Goldman-Hodgkin-Katz theory, 88–91

Henderson equation, 60
  use for calculation of liquid-junction potentials, 64–73
Heterogeneous-site glasses, 351–362, 369–371
Heterogeneous solid-state membranes, 165, 167
Hofmeister lyophilic series, 213, 223, 296
Homogeneous membranes, 56, 76, 93, 168
Homogeneous-site glasses, 343–351, 366–369
Horovitz equation, 156, 344
Hulanicki-Lewenstam theory, 193–198
Hydrated glass layer, 362, 384
Hydration energies, correlation with selectivities, 223–225
Hydration number, 15
Hydration theory, 14–18
Hydrogen bonds, effect on selectivity of neutral carriers, 269
Hyperbolic current-voltage curve, 133, 138

Image force, 116
Immobile anions in neutral carrier membranes, 277–280
Impedance theory, 377
Indicator electrode, 2
Inhibitors, 411
Interactions, attractive and repulsive, for cation-carrier complexes, 318–322
Interface, membrane/solution, model of, 29–32, 35–39, 93, 123
Interfacial kinetics, 37–39, 123, 138, 142, 377–382
Interfacial potential, *see* Boundary potential
Interstitial ions, 171, 360, 363
Iodide electrode, calibration curve, 179
  detection limit, 183
  precipitate-based, 165
  selectivity, 189, 197, 206

428

Ion-exchange, constants, 222, 227, 357
    equilibria, 188, 228, 235, 248, 280, 288, 341
Ion-exchanger membranes,
    dynamic response behavior, 379–387, 392–396
    expressions for potential and selectivity, 50, 57, 77–83, 154–157
    ion-transport behavior, 96–102, 107, 109
    liquid, 211–258, 309–315, 361–371
    requirements for electrode application, 166
    solid, 190, 340–361
Ion pumping, 28, 104
Ion transport, see Membrane transport
Ionic strength, 14
Ionogenic groups in glass, 351
Ionophores, see Charged carriers and Neutral carriers
Irreversible behavior of electrodes, 38, 192, 326
Irreversible thermodynamics, 47, 87, 211
ISFET (ion-selective field effect transistor), 9
Isosteric complexes, 145, 228
Isothermal transport, 29

Jyo-Ishibashi theory, 247–254

Kinetic domain, 138, 142
Kinetic parameter, 142–144
Kinetics, of enzyme-substrate reactions, 407
    of interfacial reactions, 37–39, 123, 377–382

Lanthanum fluoride electrode, 165–167
Läuger-Stark bilayer model, 131–134
Leached ions, 6, 172, 178, 183, 203, 363
Lewis-Sargent equation, 55, 68
Lifetime of liquid-membrane electrodes, 324–326
Ligand thickness, 321–323
Ligands, forming complexes with cations, see Charged carriers and Neutral carriers
    forming complexes with silver ion, 198–206
Lipids, see Bilayer lipid membranes
Lipophilicity of neutral carriers, 324–326
Liquid-junction potential see also Diffusion potential,
    calculations, 64–73
    definition, 4
    measured values, 66, 69
    possibilities for minimization, 68–73
Liquid-membrane electrodes, based on liquid ion-exchangers, 211–258; see also Ion-exchanger membranes
    dynamic response behavior, 382–387, 394–396

membrane materials and observed selectivities, 212–218
    theories, apparent selectivity behavior, 246–258
    potential response and selectivity, 219–236
    Sandblom–Eisenman–Walker theory and its extensions, 236–246
Liquid–membrane electrodes based on neutral carriers, see Neutral carrier membrane electrodes
Lithia glasses, 337
Lithium electrodes, based on glass membrane, 338
    selectivity, 338
    based on neutral carriers, 270
    precision, 7
    selectivity, 272
    slope of calibration curve, 275
    transference numbers for, 275
Loligo axon, 91

MacInnes convention, 15, 32
Macroheterobicyclic ligands, 268
Macrotetrolide antibiotics, 114, 265
Magnesium ions, interference in calcium electrodes, 218, 272, 396
    sensor for, 213, 217
    selectivity, 233
    transport across bulk membranes, 106
Mechanism, of carrier-mediated ion transport through membranes, 125
    of cation transport in glass membranes, 363–366
    of dissolution of solid-state membrane materials, 171, 381
    enzyme-substrate reactions, 407
    of permselectivity in neutral carrier membranes, 274–285
Membranes, definition of, 27
    model assumptions for, 30–33
    phenomena on, 27–29
Membrane conductance, ohmic, 102
    zero-current, 132, 141
Membrane materials for
    enzyme electrodes, 406
    gas sensor, 406
    glass electrodes, 337–339
    liquid ion-exchanger membrane electrodes, 212–218
    neutral carrier membrane electrodes, 268–274
    solid-sate membrane electrodes, 166–168
    summary, 8–13
Membrane potential, zero-current, definition, 2–4, 29, 30

Membrane potential, zero current (cont.)
  general result, 110
  practical solutions, 75–85
    Ciani–Eisenman–Krasne, 141
    Eisenman, 350
    Goldman–Hodgkin–Katz, 91
    ideally selective membranes, 4, 76
    permselective membranes, 77–83, 147
    from Planck's theory, 84
    Teorell–Meyer–Sievers, 82
  special results for
    bilayer lipid membranes, 141
    glass membranes, 350
    liquid ion-exchanger membranes, 221, 226, 241, 242
    neutral carrier membranes, 282, 287, 306, 311
    solid-state membranes, 168
  summary, 156–159
Membrane transport; see also Flux equation
  across bilayer membranes, free and carrier-mediates, 113–149
    comparison with bulk membranes, 143–149
    general formulation, 115–131
    model by Ciani, Eisenman, and Krasne, 134–143
    model by Läuger and Stark, 131–134
  across bulk membranes, 91–111
    electrical properties and transport selectivity, 101–111
    electrodialytic transport across neutral carrier membranes, 274–280
    transport mechanisms in glass membranes, 363–366
  classical concepts of, 87–111
    basic flux equation, 87
    Goldman–Hodgkin–Katz approximation, 88–91
    Schlögl's theory and its applications, 94–101
    for symmetrical membrane cells, 91–94
  summary, 159–161
Mercuric ions, response of silver halide electrodes to, 183–186
Michaelis constant, 407
Michaelis–Menten mechanism, 407
Microelectrodes, several aspects of, 237, 256, 309, 378, 392
Microporous membranes, 406
Miniaturization of ion sensors, 9
Mixed adsorption isotherm, 188–192, 347
Mixed solution method, 8, 192
Mobile-site membranes, see Liquid membrane electrodes
Mobilities, definition of parameters, 45

  in glass membranes, 364–366
  mean, characteristic of valency classes, 53–54, 57, 84, 155, 158, 284, 306
  values for aqueous solutions, 67
Molecular basis of ion selectivity, 313–320
Monactin, 114, 139, 263
Monensin, 229
Monovalent/divalent ion selectivity, 79–81, 107, 235, 291–295, 315–323, 326–328
Morf–Wuhrmann–Simon bulk membrane model, 91–94, 143–149
Müller's equation, 382

n-Type description of activities, 190, 341
Nernst–Einstein relation, 88
Nernst–Planck description of ion transport, 87–120
Nernst's solution for the diffusion potential, 51
Nernstian diffusion layer, 194, 249, 382
Nernstian response, 5, 76
  deviations from, 173–186, 204, 251–255, 296–315, 346, 352, 403–405
Nerve cell, electrical behavior of, 89–91
Neutral carriers, structures, 114, 265–267
  for alkali metal ions, 114, 139, 264, 268
  for alkaline-earth metal ions, 106, 114, 140, 268
  with chiral recognition, 105, 268
  design features, 322–330
  lipophilicity, 325
  molecular basis of ion selectivity, 315–322
  n.m.r-studies of the extraction behavior, 298, 299
  requirements for ionophoric behavior, 143, 269, 303
  selectivities observed for bilayer membranes, correlation with bulk-membrane selectivities, 149
  correlation with kinetic parameters, 144
  selectivities observed for bulk membranes, correlation between potentiometric and transport selectivities, 104–108
  correlation with complex stabilities, 290
  influence of the membrane solvent, 290, 295
  influence of molecular parameters, 323, 327, 329
  simulation of active or coupled transport, 28, 106
  theory of carrier-mediated ion transport across bilayer membranes, 113–149
  for transition metal ions, 268
  for uranyl ion, 268
Neutral carrier membrane electrodes, 264–330
  application in enzyme electrodes, 407

dynamic response behavior, 387−393, 398
with incorporated ion-exchangers, 309−316
membrane materials, 270
reported selectivities, 272, 316
theories and fundamentals,
    anion effects in neutral carrier mem-
        brane electrodes, 296−315
    cation selectivity, 285−295
    mechanism of permselectivity, 274−285
    molecular aspects, 315−330
Nicolsky, "first variant ion-exchange theory",
    351
Nicolsky and Shults, "second variant ion-ex-
    change theory", 361
Nicolsky equation, 8, 77, 79, 147, 191, 221,
    232, 247, 288, 293, 344
    deviations from, 79, 191−193, 228,
        235−237, 291−293, 345−362
Nigericin, 229
Nitrate electrodes, 214−216
Nonactin, 114, 144, 149, 265, 299

Ohmic behavior, 93, 146, 276
Osmotic coefficient, 18
Oxygen electrode, 402
Overpotential, 38, 124, 128

Passive transport, 28, 90
Patchwork coating, 191−198
Perchlorate electrodes, 214−216
    calibration plots for interfering ions, 254
    selectivity, 227, 254
Permeability, 90, 140, 160
Permeability ratios, definition, 141−143, 147
    observed values, 91, 144, 149
Permeability selectivity, 27
Permselective membranes, boundary potential,
    41
    diffusion potential, 50, 56−60
    membrane potential, emf, and selectivity,
        76−83
    transport behavior, 91−101, 107, 143−149
Permselectivity, loss of, 83, 109, 296
    measure of, 82, 107
    mechanism of in neutral carrier membranes,
        274−285
    requirements for, 83
pH-convention, 16
pH-electrodes, 337−338
    alkali error, 344
    selectivity, 338
pH-gradient as driving force for cation counter-
    transport, 106
pH-interference in different electrodes, 218,
    237, 272, 316, 348
pH-response of sodium aluminosilicate glasses,
    355

Phase boundary, see Interface
o-Phenanthroline, substituted, membranes based
    on iron(II)- and nickel(II)-complexes of,
    213−215, 227
Phenomenological coefficients, 87
Phenylethylammonium sensors,
    selectivity between (R) and (S) enantiomer,
    105
    transference number and slope, 276
Planck theory and its applications,
    diffusion potential, 52−56
    membrane potential, 85−86
    numerical values of liquid-junction poten-
        tials, 64−72
    response of neutral carrier membranes,
        283−285, 306−313
Plasticizers and polymers used for liquid mem-
    brane electrodes, 269
Porous membranes, 42, 81, 211; see also Perm-
    selective membranes
Potassium electrodes, glass membrane, 338
    liquid ion-exchanger membrane, 214
    neutral carrier membrane, 270, 290
        anion interference, 304
        dynamic behavior, 390−393, 398
        precision, 7
        selectivity, 272, 290
        slope of calibration curve, 275−276
        theory of potential response, 281−285
        transference number for, 275−276
Potassium transport, across bilayer membranes,
    139
    across solvent polymeric membranes,
        278−280
Potential, chemical, 35, 46
    chemical standard, 35−36, 116
    electrochemical, 35, 45
    electrostatic, 30, 35, 116
Potential dips, 233−237, 243
Precipitate-based membranes, 165, 167
Precision of emf-measurements on ion-selective
    electrode cells, 5, 7
Pressed-pellet membranes, 167
Proton carrier, 106

Rate constants, diffusional, 408, 412
    of enzymatic reactions, 407
    of interfacial transfer reactions, 37, 124,
        131, 380
    of membrane-internal translocation, 119,
        131, 145
RC time constants, 377−379
Reactions; see also Dissolution processes and
    Ion-exchange equilibria
    in enzyme membranes, 406, 407
    in gas sensors, 402, 403
Reaction order, 380, 411

Reaction-rate theory, 37, 87, 121, 123
Rectification, 131
Reduced electrical potential differences, 117, 124
Reduced free energy function, 116
Reduced mass transference number, 47
Reference electrode, 2–3, 64, 71–73, 403
Reference electrolyte, 2–3, 64
Reference potential, 4
Regular-solution theory, 190, 341
Relaxation, diffusional, of concentration polarization, 383, 412
 electrical, 377–379
Resistances, electrical, of bilayer membranes, 132
 of glass membranes, 363
 of neutral carrier membranes, 278, 309
 of solid-state membranes, 166
 in theoretical descriptions of electrical relaxation, 377–379
Response time; *see also* Dynamic response behavior and Time constants
 reduction of, 384, 392, 405
 values, 386, 387, 393, 405

Salt bridge, 2, 4, 64
 electrolytes, 66–72
Sandblom–Eisenman–Walker theory, 236–243
 extensions, 243–246, 368–371
Saturation of current, 131, 134, 138, 146
Schlögl's theory, 58–60, 94–96
 applications, 72, 96–101
Selectivities, of bilayer lipid membranes, 141–143
  correlation with bulk-membrane selectivities, 149
  correlation with kinetic parameters, 144
 of enzyme electrodes, 410–412
 of glass electrodes, 344–362, 368–371
  reported selectivity coefficients, 338
  selectivity parameters of sodium aluminosilicate glasses, 357
 of liquid ion-exchanger membrane electrodes, 221–258
  apparent selectivity behavior, 253
  correlation with hydration energies, 224, 225
  correlation with ion-exchange constants, 222, 227
  reported selectivity coefficients, 218, 222, 227, 257
 of neutral carrier membrane electrodes, 285–295
  correlation with complex stabilities, 290
  effect of incorporated ion-exchange sites, 316

 influence of membrane solvent, 290, 295
 influence of molecular parameters, 323, 327, 329
 reported selectivity coefficients, 272
 of solid-state membrane electrodes (silver halides), 186–206
  apparent selectivity behavior, 197
  reported selectivity coefficients, 165, 189, 197, 206
 of transport across bulk membranes, 101–111, 147
  correlation with potentiometric selectivities, 104–108
Selectivity coefficients, definition according to IUPAC, 8
 apparent, 191–198, 246–258
 general formulation, 78–81, 147
 summary, 156–159
Selectivity sequences, 12, 106, 195, 201, 214–217, 225
Separate solution method, 8
Separation of ions, 29, 103, 106, 269, 275
Silica group, 351, 354
 ion-exchange on, 357
Silicate glass, pH-response of, 355
Silver complexes, stability constants, 174
Silver compound membranes, 165–206
Silver electrodes, glass, 338
 solid-state, 173–183
Single-crystal membranes, 165–167
Sintered membranes, 167
Site activity or concentration, 41, 57, 93–96, 145
Slope of emf-response function, 5, 204, 252, 275
Sluggish response, 384, 394–397
Sodium electrodes, based on glass membrane, 338
 selectivity, 338
 based on neutral carriers, 270
 anion interference, 305, 314
 precision, 7
 selectivity, 272
 slope of calibration curve and transference number, 275
Sodium ions, interference in different electrodes, 218, 222, 272, 273, 338, 355
 transport across bulk membranes, 104
Solid contact, replacing the internal solution, 2, 166, 169
Solid-state membrane electrodes, 165–206
 application in gas sensors, 403
 dynamic response behavior, 381
 membrane materials, 166–168
 reported selectivities, 165, 189, 197, 206
 response curves, 176–185

theoretical treatment of silver compound membranes, basic aspects, 168–171
anion selectivity, 186–198
potential response and detection limit, 171–183
response to different cations, 183–186
response to different ligands, 198–206
Solubility products of silver salts, 174
Solvent polymeric membranes, 113, 124, 149, 277, 279; see also Bulk membranes and Liquid-membrane electrodes
Solvent used in liquid membranes, 215–217, 270
influence on the ion selectivity, 221–225, 233, 288–290, 294, 303–305, 320–322
influence on the response time, 391–393
Space-charge regions, 377
Space-charge theory of permselectivity, 144, 278, 297
Standard potential of electrode cells, 5, 169, 357
Steady-state assumption, 52, 89, 115, 126, 238, 299, 309, 407; see also Nernstian diffusion layer
Steady-state potential, extrapolation of, 398
Stephanova and Shults, ion-exchange theory, 366–369
Stepwise response, 351–362
Steric contributions to ion selectivity, 320
Stirring rate, effects of, 172, 255, 384, 391, 392
Stokes–Robinson–Bates convention, 16–19
Streaming potential, 48
Strontium electrode, 270
selectivity, 272
Structure-selectivity relationships, 315–330
Sulfide electrode, 180–183
response to cyanide, 204
Summary of fundamental relationships, 153–161
of membrane materials and selectivity-determining principles, 8–13
Surface concentrations, 126, 131
Surface layers, 187–198, 362–363
Surface potentials of lipid bilayers, 124, 133
Symmetrical membrane cells,
electrical properties, 102–109, 131–134, 138, 146
simple theoretical model, 91–94
Synthetic carriers, 114, 265–268; see also Neutral carriers

Teorell equation, 57, 109
Teorell–Meyer–Sievers theory, 27, 81–83, 281–283, 291, 297–306
Tetranactin, 114, 144, 149, 265
Tetraphenylborate and substituted homologs, use as membrane components, 214, 216, 222–224, 309–316, 392, 393
Thin membranes or Thick membranes, see Bilayer lipid membranes or Bulk membranes
Time constants of sensors, 376–396, 405, 412
Time-dependence, exponential, 376–380, 382, 405, 412
different type, 381
square-root, 378, 387–389, 394, 398
Titration, 186, 273
Transfer, coefficient, 37, 124
Transfer, free energy of, 288, 317, 322
Transference numbers or transport numbers, definition, 28, 47
experimental, 104, 105, 108, 275, 280
formulation, 103–111, 228, 366–369
integral, 53, 86
reduced, 47
Transient response, 395, 397
Transmembrane potential, 127, 141
Transport, see Membrane transport
Transport equation, see Flux equation
Transport phenomena, 28
Trinactin, 114, 144, 149, 265
Tubocurarin sensor, 214

Uphill transport, 28
Uranyl electrode, 268
response mechanism, 308
Urea sensors, 406
calibration curves, 410
Ussing's flux equation, 90

Vacancies, 360, 365
Valency classes, 50, 94
Valinomycin, 105, 114, 139, 147–149, 265, 278–285, 290, 304, 390–393, 398
homolog of, 144
Viscosity of membrane phase, role of, 255, 392

Warburg finite-diffusion process, 378
Water, effects in liquid or glass membranes, 280, 362
Water hardness electrode, 232